AF598058

# THE AMERICAN AGRICULTURAL PRESS

## 1819–1860

---

COLUMBIA UNIVERSITY STUDIES
IN THE HISTORY OF
AMERICAN AGRICULTURE
NUMBER 8

JOHN STUART SKINNER

A copy of a portrait in oil, painted in 1825 by Joseph Wood. Original in possession of Mrs. Edythe M. Brosius, Washington, D. C.

# THE AMERICAN AGRICULTURAL PRESS 1819-1860

*Albert Lowther Demaree*

*New York: Morningside Heights*
COLUMBIA UNIVERSITY PRESS
1941

FOREIGN AGENTS: Oxford University Press, Humphrey Milford, Amen House, London, E. C. 4, England, and B. I. Building, Nicol Road, Bombay, India; Maruzen Company, Ltd., 6 Nihonbashi, Tori-Nichome, Tokyo, Japan

---

MANUFACTURED IN THE UNITED STATES OF AMERICA

To

H. J. D.

# COLUMBIA UNIVERSITY STUDIES IN THE HISTORY OF AMERICAN AGRICULTURE

# EDITORS' FOREWORD

THE LAST QUARTER of a century has witnessed a growing interest in agricultural history in both the Old World and the New. In this country this interest has been manifested not only by the founding of the Agricultural History Society but by the appearance of an increasing number of publications dealing with various aspects of our agricultural development. One has only to examine U. B. Phillips' *Life and Labor in the Old South,* Avery O. Craven's *Soil Exhaustion as a Factor in the Agricultural History of Virginia and Maryland, 1606–1860,* Herbert A. Kellar's *Solon Robinson, Pioneer and Agriculturist,* and L. C. Gray's *History of Agriculture in the Southern United States to 1860*—to mention only a few—to appreciate the nature of the contributions made to the history of agriculture in the United States in recent years.

To this growing literature Dr. Albert Demaree of Dartmouth College now adds a scholarly and entertaining volume on the history of farm journalism during the ante-bellum period. The preparation of this volume has necessitated years of extensive research and travel, and the author has shown marked skill in handling a mass of details which did not easily lend themselves to organization.

In the pages which follow, the reader will find an accurate cross-section of American rural life, in its varied aspects, during the first half of the nineteenth century as portrayed by the farm journals of the time. The value of the volume is greatly enhanced by the inclusion in Part II of selected articles taken verbatim from the farm journals and the brief sketches in Part III of the more important journals which appeared during these years. Dr. Demaree has placed every student of American social and agricultural history in his debt.

Harry J. Carman
Rexford G. Tugwell

*October 22, 1940*

# PREFACE

PRIOR to the Civil War, the United States was predominately an agricultural nation. Indeed, more than eighty percent of our people, during the period from 1819 to 1860 wrested their living from the soil. In this interval, there were published in the United States well over four hundred different periodicals devoted primarily to agriculture and its related interests. Fortunately many of these agricultural journals have been preserved, although they are scattered throughout the libraries of the United States and remain, to a very large extent, untouched from year to year. As a medium of exchange for the ideas of this eighty percent of the population and as a rich depository of social and economic history of their times, these periodicals are unrivaled. Strangely enough, they have been almost entirely neglected by historians, with the exception of students of agricultural history.

So far as the writer knows, no history of the agricultural press in the United States has been written. Nor does this work on agricultural journals attempt such a task. It is well to point out at once the purpose, plan, and restricted scope of this study.

The primary purpose of the present work is to give the reader a description of the general content of these journals and to tell in part the story of rural life as seen through them. Furthermore, it attempts to point out the objectives sought by the editors and to describe the methods used to gain these ends. Approximately one hundred agricultural periodicals published in all parts of the United States have been selected for investigation. Sixteen of these have been set aside for intensive research.

The plan of presentation should be noted. Part I contains material of a general nature drawn indiscriminately from thousands of volumes. The great similarity of these publications permits a general collective treatment. The story of the individual paper, with one exception, is reserved for Part III. In order to familiarize the reader with the general set-up of an agricultural journal, one particular paper has been chosen for special study. John Skinner's *American Farmer*, the first important agricultural periodical, seems to be representative. Furthermore, this paper had the distinction of serving as a model for succeeding publications. Chapter Two is devoted to this important journal which was terminated in 1834.[1] The remaining chapters describe the subsequent change in content and emphasis of the agricultural press to 1860. Following the description of the *American Farmer*, the remainder of Part I is devoted to a study of the farm press as a whole. Among the general subjects selected for treatment are: the editors, their problems and methods; the outstanding policies of the journals; the special feature sections; advertising; poetry and the agricultural fair.

Part II consists of twenty-eight selected articles which furnish the real flavor of these periodicals, and further serve to illustrate their possibilities as sources to students of social and economic history. In general the selections are short; they represent and amplify the material discussed in Part I. For example, in the chapter on the agricultural fair, reference is made

[1] This paper was later reëstablished under another title.

to three important institutions, namely; the plowing match, the farmers' informal evening discussion, and the annual agricultural address. In each instance therefore, the reader is referred to Part II. Here he finds a concise report of an actual plowing match, a brief summary of precisely what was said at a particular evening discussion, and a portion of Lincoln's little known agricultural address before the Wisconsin State Agricultural Fair in 1859. Part II also includes an account of an agricultural tour made by Solon Robinson; a visit to ex-President Van Buren's farm; a meeting of an agricultural club; a typical article by Frances D. Gage, one of the many women feature writers for the periodicals during the fifties; a characteristic editorial on the depression of 1857; letters dealing with the moral training and care of slaves and so on.

In Part III, sixteen agricultural periodicals have been chosen for special study. This group does not necessarily consist of the sixteen most important journals of the period, but it does contain important and representative papers of the better type drawn from practically every section of the United States. An attempt is made to reveal the "personality" of each journal—which, of course, depends upon the editor—special emphasis of the journal, its geographical location, and other factors. These surveys include information such as: reason for initiation, type of editor, causes for success or failure, circulation figures, and general influence. A somewhat detailed knowledge of the important journals of the period is necessary to the complete understanding of the great influence of the farm press, which Edmund Ruffin claimed was the most inmportant factor in the agricultural revolution.

Restrictions in the scope of this volume should be carefully noted. An agricultural journal, as understood in the following pages, is a periodical published to appeal primarily to the general farmer. Included in these farm papers from the beginning were articles on all phases of rural interest—horticulture, stock raising, poultry, farm mechanics, dairying, bee culture, and the like. Toward the latter part of the period these subjects were frequently treated in specific departments, often conducted by an authority in the field. As rural activity became more and more specialized, journals devoted primarily to particular interests were inaugurated. Although these specialized papers carried important agricultural material, they are arbitrarily excluded from the present study. This investigation has been restricted to the period before the Civil War, as it is a convenient and logical unit for study. No specific mention is made of the depositories of the journals; this information is available in the *Union List of Serials.*

The difficulties involved in this field of research are suggested in L. H. Bailey's *Cyclopedia of American Agriculture* (IV, 78) which states, "The history of agricultural journalism in this country is also more complex and of earlier origin than most persons are aware. This history is difficult to follow, also, because the same name is sometimes used at different times for wholly different journals, from the practice of dating back to a former publication of the same name but which may not have been an ancestor, and from change of name in the same periodical." Possibly the most distressing problem faced by an investigator in this field of research is the inaccessibility of these periodicals. No single collection contains more than a fractional part of all the journals published during the period. These volumes are found scattered in libraries in all parts of the United States. In many instances, copies of important journals are not known to be extant. No bibliography of agricultural journals has ever been published, although the Stuntz List is in the process of publication by the United States Department of Agriculture.

Most of the research for this volume was pursued in the library of the United States Department of Agriculture, which contains the largest collection of American farm periodicals in the world. The author is deeply grateful to the staff of this institution for its helpful coöperation. Appreciation is also expressed to the following libraries, where additional research was conducted: Library of Congress; Baker Memorial Library, Dartmouth College; New Hampshire State Historical Society Library, Concord; New Hampshire State Library, Concord; Boston Public Library; Widener Library, Harvard University; American Antiquarian Society, Worcester, Massachusetts; New York Public Library; New York State Library, Albany; Library of Columbia University; New York Historical Society Library, New York City; Rutgers University Library; Philadelphia Free Library; Pennsylvania State Library, Harrisburg; State College Library, State College, Pennsylvania; University of Chicago Library; John Crerar Library, Chicago; and the McCormick Historical Association Library, Chicago. Through interlibrary loans, Duke University, the University of Wisconsin, and other institutions of learning have aided in the preparation of this work.

For permission to quote from works published by them, I am indebted to D. Appleton-Century Company (E. Douglas Branch, *The Sentimental Years*) and to The Macmillan Company (L. H. Bailey, ed., *Cyclopedia of American Agriculture*).

Several individuals have contributed invaluable assistance to the author. Herbert A. Kellar of the McCormick Historical Association has given generously of his great fund of knowledge upon this subject. Appreciation for suggestions and criticisms is expressed to Everett E. Edwards of the Department of Agriculture, to Professors Harry J. Carman and John A. Krout of Columbia University, and to Wayne E. Stevens and Randall Waterman of Dartmouth College. In her editorial capacity, Miss Matilda L. Berg of Columbia University Press has been most coöperative. Thanks are extended to Ellamae Jackson and Mary Elizabeth Vance for verifying certain source material. Finally, to my wife, Helen Jackson Demaree, deep gratitude is expressed for her helpful and intelligent coöperation.

A. L. Demaree

*Hanover, New Hampshire*
*October, 1940*

# CONTENTS

## Part III

## SKETCHES OF CERTAIN IMPORTANT JOURNALS, 1819–1860

## ILLUSTRATIONS

# THE AMERICAN AGRICULTURAL PRESS

## 1819–1860

# Chapter I

## EARLY AMERICAN AGRICULTURAL LITERATURE

*The history of agriculture shows a conservatism probably unequaled in any phase of human activity* [1]

THE SOCIAL and economic structure of the United States at the close of the eighteenth century was predominantly agricultural. Each farming community was largely self-sufficient, and the methods of tillage were exploitative in the extreme. The lack of outside markets for surpluses contributed greatly to the continuation of this situation, and the abundance of what seemed to be an unlimited amount of virgin land permitted this condition to continue through two centuries. By this time, however, the generally unprogressive nature of agriculture as an occupation began to challenge the attention of the more discerning leaders.

At the turn of the century, a number of progressive farmers, among whom Washington, Jefferson, and Madison were representative, were becoming vitally interested in the new agri-

[1] Cook, "The American Origin of Agriculture," *Popular Science Monthly*, LXI (Oct., 1902), 492.

culture of Great Britain.[2] In his correspondence with Arthur Young, the foremost English agricultural writer, Washington was hailed as "brother farmer"; at home he had established a reputation as a progressive agriculturalist.[3] In his methodical way, Washington conducted extensive experiments to check erosion, improve plowing, determine the relative value of different fertilizers, the best rotation of crops, and the usefulness of crops new to this country.[4] He, as well as other leaders, read extensively and in a wide correspondence exchanged ideas and observations about many aspects of agriculture. Though such activity greatly enhanced agricultural investigation, it had little immediate effect on the common farmer.[5]

The storehouse of modern, scientific research was largely closed to the inquiring farmer of the period. He was ignorant of the principles of animal and plant breeding. He was unacquainted with the principles of plant nutrition, and with the chemical composition of plant food. He had little or no understanding of why tillage makes the plant thrive. He was not familiar with methods of enriching the soil. He lacked the advantages of agricultural schools, experiment stations, and farm papers.[6] His agricultural implements were crude and in-

[2] Gray, *History of Agriculture in the Southern United States to 1860*, II, 612, 779.

Thomas Jefferson was the first in America to design a moldboard for a plow based on true mathematical principles. Bidwell and Falconer, *History of Agriculture in the Northern United States, 1620–1860*, p. 208.

[3] Craven, *Soil Exhaustion as a Factor in the Agricultural History of Virginia and Maryland, 1606–1860*, p. 87; Wiest, *Agricultural Organization in the United States*, p. 22.

[4] Haworth, *George Washington, Country Gentleman*, pp. 71 ff.; Kirkland, *A History of American Economic Life*, pp. 197–98. Many of Washington's agricultural experiments are described in his *Diaries*. See particularly *The Diaries of George Washington, 1748–1799*, III, 4, 178, 186, 259, 328, 330, 377.

[5] Gray, *op. cit.*, II, 779.

[6] Bailey, *Cyclopedia of American Agriculture*, IV, 362.

efficient. Years were to elapse before steel plows, horse rakes, mowers, reapers, and other important labor-saving devices were at his disposal.

In addition, he was bound by a superstitious dependence upon the phases of the moon as a factor in farm procedure; this was generally accepted as agricultural science. One writer said, "We plant, we sow, we reap and mow; we fell trees, we make shingles, we roof our houses, secure bacon, make fences, spread manure, when the moon is auspicious. If we are ready before her ladyship, we wait the happy moment when her aspect shall say, proceed." [7]

The farmer's caution and skepticism toward innovations also drastically retarded progress. An excellent illustration of this attitude was displayed in an address by James M. Garnett to the Frederick Agricultural Society early in the nineteenth century. The first Cary plows sent into his neighborhood, he said, remained lying in the Tavern stable, "objects of doubt and cunning suspicion," for about a year before anyone could be induced to give them a trial. He added that the farmer believed nothing he heard, but required to see and to feel before he gave credit to what he was told, and rarely then acted immediately upon his belief.[8]

The situation at the turn of the century is characteristically depicted by an English traveler: "Land in America affords little pleasure or profit. . . . Virginia is the Southern limit of my information in America, beyond it inquiries were unnecessary, because it appears as if agriculture had already arrived at its lowest state of degradation." [9]

A dozen years later, a Virginian, speaking of the entire coun-

[7] *American Farmer*, I (May 28, 1819), 68.

[8] *Ibid.*, IV (Dec. 6, 1822), 291.

[9] Strickland, *Observations on the Agriculture of the United States of America*, pp. 26, 45.

try, called for general agricultural reform: "Let us boldly face the fact. Our country is nearly ruined. We certainly have drawn out of the earth three fourths of the vegetable matter it contained, within reach of the plough. . . . Forbear, oh forbear matricide, not for futurity, not for God's sake, but for your own sake." [10]

The agricultural books available to the farmer before 1819 were few in number and difficult to obtain. Jared Eliot's *Essays upon Field-Husbandry in New England,* published in Boston in 1760, was the first important work of its kind in North America.[11] The author, widely traveled, was also a minister, physician, botanist, and farmer.[12] In the preface he states that the book is "not an Account of what we do in our present Husbandry, but rather what we might do, to our Advantage." [13] It contains, however, many comments on prevailing practices, based on wide correspondence and personal acquaintance with the habits of farmers in that region.

During the interval between the publication of Eliot's volume and the appearance of the *Arator* in 1813, scarcely more than a half-dozen important agricultural works were issued in this country.[14] John Taylor of Caroline, the first great

[10] Taylor, *Arator*, pp. 1, 82, 84.

[11] Bidwell and Falconer, *op. cit.*, p. 458. The six essays were published separately between 1748 and 1759.

[12] True, "Jared Eliot, Minister, Physician, Farmer," pp. 185 ff.

[13] Eliot, *Essays upon Field-Husbandry in New England*, Preface.

[14] A. C. True lists the following agricultural books published during this period: Samuel Deane, *New England Farmer or Georgical Dictionary*, 1790; John Beale Bordley, *Essays and Notes on Husbandry and Rural Affairs*, 1799; John A. Binns, *A Treatise on Practical Farming*, 1803; Job Roberts, *The Pennsylvania Farmer*, 1804; John Gardiner and Daniel Hepburn, *The American Gardener*, 1804. A. C. True, *A History of Agricultural Education in the United States, 1785–1925*, pp. 29–30. Bordley, an Englishman, brought to America a wide knowledge of the new agriculture; his experiments in this country extended over a varied field. He corresponded with

Southern reformer of the period, was the author of the *Arator*.[15] Edmund Ruffin later wrote that this book "was the first original agricultural work (worthy to be so called) which had ever been published . . . in the southern states; and it appeared at a time when agricultural improvement was still neglected by the men of intelligence and wealth whose interests were almost exclusively agricultural." [16] In the opinion of John Adams, no Northern writer had equaled this agricultural treatise.[17] Taylor not only advocated deep plowing, but laid emphasis on a "four-field system of rotation, bedding, composting and handling of manure, and the restoration of worn-out lands by inclosure without grazing." [18] This popular publication was issued in at least seven editions.[19]

Americans of the pre-farm-journal period relied largely on English contributions.[20] In 1816, when John Taylor was requested to list books valuable for an agricultural library, the only American works he recommended were the volumes of

---

such agricultural leaders as Washington and published a number of pamphlets and treatises. Gray, *op. cit.*, II, 612–13.

Percy W. Bidwell lists more than thirty works on agriculture published during this period in his "Rural Economy in New England at the Beginning of the Nineteenth Century," pp. 392–93.

[15] Craven, "The Agricultural Reformers of the Ante-Bellum South," p. 305.

[16] *Farmers' Register*, VIII (Dec. 31, 1840), 703. So impressed was Edmund Ruffin with the value of these agricultural essays that when this volume was no longer in print, he reprinted it in the *Farmers' Register*, Vol. VIII, No. 12.

[17] *American Farmer*, II (June 16, 1820), 94.

[18] Gray, *op. cit.*, II, 780.

[19] Simms, *Life of John Taylor*, p. 148; *Farmers' Register*, VIII (Dec. 31, 1840), 703 ff.

[20] It is surprising to learn how much such men of antiquity as Cato, Varro, and Pliny were relied upon as agricultural authorities during the eighteenth century. Bailey, *Cyclopedia of American Agriculture*, IV, 379; Hedrick, *A History of Agriculture in the State of New York*, p. 323.

the Philadelphia Agricultural Society. He added, "Of the agricultural books which I have, most have been compounded from theory, and have tended chiefly to prove that fine writers may be bad farmers. Arthur Young alone seems to me to occupy the station among agriculturists, which Bacon does among philosophers." [21] The available agricultural writings at this early period were of little practical value to the large planter. To the dirt farmer, who had a horror of "book farming," they were almost unknown. Even as late as 1835 the editor of the *New England Farmer* stated that the number of American books on agriculture was "very limited." [22]

The publications of the agricultural societies were among the earliest of our agricultural literature. The Philadelphia Society for Promoting Agriculture, which was established in 1785, was the first of its kind.[23] These early organizations, now generally referred to as the "literary" or "learned" societies, were pioneers in the great task of agricultural education. In their *Memoirs* and *Transactions* they published results of experiments in this country and accounts of the best practices abroad. They offered premiums for improvements in agriculture. Strange as it may seem, the initiative and direction of these organizations came from professional and business men, whose main interests were not agricultural. Naturally the chief benefits accrued to these gentlemen farmers. In reaching the "dirt

[21] *American Farmer*, II (June 16, 1820), 19.

In answer to a request for a list of books on agriculture, Jefferson, in 1817, enumerated more than fifty, including works in French, Italian, and English. *American Farmer*, II (June 16, 1820), 94.

The books recommended in the *American Farmer* in 1819 included Home's *Principles of Vegetation*, Darwin's *Phytologia*, Hunter's *Georgical Essays*, Anderson's *Essays*, Lord Dundonald's *Connection of Agriculture with Chemistry*, Davy's *Agricultural Chemistry*. *American Farmer*, I (July 2, 1819), 105.

[22] *New England Farmer*, XIII (July 1, 1835), 402.

[23] Bidwell and Falconer, *op. cit.*, p. 184.

farmer" and meeting his problems, these societies and their publications were failures.[24] "The improvements proposed fell almost dead upon the people, who rejected 'book farming' as impertinent and useless." [25]

More successful were the agricultural societies organized on the Berkshire plan, which spread rapidly after 1811.[26] Their main feature was the annual fair, which had greater appeal to the practical, working farmer than theoretical science. Agricultural journals often served as organs for publishing transactions and activities promoted by the societies.[27]

Another aid of much greater interest to the dirt farmer than the literature of the agricultural societies was the almanac. Except for the Bible, it was probably the most universally read publication in the American farm home during the eighteenth century.[28] Although intended primarily to record astronomical data, it readily became a depository for all manner of miscellaneous rural information. Frequently verse, jokes, anecdotes, riddles, essays, "cures," recipes, and interesting bits of general information appeared. Normally the schedule of local

[24] *Ibid.*, pp. 185–86.

[25] Flint, *Eighty Years' Progress of the United States*, p. 25.

The investigations made by these early societies were carefully and well conducted. The kind of topics discussed may be seen in the *Transactions* of the New York Agricultural Society (1792–1799) of which Robert R. Livingston was president. They included "Calcarious and Gypsious Earths," "Manufacturing Paper," "Experiments on Manures," "Observations on the Hessian Fly," "On the Raising of Red Clover Seed," "Domesticating the Elk and the Moose," "Experiments on Wheat, Clover and Lucerne," "Silk Worms," and "Improvements on the Steam Engine."

[26] A brief discussion of the Berkshire societies may be found on pp. 198 ff.

[27] The Columbian Agricultural Society (Georgetown, District of Columbia) set a precedent in 1810 when it chose as its organ the *Agricultural Museum*, the first agricultural journal in this country. *Agricultural Museum*, I (July 4, 1810), 3.

[28] Woodward, *The Development of Agriculture in New Jersey, 1640–1880*, p. 64.

courts and the time table for stages throughout the vicinity were noted. Often these almanacs contained blank pages which the reader used for farm and family records or for diaries and account books.

The almanac was prepared for rural communities, in which it took the place of the newspaper.[29] Its adaptation to the farmer's needs is clearly seen, for he was totally dependent upon climate and weather conditions and his round of farm duties had to keep pace with the seasons.[30] It is not surprising that bits of agricultural information appeared in the earliest almanacs, or that after 1750 this tendency showed a rather constant development.[31]

In his preface to the *Farmer's Almanack* for 1794, Robert B. Thomas welcomed agricultural observations in this wise:

> My precepts and observations on agriculture, I have the vanity to believe, have been approved of by farmers in general. Agriculture affords an ample field in this Country for the ingenious to expaciate upon, in which improvements are making every day; and as my greatest ambition is to make myself useful to the community in this way, 'tis my sincere wish that men of experience and observation in agriculture, would be kind enough to forward me such hints towards improvement, as are capable of being rendered serviceable and of general utility to the public.[32]

An important section of many almanacs was a Farmer's Calendar. This monthly guide included directions for work appropriate to the successive days of the month, with occasional moral and religious observations thrown in. The great reliance placed upon the moon for the regulation of farm activity gave a kind

[29] Hedrick, *op. cit.*, p. 311. [30] Woodward, *op. cit.*, p. 64.

[31] *Ibid.*, p. 65.

[32] A summary of the agricultural material found in the *Farmers' Almanack* may be found in *The Old Farmer and His Almanack* by George Lyman Kittredge.

of "astrological background" to agriculture. No doubt, a great deal of what they taught was educationally constructive even if some of it was scientifically and economically unsound.[33] Among other farm articles carried in the *Virginia and North Carolina Almanack* for 1800 were "Culture of Cotton," "Advice to Farmers," and "Useful Directions for the Management of Peach Trees."

Almanacks were highly respected depositories for agricultural information, as is shown by the action of the Philadelphia Agricultural Society in 1816. In that year it was resolved: "That the distribution of agricultural information, and the proposition of questions calculated to excite observations and inquiry, as well as to induce the farmers of our country to communicate their experience in husbandry, through the medium of an Almanack, would be productive of important benefit to society." They further resolved to pledge the society's patronage to *The Agricultural Almanack,* published by Solomon W. Conrad of Philadelphia, and to furnish the editor with appropriate agricultural articles.[34]

An occasional article on agriculture appeared in the newspapers of this early period.[35] Now and then a newspaper announced a policy somewhat similar to that advanced by the editor of the *New Jersey Gazette,* who wrote in 1777: "Proposals for Improvements in AGRICULTURE, and particularly in the culture of HEMP and FLAX, will be inserted with Pleasure and Alacrity." [36] Sometimes a column headed "Agricultural"

[33] Woodward, *op. cit.,* p. 79.

[34] *The Agricultural Almanack* for 1818. (Phila.)

[35] For example see the *Maryland Journal and Baltimore Advertiser* (1790); the Annapolis *Maryland Gazette* (1790, 1795, 1796, 1797, 1798); *Richmond Enquirer* (1804, 1805, 1806, 1816, 1818); Fredericksburg *Virginia Herald* (1815, 1816, 1817).

[36] Quoted from Woodward *op. cit.,* p. 80. Woodward gives a number of examples of agricultural material found in New Jersey newspapers.

or "Rural Concerns" appeared. There was no systematic treatment of the subject, however, and the material, sparse as it was, seems to have been selected with little attempt to meet the real needs of its readers.

More important as a medium for the dissemination of agricultural information were the early magazines. Mathew Carey's *American Museum*, the *Columbian Magazine*, *Niles Weekly Register*, and many others carried a great number of articles on all phases of this subject. The agricultural societies often furnished the editors with useful essays, summaries of their experiments, and the announcements of prizes offered by these bodies. Among the articles in their pages were many like the following: experiments on the culture of wheat; the cultivation of potatoes; new methods of treating hemp; the Hessian fly; and the cultivation of beets. Considerable material that later found its way into the agricultural journals was published here. *Niles Weekly Register* welcomed the *American Farmer* in 1819 as a publication long needed by the agriculturists of this country.[37]

Far more important to the farmer than books, agricultural society publications, almanacs, newspapers and magazines, were the farm journals which had their inception in 1819. Written primarily for the farmer, these were, in the main, devoted to farming, stock raising, horticulture and allied topics. The editors attempted to obtain information from every source, on every branch of husbandry, and the journals became veritable clearinghouses for agricultural information. They varied in size, price, and frequency of publication, and each emphasized the interests peculiar to its locality. A wide variety of extraneous material, including "poetry," jokes, anecdotes, news, "cures," and items of interest to women and children,

[37] *Niles Weekly Register*, n.s., IV (April 3, 1819), 108.

gradually found its way into their pages. In this brief review, only certain notable periodicals are mentioned.

The birthdate of American agricultural journalism is usually given as April 2, 1819, when John S. Skinner inaugurated the *American Farmer* (1819–97) in the city of Baltimore.[38] This famous periodical was followed two months later by the *Plough Boy* (1819–23) in Albany, with Solomon Southwick as editor. As Southwick had little interest in agriculture and no practical farm experience, the paper was not a powerful influence for agricultural improvement, and lasted only four years. Much more important was the *New England Farmer* (1822–46) established in Boston under the editorship of the eccentric but able literary figure, Thomas Green Fessenden. Reared on a farm, Fessenden soon proved himself a competent agricultural editor and the journal enjoyed a long and influential career. The *New-York Farmer* (1828–39), launched in the city from which it took its name, was issued under the patronage of the New York Horticultural Society. This journal naturally emphasized horticulture; it was also one of the first periodicals to use illustrations extensively.

Commencing in the thirties, journals sprang up rapidly in many parts of the country. Few of them enjoyed a long life, while the vast majority succumbed after a year or two of publication. The *Genesee Farmer* (1831–39) was inaugurated by Luther Tucker at Rochester, New York, and rapidly assumed an important role. In its third year of publication, Jesse Buel became the assistant editor. Buel resigned to edit the *Cultivator* (1834–65), which had been established at Albany by the New York Agricultural Society, and presently this periodical passed

[38] The short-lived *Agricultural Museum* (1810–1812) was probably the first American agricultural journal. For a brief survey of this early periodical see p. 23.

into his hands. At Buel's death in 1839, Tucker purchased the *Cultivator*—now quite generally considered the leading periodical in the country—and with it united the *Genesee Farmer*, retaining the title *Cultivator*, and continuing publication at Albany.[39] In 1853 Tucker also issued the *Country Gentleman* (1854– ), a paper with a distinctly "modern" appearance and a wide range of interests, including more general news items and literary material. Tucker was regarded as the model and leader of agricultural journalists, and did more perhaps to promote the literature of agriculture than any of his contemporaries.[40]

*The Maine Farmer* (1833–1924) was one of the most interesting of the hundreds of journals inaugurated in the Northeast before 1860. It began its long life with Ezekiel Holmes as editor and contained under his able direction until his death thirty-two years later. The *Farmers' Cabinet* (1836–48) of Philadelphia was notable for its articles on stock breeding, but its most famous editor, James Pedder, lacked something of Holmes' editorial leadership. Ex-governor Isaac Hill of New Hampshire tried his hand at agricultural journalism with the *Farmer's Monthly Visitor* (1839–49) at Concord. While Hill claimed no profound knowledge of agriculture, his journal was highly regarded in New England. Of far greater influence was the *American Agriculturist* (1842– ), founded in New York by A. B. and R. L. Allen, brothers, engaged in the livestock business. Suited to all latitudes of the United States, this publication was destined to become one of the most widely circulated and influential journals of the period.[41] *Moore's*

[39] Tucker also established the *Horticulturist* (1846–75) at Albany.

[40] *Dictionary of American Biography*, XIX, 35–36; *Genesee Farmer*, ser. 2, XVI (June 1855), 194; *Cultivator and Country Gentleman*, XXXVIII (Feb. 20, 1873), 117.

[41] A German edition was inaugurated in 1858.

*Rural New-Yorker* (1849– ) was established under the guidance of D. D. T. Moore, at Rochester, and the *Working Farmer* (1849–75) with James J. Mapes as editor, at New York City. The former, with the largest circulation of any farm journal, was notable for its interesting miscellaneous material appealing to all members of the family, and the latter for its "scientific" approach to agriculture with special emphasis on agricultural chemistry.

The Southern states, meanwhile, furnished a number of important journals. The *Southern Agriculturist* (1828–46), especially adapted to sectional needs, was first issued at Charleston, South Carolina, with J. D. Legare as editor. Five years later Edmund Ruffin, the great Southern agricultural leader, inaugurated the *Farmers' Register* (1833–42) at Shellbanks, Virginia. This publication which avoided the "lighter side" stressed by so many farm papers, was the foremost agricultural journal of the South.[42] The short-lived *Franklin Farmer* (1837–40) of Lexington, Kentucky, catered to the growing interest in the development of fine breeds in the Blue Grass State and became justly famous for its articles on pedigreed stock. The *Agriculturist* (1840–45), inaugurated at Nashville, was the organ of the agricultural societies of Tennessee. Its most important editor, Tolbert Fanning, founded on his farm an agricultural school that became the forerunner of Franklin College (opened in 1845). Volume six of the *Agriculturist* included among its editors the faculty of this college, which emphasized the teaching of manual training and agriculture. The *Southern Planter* (1841– ), containing subject matter peculiarly applicable to "Southern soil, climate and institutions," commenced at Richmond, with C. T. Botts as editor. Still far-

[42] Craven, "The Agricultural Reformers of the Ante-Bellum South," p. 310.

ther south, the *Southern Cultivator* (1843–1935) of Augusta, Georgia, had as editors during its early years such well-known men as James Camak, Daniel Lee, and D. Redmond. The *American Cotton Planter* (1853–61) of Montgomery, Alabama, while emphasizing cotton culture, nevertheless urged a change in plantation economy which included growing more grain, raising more stock, and introducing the manufacture of cotton in the South. Dr. N. B. Cloud was its first editor.

One of the first important Western journals was the *Western Farmer and Gardener* [43] (1839–45), published at Cincinnati. This magazine had the support of a wealthy and socially prominent group in Ohio and was famous for its articles on horticulture and livestock. The names of Taft and Longworth appear frequently in its pages. More widely known was the *Prairie Farmer* (1840– ) of Chicago, a journal devoted to "agriculture and education." [44] Written in a style designed to appeal to the dirt farmer and selling for a low price, this periodical soon became one of the most popular of the Western journals. Farther south was the *South-Western Farmer* (1842–45?) of Raymond, Mississippi. This journal was founded, as the editor explained, because other publications did not deal with "subjects most immediately interesting to Mississippians," especially in regard to Negro economy.[45] M. W. Philips, a prolific agricultural writer, was its most notable editor. The *Michigan Farmer* (1843– ) at Jackson was the first important farm periodical in Michigan. The editor was desirous of

[43] It was called *Western Farmer* during its first year. Another *Western Farmer and Gardener* was inaugurated at Indianapolis in 1845, although its title was *Indiana Farmer and Gardener* for the first year. Henry Ward Beecher was the most prominent editor of the Indianapolis journal, which continued until 1847 and perhaps longer.

[44] This journal, founded by the Union Agricultural Society of Chicago, was called the *Union Agriculturist* for its first two years.

[45] *South-Western Farmer*, I (March 11, 1842), 1.

introducing advanced agricultural practices and improving the cultivation of the fertile soil of the new country, and wished also "to elevate the standing and ennoble the character of the Western agriculturist." Two important Ohio journals were the *Ohio Cultivator* (1845–64) and the *Ohio Farmer* (1852– ). The former, founded at Columbus, was edited by the veteran editor of the East, M. B. Bateham, and the latter, still a popular paper today, was launched at Cleveland under the editorship of Thomas Brown. The *Wisconsin Farmer* (1849–74) at Racine was edited by Mark Miller. The first American College journal of agriculture was the *Cincinnatus* (1856–61), edited by the faculty of Farmers' College located near Cincinnati.[46] The Far West was ably represented before the close of the period by the *California Farmer* (1854–84) and the *Oregon Farmer* (1858–61?).

This hasty survey will give the reader only a sampling of the huge number of journals published before 1860. It should be borne in mind that the farm papers just mentioned are merely a selected group of representative periodicals and that this group does not include all the important and influential journals published in these years.

It is difficult to estimate the number of agricultural journals initiated during the ante-bellum period. At the end of the fourth decade, more than thirty periodicals were in circulation, having, perhaps, 100,000 readers.[47] On the eve of the Civil War, it was estimated that there were between fifty and sixty active journals with a circulation of more than a quarter of a million.[48] The vast majority of farm papers that sprang up like

[46] Bailey, *Cyclopedia of American Agriculture*, IV, 373–74.

[47] *Cultivator*, V (March, 1838), 29; VI (June, 1839), 67. In 1852 it was estimated that there were 43 journals in existence. *Journal of the United States Agricultural Society*, I, No. 2 (1852–53), 144.

[48] *Farmer and Planter*, X (Jan., 1859), 223; XI (Jan., 1860), 20; XI

mushrooms in all parts of the country were short-lived and continued for less than three years. The editor of the *Country Gentleman* calculated that the total number of journals initiated from 1829 to 1859 was about 300.[49] The present writer would place the figure at well over 400 for the entire period.[50] In 1859 the editor of the *New England Farmer* said that scarcely a week passed without the appearance of a new agricultural paper.[51]

(Nov., 1860), 330; *Country Gentleman*, XIII (Jan. 6, 1859), 9. Cf. Joseph C. Kennedy, *Preliminary Report of the Eighth Census, 1860*, p. 100.

This circulation figure seems very conservative. While data of this type is difficult to obtain, nevertheless, a glance at the few available records would lead one to raise this figure to 350,000. For example, the *American Agriculturist* printed 80,000 copies for November, 1859; the *Genesee Farmer*, after issuing 30,000 papers for February, 1859, was required to print another edition; and *Moore's Rural New-Yorker* announced in December, 1860, that their next issue would be 70,000. The circulation of the *Ohio Farmer* for 1859 was 15,000. See *American Agriculturist*, XVIII (Nov., 1859), 349; *Genesee Farmer*, XX (March, 1859), 94; *Moore's Rural New-Yorker*, XI (Dec. 22, 1860), 41; *Ohio Farmer*, VIII (Dec. 10, 1859), 396. It is true, however, that these particular journals were leaders with respect to circulation and the great majority of periodicals had less than 4,000 subscribers.

[49] *Country Gentleman*, XIII (Jan. 13, 1859), 26. (250 of these journals were inactive in 1859, 50 still in circulation.)

[50] Gray, in his *History of Agriculture in the Southern United States to 1860* (II, 788), estimates that not less than 100 agricultural journals were initiated in the South during the period. As this section inaugurated relatively few periodicals, a total of more than 400 for the entire country seems reasonable. See also *Southern Cultivator*, IX (March, 1851), 40; *Farmer and Planter*, XI (Jan., 1860), 20.

A number of years ago, Stephen Conrad Stuntz, a bibliographer in the Bureau of Soils of the United States Department of Agriculture, began a compilation of agricultural journals in the United States. His manuscript was left incomplete at the time of his death (1918) and has since been deposited in the Library of the Department of Agriculture. While in many instances it is difficult to ascertain the precise nature of the magazine because of incomplete notations in his records, nevertheless his slips indicate that well over 300 journals devoted principally to agriculture were initiated during the pre-Civil War period. The Stuntz List is in process of publication.

[51] *New England Farmer*, n.s., XI (Nov., 1859), 501.

The *American Farmer* was the first continuous, successful agricultural periodical in the United States and served as a model for hundreds of journals that succeeded it. For this reason it has seemed worth while to make a case study of this so-called typical paper and to examine it in detail.[52] The following chapter is devoted to this pioneer of American farm journalism.

[52] Certainly, no individual journal may be called typical and consequently no journal may be called typical for the entire period (1819–1860). The *American Farmer* has been chosen as fairly representative of the early years. The changes in both content and emphasis which characterized the agricultural press following the termination of the *American Farmer* in 1834 are described in the succeeding chapters.

# Part I

## THE AMERICAN AGRICULTURAL PRESS 1819–1860

# Chapter II

## AMERICA'S PIONEER FARM JOURNAL

*I know of no pursuit in which more real and important service can be rendered to any Country, than by improving its agriculture* [1]

THE PIONEER AND FATHER of American farm journalism was the *American Farmer,* established by John Stuart Skinner on April 2, 1819, in the city of Baltimore.[2] In the first num-

[1] Washington to Sir John Sinclair, July 20, 1794. *Letters from George Washington to Sir John Sinclair,* p. 22.

[2] The first strictly agricultural journal in the United States was the short-lived *Agricultural Museum,* the first number of which appeared on July 4, 1810. It was published at Georgetown, District of Columbia, under the editorship of the Reverend David Wiley and was the organ of the Columbian Agricultural Society. The editor was a man of great public spirit, energy, and remarkable versatility. While he was principal and librarian of the Columbian Academy in Georgetown, he was at the same time "the superintendent of a turnpike, the postmaster, a merchant, a miller, a preacher, the editor of an agricultural paper, and the secretary of an agricultural society." He became mayor of Georgetown the year after he began the *Agricultural Museum.* Copies of this rare journal are available in the U. S. Department of Agriculture Library, including May, 1812, which, as far as is known, was the last number issued. The journal was a small octavo in size, each number containing 32 pages. During the first year, it was issued semi-monthly, after which it was converted into a monthly. The subscription price was $2.50. No advertisements were carried and circulation figures are not available. In general content, the *Agricultural Museum* was not unlike the periodicals that came later; however, there is no evidence that it influ-

ber, which was held over a day lest the paper be taken for an April Fool joke, Skinner outlined his purpose:

> The great aim, and the chief pride, of the *American Farmer*, will be, to collect information from every source, on every branch of Husbandry, thus to enable the reader to study the various systems which experience has proved to be the best, under given circumstances.[3]

This publication was adapted to all varieties of soil and climate in the United States.

The *American Farmer*, like hundreds of journals that followed it, reflected the personality, experience, and interests of the editor. Skinner was born February 22, 1788, on a farm in Maryland, where he spent his boyhood. Later he was trained for law and admitted to the bar, but a new sphere of action was opened to him when the War of 1812 broke out. President Madison made him inspector of European mail at Annapolis and a short time later appointed him agent for prisoners of war. In 1813 his headquarters were removed to Baltimore, where he was soon commissioned a purser in the navy.[4] During these years of government service, Skinner made many warm friends among the British and American naval officers; friends who were later very helpful to him in his work as editor.[5] In 1816 he became postmaster at Baltimore, a position he retained until 1837.

---

enced them in any way. For a detailed account of this journal, see Claribel R. Barnett, "The Agricultural Museum; An Early American Agricultural Periodical," *Agricultural History*, II (April, 1928), 99 ff.

[3] *American Farmer*, I (April 2, 1819), 5.

[4] At the approach of the British forces upon Washington in August, 1814, it was young Skinner who played the part of Paul Revere by riding 90 miles in the night to announce their coming. Poore, "Biographical Notice of John S. Skinner," p. 3.

[5] While in Baltimore, September 13–14, 1814, Skinner, accompanied by Francis Scott Key, visited Admiral Cockburn to negotiate the exchange of

# AMERICAN FARMER.

RURAL ECONOMY, INTERNAL IMPROVEMENTS, NEWS, PRICES CURRENT.

"O fortunatos nimium sua si bona norint
"Agricolas. . . . . . Virg.

Vol. I. BALTIMORE, FRIDAY, APRIL 2, 1819. Num. 1.


## AGRICULTURE.

### The Ruta Baga or Swedish Turnip.

☞ THE high commendations bestowed upon the *Ruta Baga*, and the decided preference given to it over other roots and vegetables, as food for live stock, by Mr. Barney, of Delaware, (the owner of the mammoth oxen lately slaughtered in this market) will naturally beget an anxiety to know more of its peculiar qualities, and to learn the best mode of cultivating and preserving it.

All these objects will be best accomplished by the perusal of a Treatise lately written by the celebrated Mr. Cobbett, whose pen communicates new life and originality to the most exhausted subjects. We have, therefore, determined to offer to our readers, all that he has said on this matter, as well in his "first" as in his "second part of a year's residence in the United States;" both of which little volumes will be found, especially his notices of agriculture, highly entertaining and instructive.

The length of his remarks, and the near approach of the season for sowing the seed, induce us to commence the publication of his Treatise on Ruta Baga in the present number. It will be continued in each one, successively, until finished.

Mr. Barney assures us, that, but for the liberal use of the Ruta Baga, in feeding the two remarkable oxen, lately sold by him in this market, a much greater quantity of Indian meal would have been consumed; and, moreover, that without the Ruta Baga, which helped to constitute that variety so necessary to sustain a *constant appetite*, it is even doubtful whether they could have been made to attain to such *extraordinary* excellence in the weight and quality of the meat. He fully concurs with Mr. Cobbett, in estimating potatoes, and other vegetables, as altogether insignificant, in comparison with the Ruta Baga; and observes, that besides their intrinsically nutritious quality, they act finely as a medicine, counteracting the astringent effect which would result from a more exclusive use of *dry food*; all which we must confess, appears very natural and worthy of consideration.

FROM COBBETT'S YEAR'S RESIDENCE.

### RUTA BAGA.

*Culture, mode of preserving, and uses of the Ruta Baga, sometimes called the Russia, and sometimes the Swedish Turnip.*

DESCRIPTION OF THE PLANT.

IT is my intention, as notified in the public papers, to put into print an account of all the experiments which I have made, and shall make, in Farming and in Gardening upon this Island. I several years ago, long before tyranny showed its present horrid front in England, formed the design of sending out, to be published in this country, a treatise on the cultivation of the root and green crops, as cattle, sheep and hog food. This design was suggested by the reading of the following passage in Mr. Chancellor Livingston's *Essay on Sheep*, which I received in 1812. After having stated the most proper means to be employed in order to keep sheep and lambs during the winter months, he adds:—"Having brought our flocks through the winter, we now come to the most critical season, that is the latter end of March and the month of April. At this time the ground being bare, the sheep will refuse to eat their hay, while the scanty picking of grass, and its purgative quality, will disable them from taking the nourishment that is necessary to keep them up. If they fall away, their wool will be injured, and the growth of their lambs will be stopped, and even many of the old sheep will be carried away by the dysentary. *To provide food for this season is very difficult. Turnip* and *Cabbage* will rot, and bran they will not eat after having been fed on it during the winter. *Potatoes*, however, and the *Swedish Turnips*, called *Ruta Baga*, may be usefully applied at this time, and so, I think, might *Parsnips* and *Carrots*. But, as few of us are in the habit of cultivating these plants to the extent which is necessary for the support of a large flock, we must *seek resources more within our reach*." And then the Chancellor proceeds to recommend the leaving the *second growth of clover uncut*, in order to produce early shoots from sheltered buds for the sheep to eat until the coming of the natural grass and the general pasture.

I was much surprised at reading this passage; having observed, when I lived in Pennsylvania, how prodigiously the root crops of every kind flourished and succeeded with only common skill and care; and, in 1815, having by that time had many crops of Ruta Baga exceeding *thirty tons*, or about *one thousand five hundred heap'd bushels to the acre*, at Botley, I formed the design of sending out to America a treatise on the culture and uses of that root, which, I was perfectly well convinced, could be raised with more ease here than in England, and, that it might be easily preserved during the whole year, if necessary, I had proved in many cases.

If Mr. Chancellor Livingston, whose public spirit is manifested fully in his excellent little work, which he modestly calls an *Essay*, could see my Ewes and Lambs and Hogs, and Cattle, at this "*critical season*" (I write on the 27th of March), with more Ruta Baga at their command than they have months to employ on it; if he could see me, who am on a poor and exhausted piece of land, and who found it covered with weeds and brambles in the month of June last; *who* found no manure and have bought none; if he could see me over-stocked, not with mouths, but with food, owing to a little care in the cultivation of this invaluable Root, he would, I am sure, have a reason to be convinced, that, if any farmer in the United States is in want of food at this pinching season of the year, the fault is neither in the soil nor in the climate.

It is, therefore, of my mode of cultivating this Root in this Island, that I mean, at present to treat; to which matter I shall add, in another Part of my work, an account of my experiments as to the Mangle Wurzel, or Scarcity Root; though, as will be seen, I deem that root, except in particular cases, of very inferior importance. The Parsnip, the Carrot, the Cabbage, are all excellent in their kind and in their uses; but, as to these, I have not yet made, upon a scale sufficiently large here, such experiments as would warrant me in speaking with any great degree of confidence. Of these and other matters I propose to treat in a future Part, which I shall probably, publish towards the latter end of the present year.

The *Ruta Baga* is a sort of Turnip well known in the state of New-York; where, under the name of the *Russia* Turnip, it is used for the table from February to July. But, as it may be more of a stranger in other parts of the country, it seems necessary to give it enough of description to enable the reader to distinguish it from every other sort of Turnip.

The leaf of every other sort of Turnip is of a *yellowish* green, while the leaf of the Ruta Baga is of a *bluish green*, like the green peas when of nearly their full size, or like the green of a young and thrifty early Yorkshire cabbage. Hence it is, I suppose, that some persons have called it the *Cabbage-Turnip*. But the characteristics the most decidedly distinctive are these: that the outside of the *bulb* of the Ruta Baga is of a greenish hue mixed, toward the top, with a colour bordering on a red; and, that the inside of the bulb, if the sort be true and pure, is of a *deep yellow*, nearly as deep as that of gold.

MODE OF SAVING AND PRESERVING THE SEED.

This is rather a nice business, and should be, by no means, executed in a negligent manner. For, on the well attending to this, much of the success depends; and, it is quite surprising how great losses are, in the end, frequently sustained by the saving, in this part of the business, of an hour's labour or attention. I, one year, lost more than half of what would have been an immense crop, by a mere piece of negligence in my bailiff as to the seed, and I caused a similar loss to a gentleman in Berkshire, who had his seed from the same parcel that mine was taken, and who had sent many miles for it, in order to have *the best in the world*.

The Ruta Baga is apt to *degenerate*, if the seed be not saved with care. We, in England, select the plants to be saved with seed. We examine well to find out those that run least into *neck* and *green*. We reject all such as approach at all towards a *whitish colour*, or which are even of a *greenish colour, towards the neck*, where there ought to be a little *reddish cast*.

Having selected the plants with great care, we take them up out of the place where they have grown, and plant them in a plot distant from every thing of the Turnip or Cabbage kind which is to bear seed. In this Island I am now, at this time, planting mine for seed (27th March), taking all our English precautions. It is probable, that they would do very well, if taken out of a *heap* to be transplanted, if well selected; but, lest this should not do well, I have kept my selected plants all the winter in the ground in my garden well covered with corn stalks and leaves from the trees; and, indeed, this is so very little a matter

TITLE PAGE

Conscious of the worn-out state of much of the land in Maryland and stimulated by the writings of John Taylor of Caroline, John Stuart Skinner [6] decided in 1819 to establish the *American Farmer*, and to use this organ to disseminate the best methods of farm practice. Aside from the fact that he spent his boyhood on a Maryland farm, there was nothing in his education that prepared him directly for the field in which he later distinguished himself.[7] Without materials or resources, except his heartfelt devotion to farming, the paper was launched.[8] It was not undertaken, Skinner said, with a view to private emolument, so much as with the wish to employ his leisure hours in a way that would render real service to his fellow citizens.[9] It soon became the principal medium of expression for those who took an active interest in agricultural improvement.

The journal was a weekly of eight quarto pages; the annual subscription rate was four dollars in advance.[10] Its ambitious

some prisoners. The British were on the point of attacking Baltimore and refused to allow the Americans to return until the battle was over. Skinner and Key watched the bombardment, and after a night of agonizing suspense, were overjoyed in the morning to see the flag still flying over Fort McHenry. After their release they went to the Fountain Inn where Key wrote "The Star Spangled Banner." Skinner, recognizing its beauty, obtained Key's permission to have it published and the song soon reëchoed throughout the land. Poore, *op. cit.*, p. 4.

[6] *Dictionary of American Biography*, XVII, 199 ff.

[7] Ogilvie, *Pioneer Agricultural Journalists*, p. 4.

Skinner later established the *American Turf Register and Sporting Magazine* (1829), the *Farmers' Library and Monthly Journal of Agriculture* (1845), the *Plough, the Loom, and the Anvil* (1848). He edited numerous agricultural books and contributed frequently to newspapers and agricultural journals. All this was done, for the most part, outside his working hours as a public official.

[8] *American Farmer*, III (March 30, 1821), 7.

[9] *Ibid.*, I (March 24, 1820), 415.

[10] As a result of the great number of copies lost and mutilated in the mail during the first year, the editor was willing, in the second year, to guarantee

heading stated that it would deal with "Rural Economy, Internal Improvements, News, Prices Current." Since a motto, at that time, was considered an indispensable part of a newspaper, Skinner solicited the aid of a clergyman well versed in classical lore. "O fortunatos nimium sua si bona norint agricolas" from Virgil was suggested and immediately approved.[11] Four months following the *Farmer's* initial appearance, a refund was offered to any subscriber who had "been disappointed in his expectations as to the value of the work." [12]

The editor devoted every spare moment to the columns of the *American Farmer*. He worked in the evening, his days being occupied with postal duties.[13] A phalanx of the most enlightened agriculturists of the country assisted him—John Taylor of Caroline, Thomas Jefferson, James Madison, John C. Calhoun, Timothy Pickering, Henry Clay, James M. Garnett, General H. A. S. Dearborn, and General John Armstrong. Among foreign contributors were Sir John Sinclair, Thomas W. Coke, Captain Basil Hall [14] of the British Navy, and General Lafayette.[15] "Purser Skinner" regarded the of-

delivery in good condition to those who paid $5 in advance, an arrangement most subscribers accepted.

[11] *Plough, the Loom, and the Anvil,* VII (July, 1854), 5. Most of the subsequent agricultural journals likewise adopted mottoes. However, quotations in English rather than Latin were customary.

[12] *American Farmer,* I (July 30, 1819), 142.

[13] *Ibid.*, II (March 31, 1820), 7.

[14] Captain Hall talked with Skinner while in Baltimore in 1828 and received his promise of coöperation in collecting information relating to America for his forthcoming volume, *Travels in North America in the Years 1827–1828* (1829). Hall later availed himself of Skinner's generosity and at the same time urged the editor to command his services in England. The Captain was a subscriber to the *Farmer*. *American Farmer,* X (Dec. 5, 1828), 303.

[15] The editor made a very favorable impression upon General Lafayette during his visit to this country in 1824 and they became close friends. Skinner served as Lafayette's business agent in the United States, a trust he retained

ficers of the United States Navy foremost among his friends and correspondents. He claimed that of "all classes, having regard to their numbers . . . the greatest *proportion* of subscribers to the Farmer," singular as it might appear, consisted of officers of the navy.[16] Frequently, they sent him valuable communications from all corners of the earth and brought him agricultural books, stock, poultry, and seeds. Especially interested in agricultural matters were Hull, Chauncey, Hambleton, Stewart, Bainbridge, Rogers, Jones, Porter, and Perry. Practical contributions from the "dirt farmer," difficult as they were to obtain, were always eagerly sought by the editor. "It is already known [he wrote] that our *first* wish is to communicate the *experience* of the sun-browned practical Farmer, in preference to the fine spun lucubrations of the Philosopher, or the calculations, however profound, of the political oeconomist." [17]

Skinner, in his first volume, warned the subscribers that there was no easy road to an understanding of the science of agriculture.

> The readers of this journal must not expect to find its pages filled with light ephemeral speculations and essays, cooked up to satisfy the ever-craving appetites of news-mongers and politicians.—The science of agriculture is difficult—the very nature of its operations laborious, and enduring through the whole year; it is to be expected therefore, that essays treating of the vegetation, growth and management of a single object will often be long and tedious.

---

through life. General Lafayette became a subscriber to the *American Farmer* in 1824 and carried the preceding volumes with him to France. He was an active correspondent of the journal in the years that followed. Skinner's eldest son, Frederick, later attended school near Paris, remaining under the paternal eye of the General with whom he spent his vacations at La Grange. *American Farmer*, VII (Sept. 23, 1825), 216; VIII (March 31, 1826), 15; IX (March 7, 1828), 408; X (Sept. 19, 1828), 215.

[16] *Ibid.*, IV (Aug. 16, 1822), 161. [17] *Ibid.*, IV (May 3, 1822), 47.

He suggested that the farmer have his paper "read *out,*" in the evening by his sons and daughters while he pondered over the content.[18]

Editorials and articles on crop rotation, deep and horizontal plowing, new fertilizers, grasses, hedging, care of cattle in winter, agricultural chemistry, horticulture, and drainage are a few of the numerous subjects that received attention. The editor's travels about the country furnished his readers with first hand information in regard to the procedure of other farmers.[19] He copied wholly or in part the more important American and English agricultural treatises and was particularly proud that the *Farmer* was the first to print Edmund Ruffin's famous *Essay on Calcareous Manures.*[20] Skinner urged the farmers to read books that would better qualify them for their occupation, and emphasized the need for the establishment of agricultural schools and the appointment of professors of agriculture to the colleges.[21] Livestock breeding, one of his hobbies, was naturally emphasized. In coöperation with the officials of the Maryland Agricultural Society he established in 1821 a stock farm about four miles from Baltimore. On this estate, known as "Maryland Tavern," they attempted the systematic improvement of various breeds of domestic animals. The fairs of the Maryland Agricultural Society were held here.[22]

From the beginning, the *American Farmer* received the en-

[18] *Ibid.*, I (Nov. 12, 1819), 265.

[19] Traveling editors and correspondents added an interesting feature to many of the later journals.

[20] *American Farmer*, III (Dec. 28, 1821), 313.

[21] The editor frequently evaluated new and important books as they appeared. Following this precedent, the journals later inaugurated book-review sections, which became an important feature.

[22] *American Farmer*, II (March 16, 1821), 403, 408; II (March 23, 1821), 416.

thusiastic coöperation of agricultural societies throughout the country.[23] After the first year of publication, the Agricultural Society of Albermarle, Virginia, presented the first volume of the *Farmer* to each of its members; [24] while, the next year, the Maryland Agricultural Society unanimously voted that:

> The American Farmer is . . . justly entitled to the patronage and support of every farmer and planter, and . . . its Editor deserves our approbation, for his judicious and zealous efforts to advance the interests of agriculture.[25]

The Agricultural Society of South Carolina asserted that it could not more effectually promote agricultural improvement, the object for which it existed, than by furthering the interests of the *American Farmer*.[26] Skinner was personally honored by many societies during his life; indeed, there were scarcely any of importance in the United States and Europe that did not confer upon him honorary membership. The *Farmer* served as an organ for many of these societies, publishing extensive accounts of their cattle shows and fairs, including the premium awards, and agricultural addresses. Prize essays on designated subjects which often appeared in the journal were also sponsored by them. The editor took an active interest in societies and fairs, especially in the Maryland society, of which he was an official. Frequently, he delivered addresses at the Fall exhibitions and was always an enthusiastic exhibitor of his fine cattle.

[23] The societies and journals worked in close harmony throughout the period before the Civil War. It is difficult to overemphasize the importance of the coöperation of these agents for agricultural advancement.

[24] Extract from the proceedings of the Agricultural Society of Albermarle, Virginia, May 8, 1820. *American Farmer*, III (1821–1822), xxii.

[25] Extract from the proceedings of the Maryland Agricultural Society, Baltimore, June 8, 1821. *Ibid.*

[26] Notice published in South Carolina papers. *Ibid.*

Skinner made every effort to encourage new and useful agricultural inventions. He desired to make known discoveries relating to agriculture and domestic economy through the *American Farmer* and was willing to illustrate them with cuts, and "even pay the expense of engraving." [27] The farmer was encouraged to take advantage of these labor-saving devices. During the twenties, for example, the cast-iron plow and the wooden horse rake received enthusiastic support from the paper.[28] The editor was often guided in his recommendations by reports from the special committees at the cattle fairs, appointed to judge and evaluate new devices.

The introduction of new plants and seeds also received a large share of Skinner's attention. In an attempt to popularize rutabaga, he printed upon the first page of the first number an essay on the vegetable by the English agricultural authority, William Cobbett, and followed it with numerous articles and correspondence upon the same subject.[29] One farmer, while admitting that "We, unlettered clod-hoppers do not understand botanical terms," was nevertheless willing after reading Cobbett's articles, to experiment with the new vegetable and report to the *American Farmer*.[30] A few years later the journal called attention to the astonishing fertilizing properties of guano and small amounts were distributed from the office. Guano had

[27] *American Farmer*, I (April 2, 1819), 6.

[28] *Ibid.*, II (Dec. 22, 1820), 312.

[29] William Cobbett (1763–1835), an English "radical" and political refugee in the United States, was at this time farming on Long Island. Immediately after his article was published in the *American Farmer*, Cobbett sent his servant, James Hammerton, post haste to Baltimore with a supply of English seeds (including rutabaga) and also copies of his *Year's Residence* and *English Grammar*. These were placed on sale by Hammerton, who appeared daily in the "Market Place." An advertisement to this effect was carried in the *Farmer*. *Dictionary of American Biography*, IV, 248; *American Farmer*, I (April 16, 1819), 23.

[30] *American Farmer*, I (May 21, 1819), 61.

been brought to Skinner from South America by Midshipman Bland of the "Franklin." [31] While this remarkable fertilizer was not at once widely used, it was later a fundamental factor in the agricultural revival in Virginia and Maryland.[32] Indeed, one of the major activities of the editor's office was the gratuitous distribution of rare seeds from all parts of the world. He found this task rather discouraging, for—while these seeds were sent to every state in the Union—few farmers were willing to report the results of their experiments for publication.[33]

One of the most effectual methods for aiding the farmer was the adoption of a system of question and answer through the medium of the *Farmer*.[34] "I wish you would obtain the most approved mode of getting out and cleaning clover seed, and publish it," wrote an inquirer.[35] Another said, "An accurate comparison of the relative merits of the different Wheat Machines now in use, in various parts of the United States, is very desirable—and we beg the favor of gentlemen acquainted with the different kinds, to give the information." [36] Requests of this sort generally elicited replies from farmers in many parts of the United States. Skinner did not hesitate to seek information directly from the authorities. Thus in 1820 Thomas Jefferson received an appeal for advice on certain seeds, which he gladly gave.[37] A little later the editor wrote Admiral David Porter:

Being under the impression that "the proper season for felling timber, with a view to its durability," has been made, within the

[31] *Ibid.*, VI (Dec. 24, 1824), 316.

[32] Craven, *Soil Exhaustion as a Factor in the Agricultural History of Virginia and Maryland, 1606–1860*, p. 150.

[33] *American Farmer*, IX (Feb. 22, 1828), 392.

[34] *Ibid.*, VI (July 9, 1824), 128; VII (Sept. 23, 1825), 216.

[35] *Ibid.*, I (Oct. 15, 1819), 231.

[36] *Ibid.*, I (Feb. 4, 1820), 359.

[37] *Ibid.*, II (May 26, 1820), 67.

sphere of your official duties . . . I am induced to ask the favor of you to furnish me with communications on that subject, for publication, in my Agricultural Journal.[38]

The Admiral responded by giving the results of the investigation made by the Navy Department. In answer to a request from the editor, Henry Clay gave "with great pleasure" an account of his experience with imported English cattle.[39]

Since the *American Farmer* was intended to appeal to the entire family, many subjects besides agriculture naturally appeared in its columns. This so-called extraneous matter, however, sometimes brought forth criticism. On the other hand "An Old Farmer" demanded still more diversified fare, objecting to so much plowing, sowing, reaping, and mowing.[40] Skinner attempted to satisfy all his subscribers and solicited "hints from any quarter." [41] To his mind, the ideal set-up consisted of about "one half, or four pages, devoted to practical Agriculture; the remainder to Internal Improvements, Rural and Domestic Economy; selections for housekeepers and female readers, and Natural History and Rural Sports." [42]

Internal improvements were dear to Skinner's heart for he felt that roads, canals, and railroads not only contributed largely to the national wealth, but expedited transportation and lowered the costs, which influenced agricultural prosperity. Numerous editorials and articles illustrated with cuts emphasized the importance of this subject.[43]

[38] *Ibid.*, III (Sept. 6, 1821), 185. [39] *Ibid.*, IV (Oct. 4, 1822), 223.

[40] *Ibid.*, VIII (Aug. 11, 1826), 168.

[41] *Ibid.*, V (Aug. 15, 1823), 168. [42] *Ibid.*, VIII (Feb. 16, 1827), 384.

[43] In 1823 Skinner "received intimations from distant subscribers," who were "friends of exceeding good judgment" that too much space was devoted to internal improvements. He promised to suspend for a time articles on this subject since it was the editor's "inclination, as well as obvious duty to consult the wishes and interests" of the subscribers generally. *American Farmer*, V (Aug. 15, 1823), 168.

Miscellaneous items including poems, domestic and foreign news, an occasional joke, a great number of "cures" for all manner of disease, articles on health, temperance, amusements, birth and death notices, filled the pages of the *American Farmer*.

Two special departments should be noted. Articles on rural sports appeared regularly, reflecting the editor's love of the out-of-doors; in 1825 a column headed "Sporting Olio" was introduced. "How much better [wrote Skinner] to repair to the fields, the woods, or to the neighbouring streams, at the close of a week of hard study or sedentary labour, and there spend the afternoon in gunning, fishing, swimming; bowling at ninepins, pitching quoits, etc., according to one's fancy and the season, than to abuse whole days in *militia mustering!* frequenting gaming-houses, whiskey drinking, etc."[44] The other feature, early introduced, was the "Ladies' Department."[45] Deploring the fact that a great number of publications in the United States were "dedicated to the amusement and business of *men*," it was decided to present in this department "subjects for the rational amusement and instruction" of feminine readers, more especially "matrons."[46] These departments included, in the main, articles copied from a great variety of sources, very little original material being available.

---

In 1825 the Maryland Society for Internal Improvements invited him to edit and publish their quarterly journal, *American Journal of Internal Improvements*. Fearful that this new interest would interfere with his postal duties, he declined. *American Farmer*, VII (April 15, 1825), 31; VII (April 29, 1825), 46.

[44] *American Farmer*, VIII (May 5, 1826), 54.

[45] The famous departments, conducted for and by women and including articles by talented special writers, did not appear until about the middle of the century.

[46] *American Farmer*, VII (Dec. 30, 1825), 328; VII (March 3, 1826), 397.

Editorial wisdom was manifested repeatedly. In the first volume the troublesome topic of politics was wisely forbidden.

Once for all, then . . . not a word of party politics will ever be allowed to enter its columns.[47]

Contributors who insisted on political references or attempted to draw out the editor received an answer such as appeared in the *Farmer* on September 3, 1819:

"CATO" surprises us, after the repeated assurances that this paper will not intermeddle in party politics.—Whatever may be our private inclinations—it is not *here* that they can or will be manifested.[48]

Other sources of discord were ever lurking in discussions innocently inaugurated. In 1825 a dispute on the relative excellence of various breeds of cattle was carried to such violent extremes that Skinner felt compelled to end the controversy. One of the disputants from whom the *American Farmer* had received many good offices and communications withdrew his name from the subscription list in a fit of anger.

We care not a farthing [wrote the editor] for the loss of one name, or one dozen, or ten dozen names, in comparison with the mortification of our journal being the cause of exciting any ill blood, or wounding any gentleman's feelings. The American Farmer was not projected with any regard to the politicks, or merely to the number of patrons it might receive—we value the length of its subscription list at a farthing rushlight, when put in competition with the satisfaction and the honour of dispensing by the light of its columns, *solid and lasting benefit to the best interests and to the most virtuous class of society*.[49]

---

[47] *Ibid.*, I (May 14, 1819), 55. This stand was generally taken by the journals that followed. Failure to live up to this rule led to serious difficulties in many instances.

[48] *Ibid.*, I (Sept. 3, 1819), 183.

[49] *Ibid.*, VII (April 15, 1825), 31.

In order to avoid personalities in the future, a strict censorship was exercised to eliminate material causing these wrangling debates.[50]

Skinner was very proud and jealous of the *American Farmer*'s good name. On one occasion through a mistake on the part of the compositor "some indecorous extracts" from an old pamphlet were allowed to enter its columns. When the editor saw the blunder, rather than offer explanations and professions of regret, he immediately ordered the numbers sent out to be returned and the whole edition destroyed. A new edition omitting the offending extracts then went to press, costing between seventy and eighty dollars.[51]

Subscribers were urged to use the periodical as a reference work and to have a complete set on hand at all times. Inducements, offered annually, to persuade them to bind the year's copies into a single volume were enthusiastically received by ninety percent of the readers.[52] Back numbers of the paper were always on sale at the office and frequently available through agents.

The business management of the *American Farmer* was wisely directed by Skinner himself. Advertisements brought in a small return; however, space was limited and such material had to be of interest to the farmer. Each insertion cost one dollar per square. In an attempt to increase the circulation, postmasters were asked to accept subscriptions, retaining ten percent of the money collected.[53] Regular agents were also employed. In 1823 nine were stationed in various parts of

[50] *Ibid.*, VIII (Sept. 8, 1826), 200. [51] *Ibid.*, II (May 12, 1820), 55.

[52] *Ibid.*, IV (June 28, 1822), 112. Binding and preserving these old periodicals was a common custom during the period preceding the Civil War.

[53] Many postmasters acted in this capacity. Other journals later followed this precedent.

the United States while one traveled about in the South.[54] Frequently sample copies were mailed to nonsubscribers.

It is difficult to estimate the profits of the journal. Skinner insisted that his expenses were high and that frequently he paid forty or fifty dollars for his engravings.[55] It is also true that the farmer was notoriously slow in making subscription payments and very often never paid them.[56] Although circulation figures are difficult to obtain, we do know that at the close of the third year, there were at least 1,500 subscribers.[57] Many times Skinner expressed pleasure in his large subscription list and claimed circulation in every state and territory in the Union.[58] In the tenth year 300 new subscribers were added.[59] The best evidence, however, of the real status of the magazine is the fact that Skinner was able to dispose of it for $20,000, a sum huge for those days.[60]

In 1829, this leader again proved himself a pioneer, when he inaugurated the *American Turf Register and Sporting Magazine*, the first of its type in the United States.[61] Fearful that his dual editorship might be construed by some, as involving neglect of his official duties as postmaster, he sold the *American Farmer* in 1830 to I. Irvine Hitchcock & Co., publishers.[62] At this time Gideon B. Smith was engaged as editor, a position

[54] *American Farmer*, IV (Feb. 14, 1823), 376.

[55] *Ibid.*, I (Nov. 26, 1819), 281.

[56] This problem presented itself to all the editors and accounts for many of their financial difficulties.

[57] *American Farmer*, III (March 22, 1822), 416.

[58] *Ibid.*, VIII (July 28, 1826), 152. [59] *Ibid.*, X (March 13, 1829), 415.

[60] Poore, *op. cit.*, p. 15. The *American Farmer* was one of the most profitable journals published before the Civil War. *Plough, the Loom, and the Anvil*, IV (Dec., 1851), 349.

[61] This magazine was more influential than any other factor of its day in improving the breed of American horses. *Dictionary of American Biography*, XVII, 199 ff.

[62] *American Farmer*, XII (Sept. 3, 1830), 198.

he held until October, 1833, when Hitchcock took over his duties.[63] Although the journal continued its high standard, there was evidence of competition from many agricultural periodicals springing up in all parts of the country.

As the American Farmer [wrote the editor early in 1833] was from the beginning devoted to *American* agriculture, embracing all sections of our extensive country, and the interests of all in its range, it was of course neither southern, nor northern, nor eastern, nor western, in its designs or partialities. The effect, therefore, of the more local publications, has been to circumscribe the patronage of the American Farmer, as each successively became established, and we are now left with a list of subscribers barely sufficient to pay expenses.[64]

On March 7, 1834, it was necessary to discontinue publication, and the editor briefly wrote:

In regard to the *wherefore* of this occurrence, suffice it to say, that it has been caused by one of those reverses of fortune which await all men, and overtake many.[65]

---

[63] *Ibid.*, XV (Oct. 4, 1833), 233. [64] *Ibid.*, XV (March 15, 1833), 1.

[65] *Farmer and Gardener*, I (May 9, 1834), 1.

Two months after the last number of the *American Farmer* had been issued (March, 1834), I. I. Hitchcock, its editor, established the *Farmer and Gardener, and Live-Stock Breeder and Manager.* He considered this the successor to the *American Farmer.* Hitchcock edited the first volume (1834–35) but sold the paper on April 28, 1835, to Sinclair and Moore. E. P. Roberts edited Volumes II, III, IV and V. (With Volume IV, the proprietorship was transferred to Roberts, Sands, and Neilson, and with Volume V to E. P. Roberts and Samuel Sands.) On May 29, 1839, (Vol. V) the title was again changed, this time to the *American Farmer, and Spirit of the Agricultural Journals of the Day,* with Samuel Sands as publisher and John S. Skinner again assuming the editorial duties. Skinner retained this post until 1841, when he became assistant postmaster general after which Samuel Sands became the editor. *American Farmer,* n.s., III (Aug. 18, 1841), 97.

For the next fifty years, Samuel Sands served as editor and intermittently as proprietor. In the latter part of this period, his son, William B. Sands, also served as editor. Frank Luther Mott, *A History of American Magazines, 1741–1850,* p. 154.

The mutative character of many of the agricultural periodicals is exem-

Without doubt, the *American Farmer*, during its fifteen years of existence, rendered yeoman's service in the many directions already indicated. While the magazine gave a highly beneficial impulse to American agriculture, it should be remembered that the enthusiasm for reform came from the gentlemen farmers and they in turn received the chief benefits. Too much in the way of actual improvement for the mass of common farmers should not be claimed. When publication was suspended in 1834, at least fifteen agricultural journals with similar objectives, had been launched.[66] Many were inaugurated during the period only to succumb after a short existence. Notwithstanding all the pressure for reform, the "dirt farmer" was difficult to reach, for he evinced general hostility toward anything which required a deviation from the old beaten path. These farmers were "unbelievers in book farming." It is also true that a great many of them, from a purely economic point of view, felt financially unable to experiment or to abandon methods that had proved tolerably satisfactory. In 1831, the editor of the *American Farmer* said that they "will neither take an agricultural paper, read it when given them, nor believe in its contents if by chance they hear it read." In fact it was his studied opinion that not one farmer in fifty regularly subscribed to a farm journal.[67] The editors contended with the prejudice of ignorance, and the obstinate and blind perseverance in bad habits. Although the great masses remained almost untouched by these leaders of reform, nevertheless a beginning had been made; the seed had been sown.

---

plified by the *American Farmer*. While it continued until 1897, it passed through twelve series in the course of which it was modified in title, size, price, and frequency of publication. During the period it changed editors and proprietors; even the place of publication was moved from Baltimore to Washington.

[66] *Cultivator*, I (May, 1834), 36.

[67] *American Farmer*, XII (Jan. 21, 1831), 359–60.

# Chapter III

## PROGRAM AND POLICIES OF THE FARM PRESS

*But underlying them [improvements in agriculture] is the agricultural press, stimulating the people to form Farmers' Clubs, and State and County Societies, for the exhibition of all farm products; and then multiplying the power of these societies for good a hundred-fold by spreading their proceedings before the community; scattering broadcast the experiments and the teachings of hundreds of our best practical farmers; publishing to the world accounts of the best stock, the best tools, and the best fruits and vegetables.*[1]

WIDE GENERALIZATIONS may be made with respect to the program and policies of the farm press. This is true despite the fact that these periodicals vary greatly, due to the particular interests and background of the editors as well as the needs of the areas in which they circulated.

The journals from their earliest days urged the farmers to accept the elementary principles advocated by agricultural leaders. The advantages of crop rotation, deep and horizontal plowing, and variety in culture were laid before their readers,[2]

[1] Editorial in *American Agriculturist*, XVI (Nov., 1857), 244.

[2] One illustration will serve to show how the farmers themselves, through

who were further advised to sustain and strengthen the soil by manuring, draining, good tillage, and root cultivation. They were admonished not to accumulate more land than could be competently cultivated, and plundering of the soil was deplored. Businesslike methods were advocated and the wisdom of keeping farm records and account books was repeatedly pointed out.[3] Ultra-conservatism and prejudice against "book-farming" were to be laid aside; the farmers were to advance with other professions in the march of progress. For further help, agricultural reading was highly recommended, the journals doing their part by reviewing relevant American and English books, and constantly pointing out the utility of a farmer's private library.[4]

---

practical experiments, often drove home with telling effects the importance of certain farm practices recommended by the editors. The relative value of deep and shallow plowing was often debated in the press. A writer in the *Prairie Farmer,* after describing how he plowed one half of his field to a depth of four inches and the other half to seven inches before planting his wheat, continues: "Now for the result; after plowing, all parts of the field received the same treatment. In cutting the wheat, I was careful to make a division and the shallow-plowed half yielded just forty-three bushels, while the deep-plowed half yielded sixty-one bushels—or a difference of six bushels per acre. Hereafter I am for deep plowing." *Prairie Farmer,* XVIII (April 1, 1858), 105.

[3] Because of the "universal neglect of book-keeping amongst agriculturists," model forms were frequently presented in the journals. Thomas Affleck (a Mississippi planter, formerly editor of the *Western Farmer and Gardener* and widely known as a contributor to the farm press, particularly in the South), designed a journal book in 1847 for use in his section. More than 5,000 of these plantation record and account books were printed by 1860 and in wide use by overseers and planters throughout the South. Wendell Holmes Stephenson, "A Quarter-Century of a Mississippi Plantation," p. 356. See *Southern Cultivator* XIII (1855), 75–76.

[4] During the twenties, the leadership of such men as John Taylor, Edmund Ruffin, James M. Garnett, Timothy Pickering, and Jesse Buel was quite generally recognized by editors of the agricultural press. Naturally, these leaders were not in complete agreement. For example, the details of crop rotation varied greatly within a given area and the authorities modified their procedure from time to time.

American agriculture made steady and substantial advances even before the Federal Department of Agriculture was established and before the land-grant colleges and great experiment stations were in operation. Progress was largely achieved through individual initiative and experiments, which were widely publicized by the agricultural press. Take, for example, the work of John Johnston, a scientific farmer, and the father of tile drainage in the United States.[5]

A native of Scotland, Johnston came to this country in 1821, purchased 112 acres near Geneva, New York, and settled down to farming. His land, consisting of heavy clay, was cold and wet and in a worn-out condition. He remembered that tiles had been used for draining in Scotland and decided to experiment. In 1835, having procured tile made by hand after a pattern obtained from his native land, Johnston began the process of drainage.[6] While this independent experimentation was characteristic of many agricultural leaders, so was the ridicule of his neighbors who said, "John Johnston is gone crazy—he is burying *crockery* in the ground." [7] Nevertheless, it was soon evident that he was raising larger crops than they were. In 1848 his neighbor John Delafield, imported from

[5] Bidwell and Falconer, *History of Agriculture in the Northern United States*, p. 319. Johnston was born April 11, 1791; he died November 24, 1880.

No life of John Johnston has been written. There is a real need for full length biographies of many rural leaders in the pre-Civil War period. The following list includes only a portion of this important group: John S. Skinner, Jesse Buel, Andrew Jackson Downing, Solon Robinson, Thomas Affleck, Martin W. Philips, John S. Wright, Daniel Lee, and Benjamin P. Johnson.

[6] Hedrick, *A History of Agriculture in the State of New York*, p. 349; *Dictionary of American Biography*, X, 142.

The draining of wet, heavy lands had been practiced for many years in this country and the journals from the first advocated this procedure. Hedrick, *op. cit.*, p. 350; *American Farmer*, II (July 21, 1820), 131; II (Dec. 8, 1820), 293 ff.; *New England Farmer*, II (Aug. 9, 1823), 9 ff.; *Southern Agriculturist*, III (Nov., 1830), 561 ff.; *Cultivator*, I (Sept., 1834), 84–85.

[7] *American Agriculturist*, XXXIII (April, 1874), 130.

England a Scragg's tile-making machine and from this time Johnston accelerated his efforts. Tile of the horse-shoe type was laid in ditches two-and-a-half feet deep and twenty feet apart and by 1856 he had laid more than 51 miles.[8]

The agricultural press played its part in giving wide publicity to the results of Johnston's experiments and served as a major agent in the popularization of this innovation. The editors observed his practices at firsthand and described them in their papers. Johnston himself traveled widely and wrote full accounts of his methods which appeared in the journals.[9] Farmers came from all parts of the country to see the work of this man who claimed that money spent in underdraining was a profitable investment. Inquiries came thick and fast, and within a few months, Johnston answered sixty-four letters pertaining to his experiments. The importance placed upon this subject by the farm press is summarized in an editorial of the *Genesee Farmer* for 1853: "Were we asked to name any single operation that would most improve American agriculture, we should unhesitatingly answer, thorough underdraining." [10] In this manner the press encouraged and promoted the advanced practices of the leaders in this occupation.

During the period under review, the journals worked hand in hand with the agricultural societies, for their objectives were identical. Indeed, many periodicals were inaugurated under the auspices of these societies and became their organs.[11] Other

[8] Hedrick, *op. cit.*, p. 349; *Dictionary of American Biography*, X, 143.

[9] One of the first of these accounts may be found on pp. 239 ff. In 1852, the New York State Agricultural Society awarded Johnston the first premium, a silver cup, for his experiments in draining. *Cultivator*, n.s., IX (Feb., 1852), 88.

[10] *Ibid.*, XIV (Aug., 1853), 235.

[11] This was true of the *Plough Boy*, the *New-York Farmer*, *Cultivator*, *Union Agriculturist*, *Agriculturist*, and the *Pennsylvania Farm Journal*.

JOHN JOHNSTON

From the *American Agriculturist*, XXXIII (April, 1874), 121.

journals not so related also served as agencies of dissemination, and the benefits of coöperation were mutual. As organs, they provided publicity for the activity of the societies, published their papers and proceedings, advertised the agricultural essay contests, and printed the winning essays. Frequently the editors were officials of the societies and took a leading part in their activities. The members of the societies, on the other hand, contributed articles and, no less important, their hearty financial support.[12] One of the chief activities of these organizations was the annual agricultural fair; at "fair-time" the contributions made by the periodicals were indispensable.[13]

Another rural institution which received the enthusiastic support of the periodicals was the farmers' club which began to assume an important role in country life during the forties. Since the merchants, mechanics, and professional men had their associations for mutual improvement, the editors urged farmers to organize clubs for the purpose of meeting socially to discuss agricultural subjects and to compare notes on farm management. These neighborhood associations, sometimes referred to as agricultural clubs, town agricultural societies, agricultural lyceums, and (in the South) planters' clubs, varied in respect to organization in different localities.[14] The meetings were conducted informally and everybody was expected to take part.

---

Much the same relationship existed between the journals and the horticultural societies.

[12] The attitude of the Massachusetts Agricultural Society, the Maryland Agricultural Society, the Agricultural Society of Albermarle, Virginia, and the Agricultural Society of South Carolina toward the *American Farmer* may be found in the *American Farmer*, III (1821–1822), Introductory page.

[13] See pp. 198 ff. for the part played by the press during the fairs.

[14] *Michigan Farmer*, IV (Jan., 1847), 162; VI (May, 1848), 151; *New England Farmer*, XI (May, 1859), 212; *Southern Agriculturist*, n.s., IV (Nov., 1844), 401.

"Speeches" were not in order and politics was banned. Many participated in these small group discussions who never would have benefited by the meetings of county or state agricultural societies.[15]

Normally, the first step in launching a farmers' club, after deciding upon its name, was the adoption of a brief constitution and the election of officers.[16] The meetings were held weekly or monthly in a local schoolhouse, or as more often happened with small clubs, at a member's home. When conveniently located, the office of the local journal was used and in that case, the editor could be counted upon to take an active part in the discussions. In any event, the secretary usually furnished a condensed report to the editor.[17] The accounts of local clubs published in the periodicals not only possessed great informational value, but served to excite interest for the organization of new clubs.

Many clubs planned in advance a schedule of subjects for discussion, such as the relative value of certain plows, the efficacy of various fertilizers, or the most profitable crops for a particular section. Others followed a general plan set up annually to meet the needs of each month relative to the farm, to the orchard, or garden. A question period was often set aside at which time farmers could present urgent problems needing immediate solution. Sometimes refreshments, such as dough-

[15] *New England Farmer*, n.s., XI (May, 1859), 212; *Farmer's Monthly Visitor*, XI (Oct. 31, 1849), 146; *Cultivator*, X (May, 1843), 77; *Southern Planter*, X (March, 1850), 92–93.

[16] *Ohio Farmer*, VII (Dec. 4, 1858), 388.

[17] *Cultivator*, X (Feb., 1843), 31; *Union Agriculturist and Western Prairie Farmer*, II (July, 1842), 64; *Prairie Farmer*, III (April, 1843), 70; *American Agriculturist*, XVII (Jan., 1858), 10; *Ohio Farmer*, VII (Dec. 4, 1858), 388; *Western Farmer and Gardener*, V (April, 1845), 202; *Working Farmer*, V (May 1, 1853), 69; V (June 1, 1853), 76–77; V (Aug. 1, 1853), 141; V (Sept. 1, 1853), 150; *Wisconsin and Iowa Farmer*, II (March, 1850), 58; *Southern Planter*, X (March, 1850), 65 ff.

nuts and apples, were served. Very often, especially when the club rotated from home to home, a "good plain substantial farmer's dinner" was in order.[18] On occasion, wives accompanied their husbands; sometimes they took part in the discussions, but more often they withdrew to another room "to discuss their own matters." [19]

These miniature agricultural societies not only awakened inquiry among farmers, promoted investigation, and furnished the means for interchange of thought, but created social harmony and good feeling. They operated against that spirit of isolation and seclusion then prevalent among rural people.[20] Other benefits naturally accrued. An entertaining and profitable phase of many club meetings was the exhibition of farm products which appeared appropriate to the subject under discussion or seemed of unusual interest. The clubs were often instructed to send committees to visit the farms of their members, offer suggestions and finally report at a subsequent meeting "the dark as well as the bright side" of their inspections.[21] Many organizations inaugurated club libraries consisting of the best agricultural journals and books, and occasionally procured lecturers to speak before their groups.[22] Sometimes, at the close of the meeting, an opportunity was given members to announce their desire to buy or sell items of interest to

[18] *Ohio Farmer*, VII (Dec. 4, 1858), 388; *Western Farmer and Gardener*, V (April, 1845), 202.

[19] *American Agriculturist*, X (Jan., 1851), 23; XIII (Oct. 18, 1854), 82. A report of a novel agricultural club in Delaware may be found on p. 242.

[20] *Cultivator*, X (Sept., 1843), 139; *New England Farmer*, XI (Jan., 1859), 14; *Country Gentleman*, II (Nov. 17, 1853), 309; *Pennsylvania Farm Journal*, I (July, 1851), 105–6; *Valley Farmer*, IX (Dec., 1857), 362–63.

[21] *Ohio Farmer*, VII (Dec. 4, 1858), 388.

[22] *Monthly Journal of Agriculture*, III (July, 1847), 36; *Cultivator*, n.s., I (Sept., 1844), 269; *Ohio Farmer*, II (May 1, 1846), 72; VII (Dec. 4, 1858), 388; *American Agriculturist*, XVII (Jan., 1858), 10; *Indiana Farmer*, VII (Dec., 1858), 293.

farmers, and later the bargains were made. As a product of farmers' clubs, temporary associations of a few members were formed for the joint purchase and use of valuable stock animals and agricultural machinery at a considerable saving of money.[23] The journals sensed the wide benefits to be gained from these organizations in rural communities and consequently played a large part in their expansion during the twenty years before the Civil War.

The introduction and popularization of new agricultural implements and machinery was one of the most constant policies advocated by the press.[24] Owing to the ignorance, prejudice and obstinacy of a large majority of farmers, this problem presented great difficulty. The farmers evinced a decided hostility toward any deviation from the beaten path and the journals constantly battled against this deep-rooted conservatism.[25] In order to break down resistance and familiarize the farmer with new agricultural inventions, the periodicals inserted cuts and descriptions at their own expense, also printing letters from farmers bearing successful testimony to the effectiveness of the new devices. Labor-saving contrivances displayed at fairs received detailed accounts from the editors. Upon being convinced of the success of an invention, they did not hesitate to

[23] *Ohio Farmer*, VII (Dec. 4, 1858), 388; *New England Farmer*, n.s., XI (Jan., 1859), 14; *American Agriculturist*, XVII (Jan., 1858), 10.

[24] Naturally less interest in new, agricultural, labor-saving inventions was displayed in the Southern journals, due to the labor situation in this area. Certain Southern editors, such as Edmund Ruffin, who seemed unable to appreciate machinery, "saw little good in any reaper." Hutchinson, *Cyrus Hall McCormick*, I, 415. The "West," on the other hand, was vitally interested in all labor-saving devices.

[25] *American Farmer*, II (March 23, 1821), 412; III (May 18, 1821), 57; XII (Jan. 21, 1831), 359; *Cultivator*, II (March, 1835), 1; *New England Farmer*, VII (Sept. 19, 1828), 65; *Genesee Farmer*, III (Feb. 16, 1833), 52; III (Nov. 2, 1833), 345; *Farmers' Register*, I (June, 1833), 62.

give it wide and fervent publicity, a form of free advertising that became known as "puffing."

The danger of puffing an unworthy object was ever present. To obviate it the agricultural societies as well as the farm journals attempted to evaluate farm inventions by appointing small committees to examine them and report their findings.[26] Such a notice as the following found in the *Cultivator* was common in the early part of the period: "In pursuance of the recommendation of the [New York] State Agricultural Society, the subscribers, a committee appointed for the purpose, will meet at the City Hall in Albany, *on the second Tuesday in July next,* at 10 o'clock A. M. to examine and test any agricultural implements which may be offered for their inspection; and to certify to the merits of such as they may find deserving of public patronage." [27] These decisions received wide publicity through the societies and journals. Later the reaper trials generally held under the auspices of an agricultural society, served somewhat the same purpose as the committee of inspection.

It is, of course, impossible to estimate exactly the influence of the agricultural press upon the general adoption of the horse rake, steel plow, seed drill, corn planter, cultivator, thresher, mower, reaper, and other agricultural machinery in the United States before 1860. The journals not only advocated and advertised such activities as the plowing match, the fair, and the reaper trials, which developed and popularized agricultural improvements, but they abounded with enthusiastic editorials, pictures, and accounts of labor-saving devices. This activity, which was carried on with missionary zeal, without doubt facilitated the adoption of new machinery

[26] *American Farmer,* II (March 2, 1821), 389; III (Dec. 14, 1821), 301; IV (Oct. 4, 1822), 222.

[27] *Cultivator,* IV (June, 1837), 61.

which played such an important part in the agricultural revolution.

Equally important with the problem of keeping the farmer awake to new developments in agriculture was the task of breaking down old superstitions and popular fallacies. The editors felt that the journals could be as usefully employed in dissipating errors as in disseminating information.[28]

Perhaps the oldest, most prevalent and tenacious superstition which bound the farmer was the well-nigh universal belief in the influence of the moon upon agriculture.[29] Indeed moon-farming was the accepted practice. It was advocated that all tap-rooted vegetables should be sowed on the "decline" of the moon and those which appeared above the surface of the ground on the "increase" of the moon. The waning period was best for pruning and setting out fruit trees and it was believed that apples rotted if picked before the moon was full.[30] Indeed, the moon had to be auspicious for successful activity—the planting and harvesting of crops, the killing of hogs, the roofing of houses and the weaning of children.[31]

The general belief in the power of the moon was as harmful

[28] *American Farmer*, XV (Aug. 2, 1833), 161; *New England Farmer*, XII (Nov., 1833), 129; *Genesee Farmer*, III (Feb. 23, 1833), 62–63; *Southern Agriculturist*, II (Feb., 1829), 65 ff.

[29] *Farmers' Register*, I (Aug., 1833), 162 ff.

Extreme power was often credited to the moon. Many thought that moonlight upon the eyes, especially in warm climates, often proved fatal to sight; that men's weight varied one or two pounds, depending on the phases of the moon; that horses were uniformly quite blind at certain times according to the moon; that there was a close relationship between the lunar phases and epilepsy and insanity; that it was safe to get a haircut only when the moon was in Leo. *Farmers' Register*, I (Aug., 1833), 162 ff.; *New England Farmer*, III (March 18, 1825), 266–67; IX (Sept. 24, 1830), 75; *Southern Agriculturist*, VIII (Dec., 1835), 648. See also Crozier, *Popular Errors about Plants*, p. 11.

[30] *American Farmer*, II (Dec. 15, 1820), 304; *New England Farmer*, I (May 24, 1823), 339.

[31] *New England Farmer*, n.s., V (July, 1853), 318.

to agriculture as any evil with which farmers and planters contended. Times without number, farmers prepared the soil with utmost care, then ignored a favorable season for planting, because it was "not the right time of the moon." [32] Against this subservience to the "monarch of the skies" the editors and other agricultural leaders waged a campaign with striking unanimity.

A heated debate between moon-believers and nonbelievers was carried on in the press, both sides presenting seemingly conclusive "facts." However, long and well conducted experiments by scientific men completely refuted this popular belief in the influence of the "goddess Luna upon sublunary things." [33] Though the editors attacked this old superstition with reason and ridicule and outstanding agricultural writers such as Solon Robinson labeled it "all moonshine," nevertheless many still adhered to the old notions.[34]

Among the many other fallacious conceptions which had to be eradicated before scientific agriculture could be introduced was the doctrine of transmutation of wheat to chess, the most debated question in the farm papers. This controversy, innocently inaugurated through a speech by David Thomas printed in the *Plough Boy* in 1820, continued at intervals for forty years, occupying more space and causing more bitterness than any controversial subject appearing in the journals.[35]

[32] *American Farmer*, II (Dec. 15, 1820), 304; XV (July 26, 1833), 161.

[33] *Farmer's Register*, I (Aug., 1833), 162 ff.; *New Genesee Farmer*, I (July, 1840), 106; *Prairie Farmer*, VIII (Jan., 1848), 16; *Ohio Cultivator*, I (Oct. 1, 1845), 150; *Cultivator*, VI (Nov., 1839), 167.

[34] *Prairie Farmer*, VII (July, 1847), 204–5; *Ohio Cultivator*, II (Sept. 15, 1846), 138; *Farmers' Cabinet*, V (June 15, 1841), 356; *New England Farmer*, n.s., XI (Nov., 1859), 494; *American Farmer*, XV (July 26, 1833), 161; *Genesee Farmer*, VII (Sept., 1846), 207; *Farmer's Monthly Visitor*, III (June 30, 1841), 95.

[35] *Plough Boy*, I (March 4, 1820), 316; *New Genesee Farmer*, V (Jan., 1844), 4. Portions of the speech by Thomas may be found on pp. 243–44.

The question at issue was simple. Farmers planted what they considered clean wheat and reaped wheat together with a weed called chess or cheat. The debate centered upon the source of this weed. On the one side, many claimed that chess (*Bromus secalinus*) was corrupt, degenerate or diseased wheat—the change being effected by unfavorable seasons, soil conditions, excessive portions of certain manures, or accidental causes. On the other side, the "anti-transmutationists" asserted that this "preposterous notion" was the result of utter ignorance or misunderstanding of the principles of botanical science. They claimed that if really clean wheat were sown, chess would not be reaped.[36]

The journals, in the main, vigorously supported the latter view, but were willing to allow both sides to be heard in their pages in order to arrive at the truth.[37] It was obvious that so long as farmers believed that wheat could turn to weed, under circumstances beyond their control, little or no effort would be made to arrive at the real solution—the elimination of the small seeds of chess from the ground as well as from the wheat before sowing.

Letters recounting experiences and experiments poured into the editors' offices from all parts of the country and at times almost overwhelmed them. Rewards ranging from a Diamond plow to $500 in cash were offered to anyone who could prove the production of a single head of chess from a grain of wheat.[38]

[36] *New-York Farmer*, IV (June, 1831), 145; IV (July, 1831), 182.

[37] They were, however, less positive on this subject than upon the moon controversy. A few outstanding editors, for a time at least, were either uncertain or took the view that wheat could turn to chess. For example note the attitude of R. L. Allen, *American Agriculturist*, II (Nov. 15, 1843), 301; Jesse Buel, *Cultivator*, n.s., II (June, 1845), 183; and Warren Isham, *Michigan Farmer*, VII (June 1, 1849), 171.

[38] *Prairie Farmer*, III (Feb., 1843), 47; *Farmer's Register*, IX (Jan., 1841), 11 ff.; *New Genesee Farmer*, II (June, 1841), 84; *Cultivator*, n.s.,

Edmund Ruffin, editor of the *Farmers' Register*, in conjunction with several farmers subjected this question to the test of rigid scientific experiment. Negative findings induced him to offer $100 for irrefutable evidence of the truth of transmutation, but this reward like all others was never collected due to lack of satisfactory proof.[39] The chess problem was debated at agricultural fairs and clubs, and many societies appointed committees for experimentation.[40] Although most of the editors and other leaders labored to eradicate this "heresy," it is likely that the majority of farmers still clung to the doctrine of transmutation until the time of the Civil War.[41]

No policy of the farm press was more urgently and persistently advanced than the demand for agricultural education in the United States. Public interest in this subject developed slowly during the early years of the nineteenth century, increased considerably after 1840, and reached a high pitch during the fifties, when concrete results were being achieved. This development coincided with the expanding popularity and influence of its sponsor.[42]

Almost every conceivable method for promoting agricultural education was encouraged by the journals. In 1822 they

---

VI (Sept., 1849), 293, and ser. 3, VII (June, 1859), 193; *American Agriculturist*, XIX (Oct., 1860), 296.

[39] *Farmers' Register*, I (July, 1833), 83 ff.; IX (Jan., 1841), 11 ff.; *New Genesee Farmer*, V (Jan., 1844), 4; *Cultivator*, ser. 3, VI (Oct., 1858), 306; *Country Gentleman*, X (Aug. 20, 1857), 128; XIV (Sept. 22, 1859), 192.

[40] *Working Farmer*, V (June 1, 1853), 77; *Cultivator*, ser. 3, VI (Oct., 1858), 306; *Country Gentleman*, XIV (Oct. 20, 1859), 249.

[41] *Country Gentleman*, XIV (July 14, 1859), 33; XIV (July 21, 1859), 45; XIV (July 28, 1859), 64; see also Darlington, *American Weeds and Useful Plants*, pp. 86–87.

The writer's contact with farmers in certain sections of the country would indicate that many farmers today cling to moon farming and are firm believers in transmutation.

[42] Woodward, *The Development of Agriculture in New Jersey*, p. 164.

welcomed the Gardiner Lyceum at Gardiner, Maine, the first exclusively agricultural school in the United States.[43] Aside from supporting institutions of this type, they recommended agricultural courses in all manner of schools.[44] The colleges were urged to appoint professors of agriculture, as Columbia College had done in 1792.[45] The use of agricultural books to serve as readers was suggested for the district schools.[46] Solon Robinson, in his effort to induce the publication of a series of elementary agricultural school books, offered as a premium, five building lots in Chicago, "one of the numerous new towns in the west." He made this offer in the pages of the *Cultivator*, stating he felt that such a series would "have a tendency

[43] Bidwell and Falconer, *History of Agriculture in the Northern United States*, p. 194; *American Farmer*, V (Oct. 3, 1823), 224; *New England Farmer*, II (Dec. 27, 1823), 172.

Doctor Ezekiel Holmes, who later edited the *Maine Farmer*, lectured on agriculture and had charge of the school farm at the Gardiner Lyceum. A. C. True, *A History of Agricultural Education in the United States, 1785–1925*, p. 36.

[44] Many editors took an active part in the agricultural schools. For example, James Pedder, editor of the *Farmers' Cabinet* was connected with the Eden-Hill Farm Institute, started in 1843 near Philadelphia; Daniel Lee, editor of the *New Genesee Farmer*, helped to found, and taught in, a school at Wheatland, N.Y.; Tolbert Fanning, editor of the *Agriculturist*, opened a school in Tennessee which later became Franklin College. (It emphasized manual training and agriculture.) James Mapes, editor of the *Working Farmer*, established the Mapes's School near Newark, N.J. *Farmers' Cabinet*, VII (Aug. 15, 1842), 25; *Agriculturist*, VI (Jan., 1845), 10; *New Genesee Farmer*, VII (Jan., 1846), 7 ff.; *Working Farmer*, III, (March 1, 1851), 24.

[45] John P. Norton, professor of agricultural chemistry at Yale (1846–1852), was perhaps the most notable in this field. He not only taught, but delivered many public lectures, contributed widely to the farm press and wrote books on agriculture. The editor of the *Cultivator* considered him "the most practical agricultural writer and thinker" of his time. Bailey, *Cyclopedia of American Agriculture*, IV, 600; *Dictionary of American Biography*, XIII, 575; *Cultivator*, n.s., IX (Oct., 1852), 329.

For Professor Norton's views on agricultural education, see pp. 245 ff.

[46] *Farmers' Register*, VI (Aug. 1, 1838), 262 ff.; *Cultivator*, n.s., I (July, 1844), 210; n.s., V (Aug., 1848), 250; *Plough, the Loom, and the Anvil*, IX (Feb., 1857), 459.

to learn American youth such things" [47] as would help them to support themselves and their families by the labor of their own hands. The need for an agricultural dictionary was met in part by the *Genesee Farmer* in 1839 when it inaugurated a "Dictionary of Terms used in Agriculture." [48] This glossary appeared in installments in the journal.

The outstanding advocate of agricultural education during the period from 1820 to 1840 was Jesse Buel. During these years he emphasized in the New York Assembly, preached from the public platform, and stressed in his articles for the press, the need for professional schools of agriculture as well as for such instruction in the district schools.[49] When he became editor of the *Cultivator* he made this journal the medium of propaganda for his ideas on the subject, and his articles were widely copied by the other periodicals. At the request of the Massachusetts Board of Education, Buel wrote a book called *The Farmer's Companion* which was used in school and rural libraries.[50] This volume, designed as a textbook and composed largely of material from the best English works, together with extracts from the files of the *Cultivator*, went through at least eleven editions.[51] During his life he laid out an educational

[47] *Cultivator*, V (Oct., 1838), 142.

The use of "learn" in the above sense was the accepted usage in the best agricultural journals of the day. For example see *Ohio Cultivator*, III (Feb. 15, 1847), 28; *Ohio Farmer*, VI (Oct. 30, 1848), 351; *Wisconsin Farmer*, VIII (May, 1856), 232; *Northwestern Farmer and Horticultural Journal*, IV (Sept., 1859), 301; *Country Gentleman*, XVI (Oct. 25, 1860), 268.

[48] *Cultivator*, V (Oct., 1838), 141–42; *Genesee Farmer*, IX (Jan. 5, 1839), 1. This glossary was continued in the *Cultivator*, VII (Jan., 1840), 11.

[49] A. C. True, *A History of Agricultural Education in the United States, 1785–1925*, p. 47.

[50] *The Farmer's Companion* (Boston, 1847), p. xxii.

[51] Bidwell and Falconer, *op. cit.*, p. 473; Carman, "Jesse Buel, Albany County Agriculturist," p. 247.

program which was in large measure adopted in the late nineteenth and early twentieth centuries.[52]

Meager success attended this type of instruction in the secondary schools however, and more and more strongly the journals demanded the establishment of agricultural colleges.[53] The attempt to obtain a portion of the Smithsonian bequest for the erection of a great national school of agriculture, in which the United States Agricultural Society took a major interest, ended in failure.[54] After this disappointment, the journals intensified their campaign for agricultural colleges endowed by the individual states.[55] The first institution of this kind was the Michigan Agricultural College, which opened its doors in 1857, thus inaugurating a new phase in agricultural education.[56] Other states soon followed this precedent.[57] Again in

---

More than a dozen agricultural textbooks were published in the United States between 1824 and 1860. Bailey, *Cyclopedia of American Agriculture*, IV, 379 ff.

[52] *Dictionary of American Biography*, III, 238.

[53] A. C. True, *A History of Agricultural Education in the United States, 1785–1925*, p. 39.

Professor John P. Norton attributed the failure of many educational projects to the total inadequacy of competent teachers. This authority on agricultural education maintained as late as 1852 "that if any six states of the Union were within the present year to make provision for the establishment of state agricultural schools, or colleges, . . . they could not find on this continent the proper corps of professors and teachers to fill them." *Cultivator*, n.s., IX (March, 1852), 107–8. See also *American Cotton Planter*, I (April, 1853), 103.

[54] U. S. Patent Office, *Annual Report: Agriculture* (1859), p. 24.

[55] *American Agriculturist*, V (April, 1846), 106; XVII (Feb., 1858), 40; *Southern Planter*, VI (March, 1846), 60; *Pennsylvania Farm Journal*, I (Feb., 1852), 325; *Michigan Farmer*, XV (June, 1857), 184; *Country Gentleman*, X (Nov. 5, 1857), 304; *Cincinnatus*, IV (Oct., 1859), 440; IV (Jan., 1859), 29; *Ohio Cultivator*, XVI (Jan. 15, 1860), 24.

Opposition to the establishment of state agricultural colleges was frequently voiced by farmers. For characteristic objections see pp. 250 ff.

[56] Bidwell and Falconer, *op. cit.*, p. 320.

[57] *Cincinnatus*, IV (Jan., 1859), 29.

JESSE BUEL

Frontispiece, The *Cultivator*, Volume VI (1839).

the late fifties, the periodicals supported and stimulated nation-wide agitation for federal land donations for similar institutions. This resulted in the important Morrill Land Grant Act. The *Wisconsin Farmer* felt that the main credit for bringing this bill before Congress should go to the journals and societies, while the *Cincinnatus* was proud that the farm press throughout the country had indorsed and advocated it with zeal and ability.[58] The editor of the *Indiana Farmer* in 1858 suggested that the United States hold a national jubilee when the Morrill Land Bill should become a law.[59]

Popular demand for state and federal aid to agriculture came slowly and for a long time lacked vigor. The farm publications quite early, however, advocated state appropriations for agricultural schools (as already described), societies and clubs, and for the establishment of state boards of agriculture.[60] Later they fought aggressively for the appointment of state chemists to analyze soils, state appropriations for agricultural and geological surveys, crop bounties, and state rewards for the eradication of insect pests.[61] After 1835 many of these demands were conceded, and a new era in the policy of state aid to agriculture began.[62]

[58] *Wisconsin Farmer*, X (Feb., 1858), 65; *Cincinnatus*, IV (April, 1859), 145.

[59] *Indiana Farmer*, VII (June, 1858), 66.

[60] *Plough Boy*, IV (Jan. 28, 1823), 266; *Cultivator*, III (Dec., 1836), 131; III (Jan., 1837), 152 ff.; IV (Sept., 1837), 113; *Union Agriculturist*, I (Oct., 1840), 5; *Ohio Cultivator*, I (Jan. 1, 1845), 5; *New England Farmer*, XII (Jan., 1860), 30.

[61] *Cultivator*, II (June, 1835), 50–51; *Yankee Farmer*, IV (Jan. 13, 1838), II; IV (Feb. 17, 1838), 50; *Genesee Farmer*, IX (Jan. 26, 1839), 25; *Maine Farmer*, VIII (Jan. 11, 1840), 10; VIII (Feb. 15, 1840), 42; VIII (Feb. 22, 1840), 50; *Ohio Cultivator*, I (Jan. 1, 1845), 5; *Valley Farmer*, IX (Nov., 1857), 325 ff.; *Ohio Farmer*, VII (Jan. 29, 1859), 36; *Working Farmer*, I (Feb., 1849), 8; *Farmers' Register*, II (Jan., 1835), 517.

[62] Bidwell and Falconer, *op. cit.*, p. 193.

The farm press vigorously supported the crusade for the establishment of a department of agriculture in Washington, with a secretary who should be a cabinet member.[63] Indeed, the *Working Farmer*, launched in 1849, was inaugurated mainly with the idea of advocating this innovation and carried such a statement upon its title page. Since 1836 the Patent Office had been responsible for the agricultural activity of the government; it was, however, a target of farm journal criticism on charges of neglect, political corruption, and incompetency.[64] One periodical maintained that a United States Department of Agriculture would furnish the "*arm to strike the blow*" for the common objects which all farmers demanded.[65] A continuous barrage of editorials and letters demanding such legislation no doubt played an important role in the final passage of the Act in 1862 creating this department.

The necessity for the introduction and popularization of new and useful plants and animals was constantly stressed by the editors, who took an active part in this work. Their offices became distribution centers for rare and valuable seeds received from all parts of the world. These seeds were sent in from farmers, agricultural societies, American consuls, naval officers, "patriotic travellers" and after 1835 from the Patent Office

[63] The Homestead Act was also under consideration during the fifties. While little mention was made of it by Eastern and Southern journals, the *Indiana Farmer* predicted that any Senator who opposed it would "politically, be a doomed man in the West." *Indiana Farmer*, VII (March, 1859), 433.

[64] *Ohio Farmer*, I (Jan. 8, 1852), 5; *Ohio Cultivator*, VIII (June 1, 1852), 168; *Cultivator*, n.s., IX (March, 1852), 120; *Journal of Agriculture*, IV (Sept., 1853), 75; *Maine Farmer*, XXIV (July 17, 1856), 117; *Plough, the Loom, and the Anvil*, IX (Sept., 1856), 129 ff.; *Working Farmer*, VIII (Dec. 1, 1856), 217; *American Agriculturist*, XVIII (April, 1859), 104; *Wisconsin Farmer*, XII (Aug. 1, 1860), 247; *Country Gentleman*, XV (Feb. 16, 1860), 112.

[65] *Journal of Agriculture*, IV (Sept., 1853), 75.

in Washington.[66] Generally these seeds were given gratis to farmers who were willing to experiment and report their findings to the editor for publication. This work was inaugurated in the early twenties by such periodicals as the *American Farmer* and the *Plough Boy*. By the late fifties seed distribution became an important function of many journals.[67] The activity of the *American Agriculturist* furnishes a good example of what the best papers were doing. In 1856 the editor sowed a small plot with sorghum or "Chinese sugar cane," a plant introduced into the United States two years earlier.[68] Impressed with his experiments and hoping that this plant would prove valuable for making sugar and possibly provide feed for cattle, the editor decided to distribute seeds on a large scale. He procured from foreign sources nearly fifteen hundred pounds, which he sent out in small packages. While the editor admitted that this activity proved valuable as an advertisement, he also took pride in claiming that he laid the foundation of widespread culture in sorghum.[69] He further boasted that be-

[66] William H. Crawford, Secretary of the Treasury, in 1819 issued a circular requesting the coöperation of American consuls in the introduction of foreign plants into the United States. This is the first official record of governmental activity in the interests of foreign plant introduction. Ryerson, "History and Significance of the Foreign Plant Introduction Work of the United States Department of Agriculture," p. 133.

H. L. Ellsworth, Commissioner of Patents, began the somewhat systematic distribution of seeds in 1836 for the United States Government, although there was no special appropriation for that purpose at the time. Wiest, *Agricultural Organization in the United States*, p. 69.

[67] *American Farmer*, I (March 24, 1820), 415; II (Jan. 19, 1821), 342; IV (April 12, 1822), 23; IV (April 19, 1822), 32; IV (May 3, 1822), 47; *Plough Boy*, I (April 15, 1820), 366.

[68] *American Agriculturist*, XV (Oct., 1856), 305; Gray, *History of Agriculture in the Southern United States to 1860*, II, 829.

[69] *American Agriculturist*, XV (Oct., 1856), 305; XX (Jan., 1861), 6. In later years, sorghum became an important product of the United States, especially in the South. Gray, *op. cit.*, II, 829.

fore 1861 about one million parcels of various kinds of seeds had been sent from his office to all parts of the country.[70]

Another important activity encouraged by the journals was the improvement of cattle breeds in the United States. Very often the editors as well as the societies imported horses, mules, and various types of cattle for scientific breeding purposes on stock farms.[71] Such well-known stock raisers as Henry Clay frequently reported the results of their experiments. The editors expressed deep appreciation to the American naval officers for their part in the introduction of foreign animals.[72] They were responsible for bringing to our shores "the fine horses of Arabia, of Barbary and Peru—the large and spirited Asses of Malta, the beautiful cattle of Tuscany, celebrated since the days of Virgil—the sheep of Spain and Barbary—the Llamas and splendid specimens of the feathered tribes from South America—the prolific swine, and fine poultry of the East Indies." [73] American consuls were likewise active.[74] The animal

[70] *American Agriculturist*, XX (Jan., 1861), 6.

[71] For a detailed account of the attempts to improve cattle breeds in the United States during the period, see Leavitt, "Attempts to Improve Cattle Breeds in the United States, 1790–1860," pp. 55 ff.

[72] *American Farmer*, III (June 29, 1821), 107; III (July 27, 1821), 141; III (Dec. 28, 1821), 320; IV (Aug. 16, 1822), 161; IV (Dec. 27, 1822), 313; IX (April 13, 1827), 32; IX (Nov. 9, 1827), 272; XII (April 23, 1830), 49; *ibid.*, n.s., I (April, 1840), 354; *New England Farmer*, III (April 22, 1825), 307.

Sir Isaac Coffin (1759–1839), an Admiral in the British Navy, although born in Boston, Mass., was particularly interested in the Massachusetts Agricultural Society and presented this organization with a number of valuable foreign animals. *New England Farmer*, III (April 1, 1825), 286; III (April 22, 1825), 306–7; *American Farmer*, V (Sept. 12, 1823), 195; X (May 30, 1828), 87; *Dictionary of American Biography*, IV, 266–67.

[73] *American Farmer*, IV (Aug., 1822), 161.

A great interest in things agricultural was evinced by American naval officers. The gallant Commodore Chauncey, for instance, hero of the War of 1812, was president of an agricultural society, and "was proud to come

displays at fairs together with accounts of them in the press did much to popularize them.

The enthusiasm for new animals and new farm products was responsible for a series of sprees in agricultural speculation that rivaled the excesses in land and canal ventures of the period. The farmer was conservative, but conservative only in regard to the things he knew and understood. It is said that professional men, noted for conservatism in their own fields, have been consistent victims of fly-by-night ventures. So the cautious farmer, always a pliant subject for lottery-ticket salesmen and patent-medicine vendors, ever watchful for a quick and easy fortune, became the prey for one grand speculation after another. In the twenties these excesses were known as hobby riding, but by the late fifties they were dignified with such terms as "manias," "crazes," and "fevers."

An early issue of the *American Farmer* pointed out the importance of continually evaluating new plants and animals, at the same time warning its readers of the dangers of blind enthusiasm. The crazes for merino sheep, mammoth bullocks, enormous hogs (utterly unfit for any purpose "but making

---

in for the premium for the second best sow" at an agricultural fair. *American Farmer*, n.s., I (May 29, 1839), 1.

An editorial summary of the activities of American naval officers in the introduction of plants and animals into the United States may be found on pp. 253–54.

[74] *Plough Boy*, I (June 5, 1819), 6; *American Farmer*, IV (Aug. 2, 1822), 152; IV (Aug. 23, 1822), 176.

Henry Perrine, long American consul at Campeche, Mexico, was especially coöperative, but particularly in the introduction of tropical plants. It was at his suggestion that Congress in 1838 established a tropical plant station in Florida. He was active in this short-lived project. During his life Perrine introduced several noteworthy plants. *Southern Agriculturist*, VIII (Sept., 1835), 457–58; XII (Nov., 1839), 583; *Dictionary of American Biography*, XIV, 480–81. See especially *Farmers' Register*, Vols. V, VI, VII and VIII.

soap"), and unusual vegetables were decried.[75] But crazes continued with increasing intensity. The "silk mania" which started in the early thirties was among the first of importance. According to these enthusiasts, the United States was destined to become a great silk producing center. The cultivation of the Chinese mulberry tree (*Morus multicaulis*) whose enormous leaves fed the silkworms, was to be the first step toward the vast fortunes to be made by rural people in all parts of the country.[76] The farmers became widely interested; trees and cuttings were purchased, prices rose fantastically, and the fever spread far and wide. The press teemed with articles about silk and mulberry tree cultivation, together with advertisements for the sale of millions of trees.[77] By 1838 the distemper had swept the entire country, "rapidly extending amongst all classes," assuming "an alarming character" and in some sections rising "to a state of raving madness." [78] Short selling and other modern Wall Street techniques were employed.[79] One writer said he heard daily reports of individuals who had made fortunes, "rising suddenly from poverty to wealth," and he added, "the storekeeper, the farmer, the mechanic, the clerk, the teacher, are dazzled with the golden vision, and rush from their useful employment into the grand speculation." [80] However, the money was made in buying and selling the buds or

[75] *American Farmer*, III (July 6, 1821), 114.

[76] *Yankee Farmer*, IV (May 12, 1838), 145; *Maine Farmer*, II (April 18, 1834), 106; *Farmers' Register*, I (Aug., 1833), 152; IV (Aug., 1836), 251; *Franklin Farmer*, III (Sept. 7, 1839), 22; *Genesee Farmer*, VIII (Sept. 8, 1838), 284; *Southern Agriculturist*, V (May, 1832), 244; *Farmers' Cabinet*, I (Aug. 1, 1836), 29.

[77] *Farmers' Cabinet*, III (Sept. 15, 1838), 72; *Genesee Farmer*, VIII (Sept. 8, 1838), 288; *Farmer and Gardener*, IV (Nov. 7, 1837), 224.

[78] *Farmers' Cabinet*, III (Oct. 15, 1838), 81; *New England Farmer*, XVII (Dec. 12, 1838), 182.

[79] Gray, *op. cit.*, II, 829.

[80] *Farmers' Cabinet*, III (Oct. 15, 1838), 81.

trees rather than in the production of silk.[81] Indeed silk production had been almost totally neglected, and the effect of high labor costs upon ultimate profits had never been sufficiently weighed.[82] In 1839 the price of trees collapsed and speculation came to an abrupt end.[83] In announcing the termination of the silk mania, the *New England Farmer* said, "It is in common parlance, flat on its back," it "has fallen suddenly like a tremendous Colossus; and it now lies sprawling with a good many under it, who are crushed by its fall." [84]

While the silk mania was noteworthy as a case of speculation with small basis of real value in the commodity concerned, the "Berkshire fever," which raged from about 1838 to 1841, was an instance of a good thing overdone.[85] Berkshire hogs became the rage, carried along on exaggerated claims.[86] Prices soared and five hundred dollars did not seem exorbitant for a pair of the animals.[87] Early in the forties, the inevitable collapse arrived and, as usual, was followed by considerable prejudice against the animal involved. The real worth of the Berkshires, however, soon overshadowed this short-lived prejudice.[88] Dur-

[81] *Cultivator,* V (Nov., 1838), 149; *Farmers' Cabinet,* III (Oct. 15, 1838), 82.

[82] A. H. Cole, "Agricultural Crazes," p. 631. Professor Cole's article deals with the most important crazes of the period.

[83] The price of young trees was 25 cents during the early days of speculation, while in 1839 it rose as high as $1.25 and sales netting $2.00 were recorded. In the year of the collapse, trees were purchased for 2 cents each. *Ibid.,* pp. 622 ff.; *Cultivator,* V (Nov., 1838), 149.

[84] *New England Farmer,* XVIII (Nov. 13, 1839), 170.

A characteristic warning against silk speculation may be found on pp. 255 ff.

[85] A. H. Cole, *op. cit.,* p. 632.

[86] *Farmers' Cabinet,* VI (Jan. 15, 1842), 186; *Western Farmer and Gardener,* III (Nov., 1841), 45.

[87] *Cultivator,* V (Jan., 1839), 187; A. H. Cole, *op. cit.,* p. 632.

[88] A. H. Cole, *op. cit.,* p. 633; *Agriculturist,* II (Nov., 1841), 262; *Southern Planter,* I (July, 1841), 111; *Farmers' Register,* X (Feb., 1842), 89.

ing approximately the same period, there was wild speculation in broom corn, Chinese tree corn, and the Rohan potato, each following the same general pattern of extravagant claims, intense excitement, soaring quotations, and final price collapse. (Professor Arthur H. Cole says no era in American agricultural history can rival that of 1830–1842 as one of stupendous speculation in new crops.[89]) During the early fifties, the "hen fever" broke out in the United States and soon swept the entire country. Prices steadily climbed, especially on imported fowl, which were in greatest demand. The *American Agriculturist* of 1850 declared that "Few are aware of the extent to which the hen fever is now raging among our amateur farmers"; the "California fever sinks into insignificance when compared with this." [90] The hysteria lasted about five years following the general course of previous distempers.

"Hen-ry," in the *New England Farmer*, portrays the wide sweep of this malady in the following lines:

THE HEN FEVER

Lawyers, and doctors, and divines,
All practice have resigned,
And to improve the breed of hens
Their talents have combined.

. . . .

" 'Tis strange, 'tis passing strange;" the world
Is ever running wild;
Some foolish scheme will captive lead
Alike the sage and child.[91]

In general, the policy adopted by the farm press in regard to these great waves of speculation tended to promote rather

[89] A. H. Cole, *op. cit.*, p. 634.
[90] *American Agriculturist*, IX (July, 1850), 229.
[91] *New England Farmer*, n.s., II (Aug. 31, 1850), 296.

than to discourage them. Of course, the editors deemed it important to have the farmer experiment with agricultural innovations, to test their value and adaptability to this country, for who could say with certainty that the introduction of a new breed of animals or a new product might not revolutionize our agriculture. "Puffs" often appeared in the press which were intended to encourage wide experimentation. Yet these very puffs sometimes furnished the initial impulse which finally precipitated a long period of speculation. The fever, once under way, fed upon itself. Thereupon, the excited farmer demanded the latest information on the subject and the journals eagerly supplied this material.[92] The editors very often issued words of caution but these warnings generally came after the contagion had reached ridiculous heights. They knew, as we do today, that applying the brakes to "prosperity" has never been popular.[93] It is also true that many articles were written by those financially interested in spreading the malady and not a few of the editors reaped a rich reward through personal interest.[94]

Now and then an editor protested against a raging fever.[95] Judge Buel of the *Cultivator* proudly "stood alone" among his contemporaries of the press in his conviction that the *Morus*

[92] *Cultivator*, II (Oct., 1835), 119; *Genesee Farmer*, IX (March 30, 1839), 101.

[93] At the height of the silk craze, the *New England Farmer* stated, "It is not for us to give any words of caution. We have not the care of the public health; but we think it advisable for those, who desire to keep well to keep out of the way of contagion." *New England Farmer*, XVII (Dec. 12, 1838), 182.

[94] *Southern Planter*, XX (March, 1860), 143; *Franklin Farmer*, III (Aug. 31, 1839), 9; *New England Farmer*, XVII (Dec. 12, 1838), 182; *Pennsylvania Farm Journal*, II (Dec., 1852), 257; *Farmer and Gardener*, I (Aug. 26, 1834), 136; IV (Oct. 24, 1837), 201.

[95] By the middle of the century, certain editors began to assume a new role. It had been their custom to prod conservative farmers to adopt new

*multicaulis* was too tender for Northern climates, and Governor Hill, of the *Farmer's Monthly Visitor*, referring to the same speculation, "studiously withheld everything that should contribute to the excitement," and "threw cold water on the project."[96] Edmund Ruffin of the *Farmers' Register*, was always suspicious of "puffery" and "over-rated novelties," and frequently took his stand against prevalent excesses.[97] As has been noted, the *American Agriculturist* attempted to eliminate speculation in "Chinese sugar cane" which raged in the late fifties by freely distributing thousands of packages of seed for experimentation.[98]

After a particular fever had subsided, it was customary to label the object of contagion a humbug. Readers were warned never to be caught again, and the dangers of overenthusiasm were reviewed. By the end of the period, however, an occasional editorial pointed out the great advantages of former hobby riding and the benefits that had resulted.[99] While the *American Agriculturist* in 1850 had "no idea of being mixed up with the miserable humbug" called the hen fever, nevertheless, a few years later they felt its effect had been profitable.[100] The editor of the *Prairie Farmer* wrote: "Improve-

ideas. As the result of a long series of manias, they began to fear that the farmer would even turn his enthusiasm for reapers into a "reaper fever." The *Prairie Farmer* warned its readers "to beware of getting crazy, even about a good thing." *Prairie Farmer*, VI (Sept., 1846), 284; *Wisconsin Farmer and Northwestern Cultivator*, VIII (April, 1856), 187; Hutchinson, *op. cit.*, I, 235.

[96] *Cultivator*, V (Jan., 1839), 185; *Farmer's Monthly Visitor*, IV (Jan. 31, 1842), 8.

[97] *Farmers' Register*, VI (April 1, 1838), 47 ff.; X (Feb. 28, 1842), 89. For Ruffin's attitude on agricultural hobbies, see pp. 258 ff.

[98] *American Agriculturist*, XV (Oct., 1856), 305; XX (Jan., 1861), 6.

[99] *Cultivator*, n.s., VIII (Sept., 1851), 291; *Prairie Farmer*, XVII (April 2, 1857), 105; *Moore's Rural New-Yorker*, IX (Nov. 20, 1858), 373.

[100] *American Agriculturist*, IX (Nov., 1850), 352; XVII (Jan., 1858), 12.

ments in agriculture seldom come as *pure* improvements. They come in epidemics, mixed with a great deal of nonsense, and with more or less of humbug, and often of rascality." [101] Lewis C. Gray of the Department of Agriculture likewise feels that, to a considerable extent, improvements in breeds of cattle in the United States, at this period, came about through a series of crazes.[102]

Although occasional reference was made to agricultural chemistry in the early part of the period, it was not emphasized. Indeed little was known about it. The works of Sir Humphry Davy, which did so much to spread the meager knowledge of the relation of chemistry to agricultural practice, were now and then quoted.[103]

A real beginning was made by Edmund Ruffin, who has been called "the most influential leader of Southern agriculture" and "the father of soil chemistry in America." [104] After years of research upon the cause and remedy for the depleted condition of the soil of Virginia, Ruffin published the results of his investigation in the *American Farmer* in 1821.[105] He concluded that the soil, once fertile, had become "acid" by harmful cultivation and thereby lost its ability to retain manures. He contented that the application of marl would correct this situation and that proper cultivation would restore the lost fertility.[106] His *Essay on Calcareous Manures,* which was repeatedly and enthusiastically quoted by the agricultural press

[101] *Prairie Farmer,* XVII (April 2, 1857), 105.

[102] Gray, *op. cit.,* II, 847.

[103] Bidwell and Falconer, *op. cit.,* p. 319; *American Farmer,* I (Feb. 18, 1820), 374; II (March 23, 1821), 414; III (Aug. 24, 1821), 173; *New England Farmer,* I (Aug. 10, 1822), 14; I (Oct. 19, 1822), 91; I (July 19, 1823), 403.

[104] Gray, *op. cit.,* II, 780; Craven, *Edmund Ruffin, Southerner,* p. 60.

[105] Craven, *Ruffin,* pp. 53 ff.; *American Farmer,* III (Dec. 28, 1821), 313.

[106] *Dictionary of American Biography,* XVI, 215.

over the following thirty years, eventually ran through five editions and by 1852 totaled 493 pages.[107] In 1895 an authority called it "the most thorough piece of work on a special agricultural subject ever published in English."[108] In 1833 Ruffin founded the *Farmers' Register*, an agricultural journal of great influence, and here continued his crusade for the use of marl and other agricultural reforms. Where Ruffin's teachings were heeded, soil deterioration ceased, emigration was checked, and land values rose. Although less than five percent of the lands of Virginia had been so treated, land values in the lower part of this state increased by more than $30,000,000 in the twenty years before 1850.[109]

A widespread interest in agricultural chemistry in the United States was awakened in 1841 with the appearance of the American edition of Justus Liebig's *Chemistry in Its Application to Agriculture and Physiology*.[110] Liebig's ideas "came as a thunderbolt out of a clear sky."[111] He maintained first, that certain mineral elements were necessary for the growth of plants and that these were supplied by the soil; second, that analyses of both the plant and the soil would determine the required treatment. Thus, constituents lacking in the soil could be furnished through the application of mineral manures.[112] Opposition to

[107] Craven, *Soil Exhaustion as a Factor in the Agricultural History of Virginia and Maryland, 1606–1860*, p. 150.

So enthusiastic was Skinner about Ruffin's first article that he issued and distributed copies of it gratuitously. *American Farmer*, III (Dec. 28, 1821), 313. The editor of the *Maine Farmer* considered it the best work on the subject that had ever appeared. *Maine Farmer*, V (May 30, 1837), 121.

[108] Cutter, *Yearbook* U.S. Dept. of Agriculture, 1895, p. 500.

[109] Craven, "The Agricultural Reformers of the Ante-Bellum South," p. 311; Gray, *op. cit.*, II, 804.

[110] Bidwell and Falconer, *op. cit.*, p. 319.

[111] McCall, "The Development of Soil Science," p. 47.

[112] Bidwell and Falconer, *op. cit.*, p. 319.

EDMUND RUFFIN

From *De Bow's Southern and Western Review*, O. S., Volume XI (October, 1851), opposite p. 349.

these "new fine spun theories" was announced by the *Southern Planter*.

Mr. Justus Liebig is no doubt a very clever gentleman and a most profound chemist, but in our opinion he knows about as much of agriculture as the horse that ploughs the ground, and there is not an old man that stands between the stilts of a plough in Virginia, that cannot tell him of facts totally at variance with his finest spun theories.[113]

Nevertheless, Edmund Ruffin claimed that Liebig's book received more "respectful attention and applause" than any other on agriculture previously published.[114] Among the many capable pioneers who appeared in the field of agricultural chemistry might be mentioned Daniel Lee, editor of the *New Genesee Farmer* and *Southern Cultivator*, John P. Norton, professor of agricultural chemistry at Yale, and Samuel W. Johnson, protegé and successor of Norton.[115] These men wrote widely

[113] *Southern Planter*, V (Jan., 1845), 23. See also *New Genesee Farmer*, III (June, 1842), 88.

[114] *Farmers' Register*, IX (Aug. 31, 1841), 459.

[115] Doctor Lee (1802–90) served as corresponding secretary of the New York Agricultural Society and, for a period, as director of the division of agriculture in the Patent Office. He was particularly interested in the science of chemistry and geology as related to agriculture. In 1846, in coöperation with General Harmon, he attempted to establish an agricultural school at Wheatland, Monroe Co., N.Y. Lee was well known as a lecturer on agriculture and from 1855 to 1862, was professor of agriculture at Georgia University. A. C. True, *A History of Agricultural Education in the United States, 1785–1925*, pp. 71–72; *Genesee Farmer*, VII (March, 1846), 57; VII (May, 1846), 126; *Southern Cultivator*, XVI (May, 1858), 145; *Southern Planter*, V (April, 1845), 75.

Professor Norton (1822–52) studied chemistry with the elder Silliman at Yale and in 1844 went abroad, where he continued his studies over the next few years at Edinburgh and Utrecht. He was appointed professor of agricultural chemistry at Yale in 1846 and continued this connection until his death. Perhaps his most important publication was a textbook, *Elements of*

for the agricultural press, and also analyzed soils, and taught and lectured in their chosen fields.

Less able figures, however, appeared in all parts of the country. Soil analysis soon took the center of the stage, especially in the Eastern journals. It now became the panacea for all the farmers' ills.[116] Unfortunately, overconfidence and meager knowledge of the subject were blended in a dangerous mixture. Advertisements of "chemists" and "professors" who were willing to analyze soil and advise treatment filled the papers. Widely advertised commercial fertilizers appeared on the market and became the fad.[117] It was all so simple. The journals, as usual, were lost in their enthusiasm.

The spree lasted more than ten years. Not till the early fifties did editors begin to realize that soil analysis had been greatly overemphasized. Too much had been promised in view of the limited knowledge concerning this highly technical subject.[118]

---

*Scientific Agriculture* (1850). *Dictionary of American Biography*, XIII, 574–75.

Professor Johnson (1830–1909) studied under Norton at Yale and continued his work with Erdmann at Leipzig and with Liebig at Munich. In 1856, he was appointed to the Yale faculty and in the same year became chemist for the Connecticut State Agricultural Society. He was author of many works and articles on agricultural chemistry. *Dictionary of American Biography*, X, 120; *Southern Planter*, XIX (Dec., 1859), 767.

[116] Soil exhaustion, naturally, was not a problem in the newly opened lands of the West. *Ohio Cultivator*, XV (May 1, 1859), 137.

[117] Professor Mapes, editor of the *Working Farmer*, was also a "consulting agricultural chemist." He was interested in the manufacture of a fertilizer called "Superphosphate of Lime" and it was said that his analyses usually showed the need for this product. *Northern Farmer*, II (Aug., 1853), 116; *Pennsylvania Farm Journal*, III (April, 1853), 3. In the vernacular of today, his colleagues labeled him a racketeer. See pp. 96–97.

Mapes's daughter, Mary Elizabeth Mapes Dodge, long editor of the *St. Nicholas* magazine and author of *Hans Brinker; or the Silver Skates*, is more widely remembered today.

[118] *Ohio Cultivator*, VIII (Dec. 1, 1852), 353 ff.; *Southern Planter*, XIII (April, 1853), 114; *Valley Farmer*, V (April, 1853), 143; *Country*

As the extravagant claims were not fulfilled, a reaction set in against this innovation as so often occurred in the case of "over-puffed" ideas. Although the chemists were, in general, on the right track, soil analysis now represented just another humbug to the farmer. The *Ohio Cultivator* felt that the professors of chemistry were at fault, in that they did not as yet sufficiently understand the science they attempted to teach in its application to practical agriculture. The *Rural American* declared that the new theories had been "weighed in the balance and found wanting."[119] The *Pennsylvania Farm Journal* thus summarized its attitude: "Agricultural chemistry may be compared to a tree that is planted, has grown, and may have blossomed, but as yet has borne no fruit."[120] The *Northern Farmer* wisely commented: "It is not our intention to *force* the agricultural community into the next century."[121] By 1860 the best agricultural chemists throughout the country quite frankly admitted that soil analysis was of "little practical utility."[122] With the dissipation of this early feverish enthusiasm, Liebig's ideas continued to be discussed, and today "the main outline of his theory still stands."[123]

Another policy supported vigorously by the agricultural press, commencing in the early forties, was the demand for improved domestic architecture and rural art generally. Be-

---

*Gentleman*, II (Aug. 4, 1853), 69; *Wisconsin and Iowa Farmer*, VI (May, 1854), 100; *Michigan Farmer*, XII (Aug., 1854), 234; *Rural American*, I (Jan. 19, 1856), 17.

See pp. 261–62 for a characteristic editorial on the reaction that swept through the press.

[119] *Ohio Cultivator*, IX (May 1, 1853), 129; *Rural American*, I (Jan. 5, 1856), 1.

[120] *Pennsylvania Farm Journal*, III (Nov., 1853), 280.

[121] *Northern Farmer*, I (Feb., 1852), 18.

[122] *Genesee Farmer*, XXI (June, 1860), 170.

[123] McCall, *op. cit.*, p. 48.

fore this time little attention had been given to this important subject; now there was a definite recognition of the need for improvement in the construction of rural houses with regard to economy, comfort, convenience, and beauty. As stated in the *Cultivator*, the "bump of design and constructiveness" was not possessed by many people and, as good plans for farm buildings were nowhere obtainable, this journal took the lead in the publication of plans and estimates for country houses, farm houses, barns, workshops, log cabins, ice houses, piggeries, fences, and gates.[124] These were enthusiastically received and so popular did they become in all the journals that rarely a volume is found among any of the periodicals of the fifties that does not contain designs, plans, elevations, or hints for prospective builders. A grateful reader of the *Genesee Farmer* claimed that plans which he found in this paper saved him more than fifty dollars.[125] Occasionally, a journal set aside a column under "Rural Architecture" for material of this nature.

Many farm periodicals and agricultural societies offered awards including diplomas, silver medals, and cash prizes for the best plans of specific buildings. These were widely used by the press and "prize plans" often made the rounds. Selections from books on architecture and items from the *Horticulturist*, a journal which emphasized this subject and was under the editorial direction of Andrew Jackson Downing and later Patrick Barry, were widely copied. Solon Robinson contributed a great variety of plans ranging from designs of fifty-dollar log cabins to drawings of his own house, "a very convenient house without a foot of waste room." [126]

Many of these designs for city, village, and country dwell-

[124] *Cultivator*, V (March, 1838), 5; VI (Nov., 1839), 164.
[125] *Genesee Farmer*, XVI (March, 1855), 80.
[126] *Cultivator*, VI (Nov., 1839), 164; X (Feb., 1843), 37.

ings today seem in bad taste, or at best quaintly peculiar. It should be borne in mind, however, that in their day they represented an effort to escape from the rigid correctness of the Greek Revival style. During the period from 1820 to 1860, classical architectural forms prevailed, and the propriety of a building was measured by the correctness of its Greek columns, porticoes, pediments, entablature, frieze, and cornice—whether these were appropriate or not.[127] "Churches and houses likewise followed the type of the temple, which became a single unconditional ideal for all classes of buildings." [128] The reaction was twofold: first, toward greater practicality; second, toward a less formal and more picturesque style. Both of these trends are evident in the designs published by the agricultural journals at this time. The popularity of the classical style began to wane toward the middle of the century, and the Gothic influence, which had been in evidence since the thirties, became more pronounced.[129] Associated with this influence was the

[127] Tallmadge, *The Story of Architecture in America*, pp. 75, 97 ff.; Carman, *Social and Economic History of the United States*, II, 366.

[128] *Encyclopaedia Britannica*, 14th ed., XV, 637.

[129] Tallmadge, *op. cit.*, pp. 114 ff.; Carman, *Social and Economic History of the United States*, II, 366; Major, *The Domestic Architecture of the Early American Republic*, pp. 24 ff.

An editorial in the *Prairie Farmer* gives a brief history of the architecture of the period. This article may be found on pp. 263 ff.

"Carpenters Gothic" (a term used to characterize much of the architecture of the fifties) is described in the *American Agriculturist* as follows: "Suddenly, a change was introduced, and like all changes brought about by inexperienced hands, with many needed reforms, it had an equal number of absurdities. 'Rural' architects sprang up all over the country. The saw and hammer were thrown by with many a clever carpenter-and-joiner, and they went to planning houses, drawing pictures, and writing books on architecture. The country had been flooded with designs, plans, and all sorts of gimcrackery in the way of farm buildings, and 'genteel' country houses, not only for retired city folks, and village people, but farmers, cottagers, and laborers —for the educated and refined, down to the rudest dweller of the backwoods. . . . Broken lines, projections, and zig-zag angles in the walls; hips,

well-known editor of the *Horticulturist*, Andrew Jackson Downing (1815–52), who stood head and shoulders above his contemporaries in the field of "rural art." He was widely known through his books as well as his articles on horticulture, landscape gardening, rural architecture, pomology, and botany, which appeared in the agricultural press.[130] Downing was one of the first to be successful in directing the attention of the American people to beautifying their gardens and dwellings.[131] His *Cottage Residences* published in 1842 was particularly influential in creating national interest in the improvement of country homes.[132]

This reformer felt that nothing so powerfully affected the tastes and habits of a family, particularly the younger members, as the home in which it lived. He insisted that buildings and their surroundings should be harmoniously related, and that the structural form should express local fitness.[133] While Downing's designs for Italian villas, rural Gothic cottages, villas in the Romanesque style, and Swiss cottages were widely circulated in the agricultural press, it was his conception of the ideal farmhouse that holds most interest for us. Emphasizing the great value and importance of what he termed truthfulness in domestic architecture, he contended that a farmhouse must, first of all, look like a farmhouse, in order to give any

---

haws, jerks, and scarps in the roofs, costing, perhaps, half as much as the whole house, and good for nothing but to invite leaks and repairs, are the prominent features of the outside." *American Agriculturist*, XVII (March, 1858), 73.

[130] Among other books published by Downing were: *The Fruits and Fruit Trees of America* (assisted by his brother Charles, 1845); *Architecture of Country Houses* (1850).

[131] *Country Gentlemen*, III (March 23, 1854), 188–89.

[132] *Dictionary of American Biography*, V, 418; Proctor, Columbia Historical Society, *Records*, XXVII (1925), 255–56.

[133] *Cultivator*, n.s., V (Jan., 1848), 9; n.s., VII (Sept., 1850), 305.

ANDREW JACKSON DOWNING

Frontispiece, *Landscape Gardening and Rural Architecture*, by Andrew Jackson Downing, Sixth Edition, 1859.

lasting satisfaction. He wrote that since "one of the highest sources of beauty in domestic architecture is derived from its embodying the best traits in character of the man or class of men for whom it is designed," the farmhouse must, therefore, "express that beauty, whatever it may be, which lies in the farmer's life." [134] The proportions of such a dwelling "should aim at ampleness, solidity, comfort, and simple domestic feeling, rather than elegance, grace, and polished symmetry." Strong, enduring materials should be used in its construction and a "certain rustic plainness" stressed.[135] He saw no place in such a plan for Grecian columns, Italian balustrades, or Gothic carved work of any sort, and cautioned farmers concerning a "profusion of gingerbread work, and fantastic and unmeaning decoration." [136] His emphasis on the "charm of simplicity" elicited numerous objections and criticisms from farmers who called his designs "uncouth and awkward," and demanded just as much "decoration" for their homes as for city dwellings.[137]

Downing was engaged in 1851 to lay out the grounds for the Capitol, the White House, and the Smithsonian Institution in Washington. While he did not live to see the completion of this project, his ideas and plans were carried into effect by his successors. His sudden and tragic death in the following year occurred when he was at the pinnacle of his great fame.[138] In attempting to save other lives on the burning steamboat, *Henry Clay*, on the Hudson River, Downing, an excellent

[134] Downing, *The Architecture of Country Houses*, pp. 137 ff., and *Rural Essays*, pp. 205 ff.

[135] *Cultivator*, n.s., VII (Sept., 1850), 305–6.

[136] Downing, *Country Houses*, p. 138; *Cultivator*, n.s., V (Jan., 1848), 9.

[137] *Ohio Cultivator*, VII (July 1, 1851), 196–97; *Country Gentleman*, III (March 2, 1854), 140.

[138] *Dictionary of American Biography*, V, 417–18; Proctor, *op. cit.*, p. 252.

swimmer, was carried to his grave. The death of no American of this period, in any walk of life, caused such a profound sense of national loss or was so deeply mourned by the agricultural press. The accounts of this tragedy in the farm papers throughout the land attest to his enormous influence and prestige. "Mr. Downing [wrote the *Southern Cultivator*] has done more than any man in America to introduce among us correct ideas of beauty and utility in Architecture, Landscape Gardening, Horticulture and Rural Art generally, and his sudden death may well be regarded as a national calamity." The *Michigan Farmer* wrote: "There seems to be a void which no man can fill"; and the *Prairie Farmer* of that day says: "His is a country's loss, and a country mourns him." The *New England Farmer* voiced the unanimous opinion of the press when it said: "We believe the death of no man in the union could be a greater loss." [139]

Two other policies supported by the press should be mentioned in passing. The first of these was internal improvements. Although in the early days of the *American Farmer* there were objections to the emphasis placed upon this subject, nevertheless the editors in general saw the great advantages for the farmer in the form of road, canal, and railroad construction, as a means toward opening new markets, speeding transportation, and reducing haulage costs.[140] Indeed, the editor of the *Kentucky Cultivator*, in 1852, changed its title to *Kentucky Cultivator and Railroad Journal*, as he felt that agricultural ad-

[139] *Southern Cultivator*, X (Sept., 1852), 273; *Michigan Farmer*, X (Sept., 1852), 274; *Prairie Farmer*, XII (Sept., 1852), 415; *New England Farmer*, n.s., IV (Sept., 1852), 414; *New Genesee Farmer*, XIII (Oct., 1852), 280; *Maine Farmer*, XX (Aug. 5, 1852), 126; *Cultivator*, n.s., IX (Sept., 1852), 297; *Wisconsin Farmer*, IV (Sept., 1852), 214; *Pennsylvania Farm Journal*, II (Oct., 1852), 209.

[140] *American Farmer*, V (Aug. 15, 1823), 168.

vancement and railroad improvement went hand in hand.[141] Many farmers opposed the railroads, some of them fearful that a train of roaring cars would frighten and kill their livestock as well as scatter dangerous sparks in all directions. One troubled farmer complained: "The noise of the steam-cars in passing is so great that at times they may be heard at the distance of two miles,—much resembling the clatter of the instrument called a horse fiddle. The consequence is that many horses are greatly alarmed; and farmers who have to cross the track . . . approach it under apprehensions of danger." [142] Others feared that the railroads would gain a monopoly of transportation with resultant abuses.[143] The editors attempted to allay these fears. Few policies advocated by the press received the same unanimous approval as the demand for improved transportation facilities.[144]

The encouragement and improvement of the common school in the United States was the other policy to be included. While less attention was devoted to this subject in Southern journals, the Northern and Western periodicals attributed much importance to it. Many of them, including the *Prairie Farmer*, the *Farmer's Monthly Visitor*, the *Michigan Farmer*, and the *Wisconsin Farmer* had special educational departments, with the state superintendent of schools or some other expert in

[141] *Kentucky Cultivator*, I (Dec., 1852), 49.

[142] *Genesee Farmer*, V (July 25, 1835), 233; *Country Gentleman*, II (Aug. 18, 1853), 104.

[143] *Genesee Farmer*, VI (Oct. 15, 1836), 332.

[144] The following references illustrate the general support of this policy by journals of widely different interests. *New England Farmer*, III (March 4, 1825), 251; *Genesee Farmer*, VI (Oct. 15, 1836), 332; *Ohio Cultivator*, III (June 15, 1847), 92; *Ohio Farmer*, I (Jan. 8, 1852), 6; *Southern Planter*, VII (Sept., 1847), 257; *Southern Cultivator*, IX (March, 1851), 40; *American Cotton Planter*, I (Jan., 1853), 25; *Oregon Farmer*, II (April 7, 1860), 137.

charge.[145] Universal public education in all its phases became the popular theme of the press. Such topics as these were discussed: the defects of our schools and their remedies; moral education in common schools; physical, mental, and social education; common-school libraries; agriculture in common schools; the need for a normal school; the employment of female teachers; school architecture, and "the common school, the bulwark of our civilization." "Herein lies the sovereign remedy for social and national evils, the safeguard of Republican Institutions, the glory of a people, and the hope of a world," said a writer in the *Wisconsin Farmer*.[146] The editor of the *Pennsylvania Farm Journal* in 1851 went so far as to say that the agricultural press was the most efficient and legitimate advocate of common-school education in the country.[147]

There were other subjects, however, which demanded the editors' attention. The quarter of a century following 1830 is characterized in American history as a period of reform. The story of most of the major reform movements may be traced in the periodicals for the editors of the agricultural press did not stand aloof on moral issues.[148] The *Country Gentleman* of 1854 in a leading editorial says: "This is an age of reform. Conventions are in session almost every day to devise some method to put a stop to the existing evils of Society, and new plans for

[145] *Union Agriculturist*, 1 (Nov., 1840), 14; I (Nov., 1841), 85; *Farmer's Monthly Visitor*, XI (June 30, 1849), 95; *Michigan Farmer*, VIII (Jan., 1850), 12; *Wisconsin Farmer*, II (Feb., 1850), 67; *Maine Farmer*, VIII (Feb., 1840), 29; *Western Farmer and Gardener*, II (Jan., 1846), 2; *Valley Farmer*, IX (Nov., 1857), 327–28; *Agriculturist*, III (Aug., 1842), 183; *South-Western Farmer*, I (April 15, 1842), 44.

Particularly influential was the *Prairie Farmer*, which was "dedicated to the cause of the common schools of Illinois." See *Prairie Farmer* survey pp. 376 ff.

[146] *Wisconsin Farmer*, I (Feb. 1, 1849), 42.

[147] *Pennsylvania Farm Journal*, I (Aug., 1851), 152.

[148] A discussion of the part played by the women in certain reform movements may be found in Chapter VII.

the preservation of health, the securing of rights, and the extinction of wrongs are urged upon the public every day." [149]

The columns of the press were filled with editorials, correspondence, and poems directed against "demon rum." [150] In the early period, temperance was strongly advocated, but gradually a goal of total abstinence was sought. However, when an editor began "riding the horse to death" he was usually warned in no uncertain terms. "Has a man [wrote a subscriber in the *New England Farmer*] after receiving subscriptions and pay in advance for a public journal any moral right to devote it to other objects than those mentioned in his prospectus? If not, why are the pages of the 'New England Farmer' and 'Horticultural Journal' filled with the productions of the Temperance advocates?" [151] In this case the editor promised to do less preaching. Many editorials appeared combating the old custom of supplying a barrel of whiskey for farm workers during the harvest season.

Tobacco in any form was viciously attacked throughout the entire period. Although in common use, it was not considered fashionable.[152] While the weed had no defenders, it was generally admitted that its use was "increasing at an alarming rate." Aside from being a "filthy," extravagant habit, it was also ruinous to health. That hundreds of thousands of people were dying each year because of lung disease, induced by tobacco smoking, was generally accepted. The need for Antitobacco Societies similar to Temperance Societies was often discussed in the pages of the press.[153]

[149] *Country Gentleman*, III (March 9, 1854), 156.

[150] Temperance poems may be found on pp. 189–90.

[151] *New England Farmer*, XII (Nov. 20, 1833), 145.

[152] Branch, *The Sentimental Years*, p. 258.

[153] *New England Farmer*, XII (Aug. 28, 1833), 54; *Maine Farmer*, II (Dec. 12, 1834), 371; *Rural New-Yorker*, VIII (March 28, 1857), 104; VIII (April 11, 1857), 120.

Although the crusade against tobacco was nation-wide, less was heard of its evils in the Southern, tobacco-raising states. To the editor of the *Maine Farmer*, the use of tobacco was "in reality as filthy and degrading as the use of rum." [154] The *Prairie Farmer* commented upon its value in killing lice, ticks, and the like, but urged its readers to keep it out of their mouths.[155] The editor of the *Genesee Farmer* was called on the carpet for writing an article on the cultivation of tobacco in answer to a question from one of his readers. In defense, he said he was personally opposed to the raising of tobacco, but felt obliged to answer any such questions directed to him.[156] Although the *Agriculturist* published articles on all phases of tobacco culture, it discouraged such cultivation in Tennessee and considered its use "one of the greatest evils of the world." [157] Quite exceptional was the praise directed to a certain chewing tobacco by Frank Ruffin, the independent editor of the *Southern Planter*. "We write with a quid in our mouth," he said, but hastily added that no man in the world should chew.[158] Gambling, particularly lotteries and land speculation, also came in for a share of the criticism in this reforming age.

While the editors were eager to assist in establishing moral reforms, they refused to enter into the muddy waters of party politics.[159] Thus the editor of the *Western Farmer and Gardener* characteristically announced: "It is part of our compact

[154] *Maine Farmer*, VI (Nov. 23, 1838), 326; VI (Nov. 13, 1838), 314.

[155] *Prairie Farmer*, XV (Aug., 1855), 263.

[156] *Genesee Farmer*, XIII (June, 1852), 193.

[157] *Agriculturist*, IV (Jan., 1843), 4.

[158] *Southern Planter*, XVIII (Nov., 1857), 643.

[159] Religious subjects were also avoided. However, the great wave of interest in spiritualism about the middle of the century forced the editors to devote considerable space to it. Millerism was almost completely neglected. The *Boston Cultivator* carried a column entitled "Moral and Religious," but this was unique.

with our subscribers not to talk politics."[160] The editors went to great lengths to avoid this troublesome topic, which had a way of turning up in the most innocent correspondence. A few paragraphs of one of Solon Robinson's letters were deleted by the editor of the *Prairie Farmer* as he said they dealt with a phase of politics and were thus not admissible in an agricultural journal.[161]

With regard to the few exceptions to this rule, the editors in most cases eventually regretted their foray into the field of political controversy. The *Plough, the Loom, and the Anvil* probably contained more political features than any journal. While the *Rural American* discussed party politics rather freely in its first volume, the editor soon announced: "*No Politics Hereafter*," explaining that he had been led to say more on politics than he thought prudent or had originally contemplated.[162] Ruffin's abandonment of the *Farmers' Register* was due in part to the opposition of those who resented his excursion into political fields.

The tariff was discussed at great length in some journals and shunned as a party issue in others. Among those who felt the subject admissible to an agricultural paper, many deemed protection a burden to the farmer, while others favored it. The attitude depended somewhat on local interests. In 1842 the columns of the *Maine Farmer* were opened to a free discussion of this subject, but the editor begged his correspondents not to "stray off, to abuse this or that political party."[163] The following year the *Agriculturist* urged farmers to give their strongest arguments pro and con.[164] The editor of the *New*

[160] *Western Farmer and Gardener*, III (May, 1842), 191.
[161] *Prairie Farmer*, III (April, 1843), 87–88.
[162] *Rural American*, II (Jan. 15, 1857), 15.
[163] *Maine Farmer*, X (Feb. 26, 1842), 30.
[164] *Agriculturist*, IV (Dec., 1843), 189.

*Genesee Farmer*, asserting that the only certain road to truth was free inquiry and open discussion, added: "We do not consider it [tariff] at all a party question, but a great national question." [165] The *Plough, the Loom, and the Anvil* took a strong stand for protection and on all occasions pointed out the evils of free trade.[166] The *Ohio Cultivator* cautiously admitted one letter on each side of the question and then hastily closed the gates, fearing that the comments would be interpreted as having a "*party*-bearing" and cause offense.[167] The *Southern Cultivator*, on the other hand, announced its policy in 1845 as follows: "But this [tariff], most unfortunately for the South, has been made a topic of party politics, and has therefore, been steadily avoided by us in the conduct of the *Cultivator*." [168] In the same year the editor of the *Southern Planter* stated, "There is a great question of the policy of affording governmental protection to manufactures, with which we have nothing to do; because, it has been made a political and party question, and this is a pool much too foul and turbid for our taste." [169]

Bearing in mind the reluctance with which the vast majority of the editors mentioned matters pertaining to party politics, the editorials on business depressions of the period are of particular interest.[170] In contrast with more recent times, the editors laid no responsibility at the door of the party in power.

[165] *New Genesee Farmer*, III (March, 1842), 40.

[166] *Plough, the Loom, and the Anvil*, IV (Dec., 1851), 350; *Ohio Cultivator*, VI (Sept. 15, 1850), 28. A Western journal said of the *Plough, Loom, and the Anvil:* "It yokes Agriculture to Protection, and drives the team together," *Prairie Farmer*, XIII (March, 1853), 124.

[167] *Ohio Cultivator*, VI (Sept. 15, 1850), 28.

[168] *Southern Cultivator*, III (Sept., 1845), 136.

[169] *Southern Planter*, V (July, 1845), 155.

[170] A characteristic editorial on the depression of 1857 may be found on pp. 266–67.

While analyzing the causes of the depression of 1837, the editor of the *Maine Farmer* said that it would "not be necessary to go into the arena of politics, or scold at this or that party, as being the remote or proximate cause of the trouble. This we leave for those who delight in such warfare." [171] In the same year the editor of the *Cultivator*, in speaking of the causes of the existing ills, said: "How far the measures of the government have contributed to the present distress, or how far it is in the power of the administration to afford relief, by the revocation of the specie circular or other means, are questions which it is not our province or intention here to discuss." Then, expressing the philosophy of the period he added: "We are sick—the disease is seated, and will run its natural course; and when the crisis has passed, the patient will regain his strength and vigor, if he is not dredged [*sic*] with quack nostrums. All that can be done safely is to make him as comfortable as possible; to watch and profit by the first symptoms of convalescence, and to guard against a relapse." [172] Normally the causes of panics were listed as: extravagance, speculation, the "prevalent impatience to get rich faster than one's neighbor," the "stampede of men of all professions and pursuits to the Western States," and the unwillingness of cultivators to remain on the farm.[173]

The discussion of slavery in its controversial, political aspects was theoretically taboo in the agricultural press. However, this subject, since it was of greater vital interest to the

[171] *Maine Farmer*, V (June 20, 1837), 145.

[172] *Cultivator*, IV (June, 1837), 61.

[173] *Maine Farmer*, V (June 20, 1837), 145; *Cultivator*, IV (June, 1837), 61–62; *Genesee Farmer*, VII (Nov. 4, 1837), 345; *Country Gentleman*, IX (May 28, 1857), 353; *Valley Farmer*, IX (Nov., 1857), 349, 352; *Rural American*, II (Nov. 2, 1857), 166; II (Dec., 1857), 177; *Prairie Farmer*, XVIII (April 29, 1858), 140.

editors of the South, appeared more frequently in their pages. After the middle of the century many of the Southern papers went through a process of evolution and more and more this institution was openly defended.[174] Nevertheless, it is surprising to find comparatively few references to this troublesome topic even in this section.[175]

A favorite device for voicing an editorial attitude lay in reviewing new books on slavery. The editor of the *Southern Agriculturist*, in commenting on J. K. Paulding's *Slavery in the United States*, said: "The abolitionist and fanatics, to whom the work is principally addressed, cannot fail of feeling its force." [176] Likewise, the *Southern Planter* considered Professor Bledsoe's *Liberty and Slavery*, which attempted to reconcile the dogma of the Declaration of Independence with property in man, as "terse, original and unanswerable." [177] Feeling that the works which constantly flowed from the abolition presses gave an untrue picture of the South and its domestic institution, the *South-Western Farmer* printed large extracts from T. C. Thornton's *An Inquiry into the History of Slavery*.[178]

One of the few journals to state its position at the outset of its career was the *Rural American* (1856) of Utica, New York, which asserted that it would oppose the extension of slavery into new states, adding however, that it would not "meddle with the question otherwise," as no possible good could come from this agitation.[179] The following from the *Boston Culti-*

[174] The transition made by the *Southern Planter*, during the fifties, is described in *Southern Planter*, C (Jan., 1940), 48 ff.

[175] A bristling defense of slavery by a Southern correspondent may be found on pp. 268 ff.

[176] *Southern Agriculturist*, IX (June, 1836), 295.

[177] *Southern Planter*, XVI (May, 1856), 148–49.

[178] *South-Western Farmer*, II (Dec. 22, 1843), 332.

[179] *Rural American*, I (Jan. 5, 1856), 8; I (Jan. 12, 1856), 14.

*vator* of 1848 was the sort of statement usually avoided in all the papers: "Slavery, of whatever species, physical, intellectual, or moral, is a curse to the world, and a foe to human happiness." [180] Occasionally an editor felt obliged to clarify a previous statement which had been challenged, and this was often done in no uncertain terms. Thus the editor of the St. Louis *Valley Farmer* wrote: "We have no sympathy with Abolitionism in any way. We think most of the trouble masters have with their slaves is occasioned by free negroes. We are no *Abolitionist*, and should consider it an insult for any person to hint that we were." [181]

A great many articles, however, appeared in the press dealing with slavery in its noncontroversial aspects. A correspondent of the *Southern Planter* wrote as follows in describing the economy of slave labor: "I do not propose, therefore, to discuss this point in its political bearings, both because you very properly reject such discussions, and because upon those points your readers and myself are no doubt perfectly agreed." [182] The acceptable topics included phases of plantation life, such as descriptions of slaves, their homes, work, care, discipline, education, moral training, food, and clothing.[183] Proper government of overseers, so difficult yet so important in good plantation management, received much attention and "rules for overseers" often made the rounds of the Southern journals.[184]

Before concluding this chapter, a consideration of the attitude of the journals respecting two basic currents of the period

[180] *Boston Cultivator*, X (July 22, 1848), 240.

[181] *Valley Farmer*, V (Dec., 1853), 436.

[182] *Southern Planter*, I (Aug., 1841), 137 ff.

[183] An editorial on the moral culture of slaves may be found on pp. 272 ff.

[184] The overseer was frequently pointed out as the most objectionable feature of slavery—the source of Negro maltreatment as well as of inexcusable soil exhaustion.

A widely copied series of rules for overseers may be found on pp. 275 ff.

is essential. They are: the Westward movement which resulted from the opening up of cheap, fertile lands in the Middle West and the movement cityward, brought about by the first phases of the industrial revolution.

The great movement Westward caused grave apprehension among the journals of the East.[185] In New England especially, the press frequently expressed alarm because of the deserted farms and the increasing evidence of successful Western competition with Eastern farm products.[186] An attempt was made to stem this Westward tide. After pointing out the great advantages of the East, a writer in the *New Genesee Farmer* paints a drab picture of the West, characteristic of hundreds appearing in the journals of the older section. "The young of both sexes foolishly give up such a birth-right, for a bare-foot existence in the western wilderness—amid privation, and toil, and sickness—where pork is the greatest luxury, and a log-house raising the most exciting recreation." [187] The attitude of the South was expressed by the editor of the *Southern Cultivator* when he called attention to the need for arresting the continual tide of emigration Westward, which was depriving the older regions of their best and most enterprising citizens.[188]

[185] *Farmer's Monthly Visitor*, III (May 31, 1841), 69; *Maine Farmer*, XI (Aug. 5, 1843), 121; XXII (June 8, 1854), 94; *New Genesee Farmer*, IV (March, 1843), 46; *New England Farmer*, XXIV (July 9, 1845), 14; *Country Gentleman*, IX (May 28, 1857), 353; *Farmers' Register*, IV (May, 1836), 10 ff.; IV (April 1, 1837), 736, 747; *Southern Agriculturist*, XII (July, 1839), 349; XII (Nov., 1839), 571 ff.; *Southern Planter*, X (Jan., 1850), 17; *Southern Cultivator*, XVII (Feb., 1859), 43; *Prairie Farmer*, XVII (May 14, 1857), 158.

[186] *New Genesee Farmer*, V (Sept., 1844), 76; *Farmers' Cabinet*, XI (Sept. 15, 1846), 63; *Cultivator*, n.s., VI (April, 1849), 109 ff.; *Union Agriculturist*, I (Nov., 1841), 85.

[187] *New Genesee Farmer*, VIII (Jan., 1847), 17.

[188] *Southern Cultivator*, I (July 5, 1843), 111.

The Southern editors realized that soil exhaustion was a fundamental cause

As the Middle West became populated and new journals were inaugurated in this region, they, on the other hand, beckoned to the farmers of the East. To the *Prairie Farmer*, it was "men of nerve and energy and ambition," not "faint-hearted, irresolute men" who caught the "western fever." [189] The *Wisconsin Farmer*, with characteristic assurance, advised the East: "Stay the Niagara, or stop the tides from coming into Boston harbor; but don't be so fool-hardy as to try to stop the most enterprising of your sons, and the most beautiful of your daughters, from coming West, by mere Fourth of July orations, or County Fair addresses." [190] However, in the late forties, as might be expected, the "western" journals joined the Eastern journals in opposition to their common enemy, the "gold fever," which was generally called the "California fever." [191] For its numerous advantages to agriculture, on the other

---

for emigration. They were convinced that improved agricultural methods could "reform" this land. Therefore the restoration of the impoverished soil became their major task. As one farmer wrote, "If any system can be devised by which the worn lands can be restored, emigration to the west" will cease. *Farmers' Register*, II (Nov., 1834), 382. See also *Farmers' Register*, III (July, 1835), 138, III (Aug., 1835), 233 ff., III (Dec., 1835), 508–9; *Farmer and Gardener*, II (Aug. 4, 1835), 106; *Southern Agriculturist*, IX (1836), ii; *Plough, the Loom, and the Anvil*, I (Sept., 1848), 175 ff.

[189] *Prairie Farmer*, XVI (Jan. 17, 1856), 10.

[190] *Wisconsin Farmer*, VIII (Sept., 1856), 397.

Nevertheless, there was sometimes a note of caution in these "western" journals, as was expressed by the editor of the *Ohio Farmer:* "Old trees do not commonly bear transplanting well, especially in distant fields." *Ohio Farmer*, VI (May 30, 1857), 86.

[191] *Ohio Cultivator*, V (Jan. 1, 1849), 13; *Michigan Farmer*, VII (March 15, 1849), 89; *Farmer's Monthly Visitor*, XI (May 31, 1849), 70; *Cultivator*, n.s., VI (April, 1849), 124.

The same hostility was shown toward the "Pike's Peak fever" in the later fifties. The *Prairie Farmer* editorialized thus: "Again is the attention of young men, old men—speculative, unstable, adventurous and discontented—being drawn to Pike's Peak." *Prairie Farmer*, XXI (March 8, 1860), 148. See also *Wisconsin Farmer*, XI (Feb., 1859), 67; *Michigan Farmer*, n.s., I (Feb. 5, 1859), 45.

hand, the Far West was championed by the *California Farmer*. This journal resented the deprecatory attitude taken by the Eastern press and boasted that its "home" possessed a perfect climate as well as soil "of the most fertile and exhaustless character." It disclaimed any desire for the emigration of "adventurers, speculators and loafers" and stated firmly that it wanted only "families of permanent settlers." [192]

The drift to the cities created another problem to be faced by the agricultural press.[193] From ancient times the city had been looked upon by the rural population as the "seat of corruption" and now this "den of iniquity" was absorbing the flower of rural youth.[194] Yet the editors knew that all too often the farmer secretly "looked down" on his occupation and that the younger generation was "oppressed with a growing sense of social inferiority to the city population." [195] That the city offered obvious material advantages could hardly be denied. Thoughtful agricultural leaders looked with distress and alarm at this disturber of the rural peace; and the city together with its occupations was lashed in the press. The *Farmer and Planter* discouraged the "disposition to flock to cities and villages, and engage in professions already filled with drones." [196] A writer in the *Farmers' Register* said, "Of lawyers, and physicians, and merchants, our country has more than enough. Not a few in these callings, are absolutely the pests of society." [197] A leading editorial in the *Indiana Farmer* characteristically said: "The farm is the natural home of man. Placed in any other

[192] *California Farmer*, I (Jan. 5, 1854), 4; III (May 31, 1855), 170.

[193] See pp. 278–79 for a characteristic warning to young men to "remain on the farm."

[194] For a more detailed discussion of this conception, see pp. 183 ff.

[195] *Cultivator*, VIII (Feb., 1841), 27; ser. 3, I (Jan. 1853), 15; Bidwell, "The Agricultural Revolution in New England," p. 700.

[196] *Farmer and Planter*, X (Dec., 1859), 371.

[197] *Farmers' Register*, IV (April 1, 1837), 739.

condition he naturally degenerates, both physically and morally, and soon acquires an inferior type." [198] Another writer cried: "Oh how fearful a thing is a city! How full of sin and sorrow." [199]

To keep the farmer contented and to fortify his morale became a major task of the agricultural press. In editorials, letters and speeches, the farmer was told and retold that he was the "chosen of God," that he belonged to "the order of real noblemen," his was "the only, the hard-earned, the true nobility." [200] The dignity and independence of a farmer's life became a favorite editorial theme. It was said that no class could "rank with him in honesty, high moral feeling, and nobleness of soul." [201] The editor of the *Country Gentleman* felt that one of the main aims of an agricultural journal should be to give the farmer a "more exalted opinion of his own calling." [202] The editor of the *Farmer and Planter* urged his readers to infuse into their children a belief that agriculture was "an honorable and fascinating occupation." [203] The *Cincinnatus* went so far as to recommend that all books that had a "tendency to render children of the farmer discontented with their lot in life, should be discarded at once" from rural libraries.[204]

While temporarily assuaging the restlessness and discontent of the farmer with fulsome flattery, the agricultural leaders knew well that a tangible remedy was imperative. It lay in

[198] *Indiana Farmer*, VII (June, 1858), 65.

[199] *Ohio Farmer*, IV (Aug. 25, 1855), 134.

[200] *Maine Farmer*, VIII (April 18, 1840), 120; *Prairie Farmer*, III (Jan., 1843), 2; *New England Farmer*, XXIV (July 9, 1845), 14; *American Agriculturist*, XIX (Aug., 1860), 234.

[201] *Maine Farmer*, VIII (April 18, 1840), 120.

See pp. 280–81 for one of the many thousands of such articles.

[202] *Country Gentleman*, XV (Jan. 5, 1860), 20.

[203] *Farmer and Planter*, X (Dec., 1859), 371.

[204] *Cincinnatus*, IV (Feb., 1859), 63.

better transportation facilities, improved methods of agriculture, wider markets, and higher profits. Higher profits would change the picture completely and the editors and other agricultural leaders worked in season and out of season to this end.[205]

[205] It is worth noting that the intensity of interest in agricultural improvements and the size of the subscription lists of the farm journals varied more or less directly with the rising and falling prices of farm commodities. *Cultivator*, n.s., I (March, 1844), 73; *Prairie Farmer*, IV (Aug., 1844), 178; VI (Feb., 1846), 51; *Southern Planter*, XVII (Oct., 1857), 637.

# Chapter IV

## THE EDITORS

*A unity of action through the whole land is what we are desirous of establishing—of drawing the whole farming community into one bond of intercourse, and exchange of information. We consider this is best done by breaking down all idea of locality, and we are gratified by being able ourselves to exemplify the fact that an agricultural paper can be so carried on as to destroy all conception of local feeling, by making its editorial department embrace a field, literally without a boundary, and alive to the interests of the whole extent of the Union.*[1]

THE MEN who guided the destinies of the agricultural journals were drawn from all walks of life and represented most of the professions. A great majority of them were experienced in the field of journalism, principally political journalism, before assuming editorial posts on agricultural papers.[2] The medical profession was also strongly represented in such men as John Hoyt (*Wisconsin Farmer*), N. B. Cloud (*American Cotton Planter*), M. W. Philips (*South-Western Farmer*),

[1] Editorial in *Western Farmer and Gardener*, III (May, 1842), 192.

[2] Examples are: John Skinner, Solomon Southwick, Jesse Buel, Luther Tucker, Thomas G. Fessenden, Isaac Hill, Thomas B. Stevenson, Ezekiel Holmes, Daniel Lee, Henry Ward Beecher, and D. D. T. Moore.

and Ezekiel Holmes (*Maine Farmer*). The ministry too made its contribution in Henry Colman and Allen Putnam (*New England Farmer*), Tolbert Fanning (*Agriculturist*), and Henry Ward Beecher (*Western Farmer and Gardener*). A number of the editors served in their state legislatures. Isaac Hill (*Farmer's Monthly Visitor*) had been governor of New Hampshire and United States Senator before inaugurating his paper. Many scientific farmers and planters who had shown special aptitude in writing gave their time wholly, or in part, to editorial work. It is interesting to note that a great many postmasters inaugurated periodicals, particularly in the early period, due in part no doubt, to the existing generous franking privileges.[3]

A vast majority of the editors had practical farming experience and most of them continued to farm on a small scale after they assumed their editorial duties. This fact is contrary, however, to the statement of a subscriber in the *Western Farmer and Gardener* which probably reflected the contemporary farmers' opinion. He said that these leaders, in general, professed a profound love for agriculture but had never "embrowned their hands at it." [4] Actually, many of them were close to the soil, received substantial incomes from their farms, published accounts of extensive and varied practical experiments conducted on these model projects, and frequently won awards at the fairs. These achievements therefore help to substantiate the claim that this group could "handle the plow as well as the pen." [5]

[3] *American Farmer*, n.s., I (April 1, 1840), 354.

[4] *Western Farmer and Gardener*, II (May 15, 1846), 150.

[5] *American Farmer*, V (Nov. 21, 1823), 274; *New England Farmer*, XXIV (July 30, 1845), 33; *New Genesee Farmer*, VI (Dec., 1845), 177; XX (Jan., 1859), 1; *Ohio Cultivator*, II (Jan. 15, 1846), 12; II (Sept. 15, 1846), 140; *American Agriculturist*, XII (March 15, 1854), 16; XIX

Just as the two decades between 1840 and 1860 have been called the period of personal journalism among the newspapers, so the years from 1819 to 1860 may be similarly labeled for the agricultural press.[6] As Greeley was to the *Tribune*, Bennett to the *Herald*, Raymond to the *Times*, so Skinner was to the *American Farmer*, Ruffin to the *Farmers' Register*, and Buel to the *Cultivator*. Indeed, the personality and ability of more than a score of these dynamic rural leaders stamped their journals with a distinct individuality. It is also evident that this leadership gave prosperity and vitality to the papers. This was demonstrated frequently by the utter collapse of a popular periodical upon the death or resignation of the editor.

The friendly and intimate relationship that existed between editor and reader is visible in many directions. Baskets of fruit and other gifts sent by subscribers were constant reminders of the esteem in which the editors were held. The recipients often voiced their thanks as did the editor of the *Northwestern Farmer and Horticultural Journal:* "Our table groans under the favors of kind, thoughtful friends." [7] Urgent requests to visit subscribers came so frequently that it was necessary to explain that physical limitations made it impossible to accept them all. Speaking of the editor's popularity, the *Ohio Farmer* declared, "Let him go where he will, from Maine to Minnesota, he will find a warm welcome from his friends who have learned to love him through the medium of his paper." [8] Very often, an illness of the editor, fully described in his paper, elicited a torrent of solicitations and suggested cures.

(Feb., 1860), 48ff.; *Michigan Farmer*, n.s., I (April 16, 1859), 125; *Working Farmer*, III (March 1, 1851), 24; *Prairie Farmer*, XX (July 12, 1860), 20.

[6] *The Cambridge History of American Literature*, II, 189.

[7] *Northwestern Farmer and Horticultural Journal*, III (Oct., 1858), 354.

[8] *Ohio Farmer*, VI (May 2, 1857), 70.

The simple directness of these leaders is seen on every hand. Skinner did not think it inappropriate to discuss with his readers in the *American Farmer* the problems experienced in rearing his sons. Nor did the editor of the *American Agriculturist* hesitate to express regret that he was forced to disappoint his Hookertown friends by his inability to attend "Sally's wedding." [9] The editor of the *New England Farmer* presented a picture of his home as a frontispiece to one number; another included in his paper plans for the home he was building, and from time to time, acknowledged suggested changes and improvements from his readers, who took a vital interest in the project.[10] An editor often called a favorite contributor by name, reminding him that he was neglecting his friends in his failure to correspond more regularly.

Likewise, a close and harmonious relationship existed between the editors themselves. They were known to each other personally as well as officially, meeting on tours of observation about the country, at conventions, and agricultural fairs. At the Fall exhibitions, making the rounds of the animal and farm-machinery displays they resumed old friendships and, later, freely quoted their colleagues. These personal contacts aided in molding more firmly the common objective; namely, the advancement of agriculture. The editors formed a kind of fraternity, bound together by no formal organization, but held by bonds of mutual interest.

The appearance of a new journal was always welcomed with a friendly greeting from other periodicals as well as an editorial "puff" in which the good qualities of the new paper and its editor were extolled. It was customary for each publication to

[9] *American Agriculturist*, XVII (Sept., 1858), 264.

[10] *New England Farmer*, XI (Jan., 1859), 1; *American Agriculturist*, XVII (Oct., 1858), 297; XVII (May, 1859), 137–38; XIX (Feb., 1860), 48 ff.

volunteer to accept subscriptions for the newcomer, and this form of mutual aid continued throughout the entire country. A general feeling of good-will prevailed. For instance, the *Franklin Farmer* (Kentucky), in boosting the time-honored Albany *Cultivator* said, "The farmer who does not subscribe for this paper, deserves to have briers and thorns grow over his place, instead of corn and wheat." [11]

In such an atmosphere of friendship it is not strange that one of the chief characteristics of the agricultural press was the ever-recurring evidence of understanding, sympathy, coöperation, and tolerance between the editors. The anxiety of the *Indiana Farmer*, lest anything "personal and acrimonious" enter its columns, was characteristic.[12] M. W. Philips, the Southern editor of the *Western Farmer and Gardener*, voiced the sentiments of his colleagues, when he urged his readers to "seize with avidity knowledge" wherever found and "be not sectional." [13] D. Redmond of the *Southern Cultivator*, at an

[11] *Franklin Farmer*, I (Nov. 16, 1837), 23.

[12] *Indiana Farmer*, VII (Sept., 1858), 175.

[13] *Western Farmer and Gardener*, III (May, 1842), 181.

There was a growing feeling on the part of agriculturists, both of the North and South, during the late fifties, that the sectional crisis was due to the "one-horse politicians" who were running the country. It was occasionally urged that farmers from all sections form a "solid phalanx" untainted with "political partisan influences" to save the nation through "agricultural statesmanship." A widely copied article in the *Southern Cultivator* (1859) suggested: "If we could now have a convention of Farmers (no politicians of any party); if we could have one assemblage of talented farmers from all parts of the Union, such a convention would do away all sectional discord, and materially and beneficially change the policy of the government." *Southern Cultivator*, XVII (Jan., 1859), 121. The editor of the *American Cotton Planter* said, "We earnestly bespeak for this communication a calm and careful perusal by all our readers. . . . It cannot, it should not, be disguised that the government of the country is in great peril and its dismemberment one of the commonplace topics of the *politicians!*" *American Cotton Planter*, XIII (June, 1859), 196. It would be interesting to speculate upon the possibility of an altered situation in 1860, had suggestions of this type received serious attention at an earlier date.

editors' meeting in New York in 1858, made a plea that there be no "jealousies, or bickering, or contests, between the individual members of the fraternity" and it struck a most responsive chord at the convention.[14]

To a remarkable degree the editors looked upon their profession from a nationalistic point of view, for theirs was a national cause. In general, and as late as 1860, they recognized no sectional antagonism.[15] This was due, in part, to certain restrictions upon the subject matter appearing in their pages. Politics, and the discussion of slavery in its controversial aspects—the major troublemakers of the period—were quite generally taboo in the journals.[16] In addition, these editors through wide

[14] *American Agriculturist,* XVII (Oct., 1858), 291; *Southern Cultivator,* XVII (Jan., 1859), 8.

[15] It is true, however, that as time went on, the North was gradually assuming national leadership in agriculture, as may be seen from the composition and direction of the two national agricultural societies inaugurated before the Civil War. The first of these, the "National Society of Agriculture" which was organized in 1841 in Washington soon succumbed because of its failure to secure the Smithson fund. However, this short-lived society had all the appearance of a "national" society in its plans for a national agricultural school, national agricultural fair, and a national agricultural journal. One of the important reasons advanced for its formation was the purpose of binding "our union in still closer and more effectual bonds." *American Agriculturist,* I (April, 1842), 4. Solon Robinson, its chief proponent, supported this project, hoping it would draw people together in "one strong bond of brotherhood," and James M. Garnett of Virginia felt it would strike a blow to the "deadly poison" of sectional jealousies. *Cultivator,* VIII (June, 1841), 94; *Farmers' Register,* IX (April 30, 1841), 249.

In 1852 the "United States Agricultural Society" was also organized in the capital and continued an active life until 1860. It is interesting to note that the initiative and leadership of the latter society were largely in the hands of Northern men and the majority of its members came from that section. *Cultivator,* n.s., IX (Aug., 1852), 284; U. S. Patent Office, *Annual Report: Agriculture* (1859), pp. 24 ff. In 1853 only 69 of the 399 members were from the South (mostly Maryland and the District of Columbia) and in 1856 only 36 out of 585 were from this section. Gray, *op. cit.,* II, 788.

[16] Kellar, *Solon Robinson,* I, 37.

travel in this country and Europe, were better able than most contemporaries to rise above local animosities. Indeed, they could with ease transfer their allegiance from one journal to another, in widely separated areas, as editorial opportunities presented themselves.

That the readers were less broadminded and in many instances sensitive along sectional lines, was evident from remarks frequently appearing in the pages of the press. A writer in the *Farmers' Register* said: "Mr. Editor, it is not enough for me, to hear that a man is a patriot—he must be a *Virginia* patriot." [17]

A local subscriber to the *Southern Cultivator* wrote the editor, Daniel Lee (who had formerly edited a Northern periodical), that although his prejudice against Northern men was very strong, nevertheless he was willing, in this instance, to receive instruction in agriculture even from a "Yankee." [18] The editor of the *Southern Agriculturist*, before describing the remarkable progress of Northern agriculture, felt called upon to impress upon his readers that he was "wholly and entirely identified with the South, by birth, education, feeling and interest." [19]

While harmony normally characterized the relationship between the editors, minor discords naturally developed. Possibly the greatest single cause for friction arose from the use of copy without due credit. Editorials such as "Thou shalt not steal," "Give unto Scissors the Things which are Scissors," and "Editorial Pilfering," appeared frequently in the journals.[20]

[17] *Farmers' Register*, III (May, 1835), 32.

[18] *Southern Cultivator*, X (Feb., 1852), 39.

[19] *Southern Agriculturist*, II (March 20, 1844), 47.

[20] *Farmers' Register*, II (Feb., 1835), 581; *Maine Farmer*, V (March 7, 1837), 25; *Franklin Farmer*, II (Oct. 20, 1838), 61; *Prairie Farmer*, III

After the editor of the *Cultivator* (Albany) called attention to the fact that nearly one quarter of the first volume of the *Ohio Cultivator* had been lifted from the Eastern paper without credit, he added: "If the editor is an old man, he ought to *know* better. . . . If he is a young man we hope he will *learn*, that 'honesty is the best policy,' the world over." [21] In calling attention to a similar indiscretion in another paper the *Indiana Farmer* said: "We suppose the editor of that paper to have been absent when that number was set up, and the compositor neglected to give us credit; we have no doubt the editor felt chagrined when he returned and saw it, and we tender our sympathies to him." [22] The *Maine Farmer*, in pleading with other editors not to "hook" their articles, added: "When you *thunder* with our *thunder*, have the goodness to label it correctly." [23] Normally an accusation by one paper was followed by a countercharge from the other and at times the wrangling continued for months.

Many of the publications struck persistent "plunderers" from their exchange lists. As a last resort against inveterate "poachers," the *American Agriculturist* felt obliged, in 1858, to copyright all material, at the same time inviting other journals to draw freely from it and promising never to take advantage of the copyright when proper credit was given.[24]

One of the few vigorous editorial combats of the period was waged between James J. Mapes of the *Working Farmer* of

---

(May, 1843), 89; *Indiana Farmer*, I (July 12, 1845), 193; *Genesee Farmer*, IX (Feb., 1848), 53; *American Agriculturist*, IX (July, 1850), 229; *New England Farmer*, n.s., IV (Oct., 1852), 477; *Country Gentleman*, IV (Aug. 31, 1854), 136; *Rural American*, I (March 22, 1856), 95.

[21] *Cultivator*, V (March, 1839), 16.

[22] *Indiana Farmer*, I (July 12, 1845), 193.

[23] *New England Farmer*, n.s., IV (Oct., 1852), 477.

[24] *American Agriculturist*, XVII (Feb., 1858), 33.

New York, and the majority of the members of the agricultural press. Mapes was not only an agricultural editor, a "Professor," a "Consulting Agricultural Chemist," but was also a manufacturer of fertilizers. According to contemporary editors, the source of his titles, "Professor" and "Chemist" were never satisfactorily divulged and caused wide speculation. The best known of Mapes's fertilizers was "Superphosphate of Lime." During the fifties, when grave doubts were arising in the minds of agricultural authorities as to the practical value of soil analysis and mineral fertilizers, Mapes, in his paper, was making extravagant claims for his manural panacea, and incidentally charging the farmers twenty-five dollars for his "letters of advice" after collecting a five-dollar fee for a soil analysis. When Professor S. W. Johnson of Yale analyzed samples of the superphosphate, he not only found them of little worth, but announced that the product which was sold commercially was far inferior to the samples that the "Professor" distributed for purposes of analysis. Mapes accused the editors, who gave this report wide publicity, of being jealous and unscientific. He considered their periodicals the "constitutional enemies of the *Working Farmer*." The terms "humbug," "fake," "charlatan," and "quack doctor" were openly applied by editors in all sections to the "man of phosphates" who, they claimed, "loved supremely the almighty dollar" and "undermined" the prestige of the agricultural press.[25]

[25] *Genesee Farmer*, XIV (July, 1853), 205 ff.; XXI (June, 1860), 170; *Country Gentleman*, I (June 2, 1853), 340; II (Dec. 22, 1853), 390; VII (Feb. 28, 1856), 137–38; IX (May 7, 1857), 304; XIV (Dec. 8, 1859), 368; *Northern Farmer*, II (Aug., 1853), 116; *Pennsylvania Farm Journal*, III (April, 1853), 3; *Prairie Farmer*, XIII (Dec., 1853), 458–59; *Valley Farmer*, V (April, 1853), 143; *Working Farmer*, VI (March 1, 1854), 4; VI (May 1, 1854), 64, 70; VI (Jan. 1, 1855), 246–47; IX (March 1, 1857), 1; *Southern Planter*, XV (Jan., 1855), 24–25; XIX (Dec., 1859),

Characteristic of the friendly tiffs that frequently occurred in the press was an occasional exchange between Henry Ward Beecher of the *Western Farmer and Gardener* and other members of the fraternity. For instance a few editorial eyes were lifted when the "college learned" preacher and editor committed the "orthographical sin" of spelling acre, a-k-e-r in his journal. The *New England Farmer*, objecting violently to this unorthodox procedure, called upon his fellow editors to "hoot at it" till it should "become obsolete." Beecher, claiming the support of Webster, answered provokingly that nothing becomes obsolete until it has been in vogue.[26] A few years later, however, when this popular editor had won a national reputation in the pulpit, his refusal to accept an honorary degree from Amherst was greeted with applause, especially in the Western journals, who felt there was "altogether too much of this A.M. and D.D. business carried on." [27]

---

767; *Ohio Cultivator*, XV (Dec. 1, 1859), 356; *Farmer and Planter*, XI (March, 1860), 82.

In 1859 Professor Johnson made the following statement: "Of all the many fraudulent and poor manures which have been from time to time imposed upon our farmers during the last four years, there is none so deserving of complete exposure, and sharp rebuke, as that series of trashy mixtures known as Mapes's Superphosphates of Lime." *Southern Planter*, XIX (Dec., 1859), 767.

However, visitors to the large Mapes's farm at Newark, N.J., were usually favorably impressed. One writer after a tour of investigation said: "I can assure you that Professor Mapes' *farming* is no *humbug*." *Prairie Farmer*, XXII (July 12, 1860), 20.

[26] *Western Farmer and Gardener*, II (May 15, 1846), 151 ff.; II (April 15, 1846), 114; *New England Farmer*, XXIV (April 22, 1846), 342.

[27] *Michigan Farmer*, n.s., II (Sept. 1, 1860), 276. The editor adds: "No sooner does a man raise himself to any kind of eminence than some interested squad of the professional fraternity are impressed with the duty of making him one of themselves, by a parchment process, that is really of not the least value. . . . We are heartily glad to see some man who views the subject as it ought to be, and hope the example will become a precedent which will be

The editors' offices were not only storehouses for all kinds of agricultural information, but were in many cases social, educational, and business centers for the farmer, where the genial editor welcomed his patrons and friends. Here farmers met informally on their visits to the city. They talked over their problems and ofttimes sowed the seed for the initiation of agricultural and horticultural organizations.[28] The rendering of services to farmers along almost any line was considered a part of the editors' duty. The editors also benefited, as they became acquainted with the farmers individually and often learned practical things from them.[29]

It was customary from the earliest days of the agricultural press for the farmers to present editors with choice or unusual fruits and vegetables as well as agricultural curiosities and monstrosities. Thus, the editors' offices were gradually converted into small museums suggesting miniature agricultural fairs. These collections consisted of rare, unusual plants and seeds, specimens of rocks, minerals, fossils, and, occasionally,

---

followed, and thus tend to reduce the extravagant vanity which is apt to inflate pedants and pedagogues full of the pomposity and arrogance which A.M. and B.A. confers; and who beyond scribbling x and y on a blackboard have neither brains, ability nor industry."

The *Wisconsin Farmer* recognized "true manliness" in Beecher's stand. *Wisconsin Farmer*, XII (Dec., 1860), 388.

[28] The following quotation illustrates the role played by the *New England Farmer* in the establishment of the Massachusetts Horticultural Society: "The office of the *Farmer* became an exchange for the discussion of all matters of interest to cultivators. It was here that the subject of forming a horticultural society was discussed; and when such a society was formed, the *Farmer* naturally became its organ, and continued to be as long as it existed." Manning, *History of the Massachusetts Horticultural Society*, p. 47.

[29] *Union Agriculturist*, I (June, 1841), 41; *New Genesee Farmer*, XI (Jan., 1850), 29; *Farmers' Register*, VI (April 1, 1838), 47; *Prairie Farmer*, XVI (Jan. 3, 1851), 2; *Southern Cultivator*, X (Feb., 1852), 59; *Southern Planter*, I (Feb., 1841), 32; I (April, 1841), 54.

needlework and other forms of household art. The walls were hung with "portraits" of blooded stock.[30] The editor of the *Franklin Farmer* outlined his plans as follows: "We propose to make our office a sort of museum of the pictures of distinguished animals and of specimens of the geology, mineralogy and botany of the State." [31] The editor of the *Ohio Cultivator* had a much wider range of interest for the "Ohio Cultivator Museum." He said, "We wish especially for good portraits, prints, paintings or casts of all kinds of Live Stock, Garden Vegetables or Flowers, drawings or models of Machines and Implements, plans and elevations of Houses, Barns, Gates." [32]

After the *American Agriculturist* in the late fifties moved to new quarters on Park Row, New York City, it probably possessed the most pretentious museum of any of the periodicals. The new location, "entirely fire-proof" was selected to secure spacious and appropriate rooms where "thousands" could view their "Perpetual Exhibition." It was their aim to make this office the headquarters of agriculture and horticulture in New York—a sort of "Agricultural Exchange." [33]

Often closely associated with the museum was a farmers' reading room in which a large file of exchange papers as well as standard agricultural works were available. Upon occasion the county agricultural society or the local farmers' club held its meeting here.[34]

[30] *American Farmer*, XIV (Dec. 21, 1832), 321; *Franklin Farmer*, I (Sept. 23, 1837), 29; *Union Agriculturist*, I (June, 1841), 41; *Dollar Farmer*, I (Oct., 1842), 50; *Agriculturist*, IV (Jan., 1843), 2; VI (Oct., 1845), 150; *Southern Cultivator*, IX (Dec., 1851), 185; *Ohio Cultivator*, VII (Sept., 1851), 266; XII (Feb. 15, 1856), 56; *California Farmer*, III (May 24, 1855), 162; *American Agriculturist*, XIX (Oct., 1860), 305; *Southern Planter*, II (Sept., 1842), 213.

[31] *Franklin Farmer*, I (Sept. 23, 1837), 29.

[32] *Ohio Cultivator*, XII (Feb. 15, 1856), 56.

[33] *American Agriculturist*, XIX (Oct., 1860), 305, 309, 320.

[34] *Plough Boy*, II (Jan. 27, 1821), 278; *New England Farmer*, XVII

OFFICE OF THE *AMERICAN AGRICULTURIST*,
No. 41 PARK ROW, NEW YORK CITY

From *American Agriculturist*, XIX (October, 1860), 304.

Since the journals were primarily designed to serve as the main clearing house of information pertaining to agriculture and rural affairs, contributions from all manner of sources were solicited.[35] Perhaps the so-called practical farmer, who as Edmund Ruffin said had a "strong aversion to writing" was the most reluctant to take up the pen.[36] The editors did all in their power to facilitate contributions from this valuable source, urging the farmers to write everyday experiences in plain words and common phrases, insisting that "holiday terms" and fine writing were unnecessary.[37] The *American Agriculturist* promised to correct any imperfections in the contributions and added: "We like to have farmers sit down and scribble upon paper, just how they do this and how they do that—write it down just as they talk it to a neighbor." [38] The *Wisconsin Farmer* suggested to contributors to "write plain, with or without punctuation, and always directly to the point as possible" while the *Southern Cultivator* counseled its readers not to have the slightest fears about spelling or diction as the editor would put their contributions in "ship shape." [39] The *Ohio Cultivator's* ambition was to make every proposition set forth

---

(Jan. 9, 1839), 214; *Union Agriculturist*, I (June, 1841), 41; *American Agriculturist*, I (Feb., 1843), 336; *Prairie Farmer*, III (April, 1843), 70; XVI (Jan. 17, 1856), 10; *New Genesee Farmer*, VI (Aug., 1845), 116; XI (Jan., 1850), 29; *Southern Cultivator*, X (Feb., 1852), 59; *California Farmer*, III (May 24, 1855), 162.

[35] Edwards, "Some Sources for Northwest History; Agricultural Periodicals," p. 407.

[36] *Farmers' Register*, I (Sept., 1833), 193; III (May, 1835), 44–45.

[37] *New England Farmer*, I (Aug. 3, 1822), 6; *Maine Farmer*, I (Jan. 28, 1833), 10; *Cultivator*, I (June, 1834), 49; *Farmer and Gardener*, V (Jan. 15, 1839), 297; *Farmers' Cabinet*, IV (April 15, 1840), 286; *Prairie Farmer*, III (Jan., 1843), 1; *Southern Cultivator*, I (April 5, 1843), 38; *Michigan Farmer*, IV (Nov., 1846), 128.

[38] *American Agriculturist*, XII (Aug. 2, 1854), 328–29.

[39] *Wisconsin Farmer*, VIII (April, 1856), 187; *Southern Cultivator*, I (June 21, 1843), 103.

in their periodical "as plain and practicable as the two handles of a plow." [40] While it is evident with passing years, that the practical farmer was less timid about making contributions, nevertheless securing these articles continued to be the editors' major task.

A second and more active group of writers consisted of what might be termed amateur contributors.[41] Though rarely compensated financially for their contributions, members of this group, generally comprising gentlemen farmers who had time for experimentation and writing, took an active part in agricultural improvement. A few chosen at random will indicate their importance. Adam Beatty (1777–1858) was trained as a lawyer and spent most of his life in Kentucky, where he was active in politics, serving in the state legislature and senate. After 1823 however, he made farming his principal business. He imported purebred livestock and became deeply engrossed in the study of agriculture, especially Kentucky agriculture. Beatty was active in the Kentucky Agricultural Society and a popular contributor to the *Kentucky Farmer*, *Dollar Farmer*, *Farmers' Register*, and other agricultural periodicals.[42] Although Caleb N. Bement (1790–1868), whose background was probably less colorful, spent his early life in New York as a printer and hotelman, in 1834 he also decided to devote himself to agricultural theory and practice. He purchased a farm near Albany called "Three Hills Farm," and subsequently

[40] *Ohio Cultivator*, XVI (Jan. 1, 1860), 1.

[41] A large group of prominent national figures contributed freely to the agricultural press, especially in its early years. In this group were James Madison, Thomas Jefferson, Henry Clay, Timothy Pickering, and Daniel Webster.

[42] *Dictionary of American Biography*, II, 99–100; *Dollar Farmer*, III (March, 1845), 129, 142; IV (Feb., 1846), 121, 125; Bailey, *Cyclopedia of American Agriculture*, IV, 556.

In 1843 Beatty published a book of essays entitled *Southern Agriculture*.

made it famous. Here he became well known as an important breeder, importer, and distributor of improved livestock and probably still better known as an inventor of several important agricultural implements. In his frequent contributions to the agricultural periodicals, Bement presented the results of his experiments at "Three Hills." [43] He also called attention to new machines, improved livestock and advanced agricultural practices which came to his attention through extensive reading and travel.[44] Unlike Beatty or Bement in training and background was Martin W. Philips (1806–89) who was born in South Carolina and received his medical degree from the University of Pennsylvania. After practicing medicine for a short time Philips turned to farming and in 1836 purchased a tract of land in Mississippi. Here at "Log Hall" he raised fruit trees and became a successful cotton planter and stock breeder. To his contemporaries he was best known, however, as a prolific contributor to the farm press, where he recounted the results of many personal experiments. Philips urged upon the South the wisdom of diversification and the advisability of accepting improved agricultural implements and the advantage of obtaining better blooded stock.[45] Among the many journals, both

[43] Bement tried his hand as editor on two occasions. In 1844 he became an editor of the *Central New York Farmer*, and in 1848 the editor of the *American Journal of Agriculture and Science* (this was Vol. VII of the *American Quarterly Journal of Agriculture and Science*). In both cases the papers continued publication for only a short period because of lack of support.

[44] *Dictionary of American Biography*, II 172–73; *Genesee Farmer*, ser. 2, XX (Feb., 1859), 49; ser. 2, XX (April, 1859), 116; ser. 2, XX (May, 1859), 147; *American Agriculturist*, II (May, 1843), 55–56; II (June, 1843), 84–85; III (July, 1844), 201–2; VI (Dec., 1847), 369–70; *Cultivator*, n.s., I (March, 1844), 101.

[45] *Dictionary of American Biography*, XIV, 537; *Cultivator*, n.s., I (April, 1844), 124–25; n.s., I Nov., 1844, 331; *American Farmer*, I (Oct., 1845), 115 ff.; *American Agriculturist*, IX (Aug., 1850), 244; IX (Oct., 1850),

in the North and the South, to which he contributed most frequently, were the *American Farmer*, *Cultivator*, *Southern Cultivator*, and *American Cotton Planter*. From 1843 to 1845 he was one of the editors of the *South-Western Farmer*, published at Raymond, Mississippi, and, as was characteristic of this group of reformers, he received no financial compensation.[46] In 1859 he became editor of the *Southern Rural Gentleman* (Grenada, Mississippi). Philips has been placed second only to Ruffin as an influential agricultural leader in the South.[47]

The third group of contributors comprised those professional agricultural writers who made their living, in part, from contributions to the farm press. Three noteworthy examples of this group are Andrew Jackson Downing, John P. Norton, and Solon Robinson. Downing,[48] long engaged in the nursery business in New York, wrote for a number of farm papers on landscape gardening, rural architecture, and horticulture and finally in 1846 became the first editor of the *Horticulturist*. John P. Norton's agricultural tours of Europe, which appeared widely in the journals, were exceedingly popular; later, while professor of agricultural chemistry at Yale, he wrote regularly for a number of periodicals, principally on agricultural chemistry. Solon Robinson was best known for his tours in all parts of the United States; these are discussed fully in succeeding pages. In addition, he wrote many articles upon a wide field of subjects, including rural architecture, agricultural societies,

---

303; *Southern Cultivator*, XIV (Feb., 1856), 52–53; *American Cotton Planter*, IV (Jan., 1856), 30.

[46] *South-Western Farmer*, II (March 17, 1843), 9; *Dictionary of American Biography*, XIV, 537.

[47] Gray, *op. cit.*, II, 781.

Dr. Philips's diary of farm operations (1840–63) has been published under the title, "Diary of a Mississippi Planter," pp. 305 ff.

[48] A more detailed consideration of Downing may be found on pp. 72–73.

agricultural education, as well as articles advocating scientific agricultural practices, suggestions for the guidance of prospective Western settlers, and accounts of his own experiments and investigations. This rapidly growing group of professional agricultural writers had come to the fore in the twenty years before the Civil War.

The last group to be mentioned included the editors themselves. The proportion of articles written by them varied greatly and depended largely upon individual ability and in many instances, upon available time, aside from other pursuits. For example, Governor Hill, who frankly admitted that he was a "mere gatherer" when compared with many of his colleagues, contributed comparatively few articles to his paper, the *Farmer's Monthly Visitor*.[49] On the other hand, Edmund Ruffin probably wrote more than half of the articles appearing in his *Farmers' Register*, while Solon Robinson also furnished a major portion of the text for the *Plow*.[50]

The editors wrote upon a great variety of subjects, depending largely upon their particular interests and experiences. As a rule they dealt with scientific agriculture, results of experiments at their farms, visits to fairs, agricultural tours, and personal views upon contemporary problems. A question and answer corner provided a popular feature. Often however, the editor was forced to appeal to his readers for satisfactory answers to knotty farm problems. Important agricultural and horticultural books were reviewed under such headings as "Book Review Section," "Literary Notices," and "Our Book Table." Frequently, selections from foreign books and excerpts from periodicals, especially those of France and England, were

[49] *Farmer's Monthly Visitor*, III (Dec. 31, 1841), 190.

[50] Swem, *An Analysis of Ruffin's Farmers' Register*, p. 42; *Dictionary of American Biography*, XVI, 215; Kellar, *Solon Robinson*, I, 30.

reprinted. In most cases the editors carried a column entitled, "Work of the Month" which successively outlined the farmers' duties in detail.[51]

Although it was always a problem to obtain satisfactory contributions, the editors experienced less and less difficulty in filling their pages with valuable original articles. In the fifties it was not uncommon for them to comment upon the "avalanche of communications" received at their offices. Nor was the statement in the *Ohio Cultivator* of 1858 unusual, when the editor said that a recent number contained only articles written expressly for that periodical.[52]

A favorite feature comprised a series of letters from the editor, a staff member, or correspondent, traveling either in this country or in Europe.[53] As first conceived, an agricultural tour provided a substitute for the written reports which farmers were reluctant to make.[54] The traveler went about the country, not as an instructor, but as a seeker of knowledge from practical husbandmen who were delighted to give information verbally.[55] These writers described agricultural conditions

[51] A representative article of this type may be found on pp. 282–83.

[52] *Ohio Cultivator*, XIV (March 15, 1858), 88; *American Agriculturist*, XVII (Feb., 1858), 60; *Valley Farmer*, XII (Jan., 1860), 11.

[53] A similar feature was popular in the newspapers of the period. *Cambridge History of American Literature*, II, 190.

[54] *American Farmer*, II (Jan. 12, 1821), 329; *New-York Farmer*, V (Jan. 26, 1832), 33; *Southern Agriculturist*, V (July, 1832), 354; *Farmers' Register*, IV (Dec. 1, 1836), 512; *Farmers' Cabinet*, IV (May 15, 1840), 297; *Southern Cabinet*, I (Jan., 1840), 13; *Southern Planter*, V (April, 1845), 75–76; *Country Gentleman*, X (July 9, 1857), 32.

Arthur Young of England was the first famous touring farming reporter. Rodney C. Loehr, "The Influence of English Agriculture on American Agriculture," p. 9.

[55] These reporters often acted as agents for the journals, sometimes as representatives of agricultural seed stores and warehouses, and occasionally, as is true of Southern reporters, became purchasing agents for machinery and stock, when in the North.

throughout the sections visited, told about the principal crops and methods of culture, and made comments on the machinery and implements which were in use. They also included statistics of the resources of the district as well as data on the exact conditions and prospects of crops for that year. These traveling correspondents, with eyes wide open, journeyed about the country on horseback, by carriage, canalboat, steamboat, and railroad train.[56]

The travelers naturally recorded interesting items on a host of topics relating to all phases of rural life. Comments were also made upon the climate, roads, travel accommodations, and living conditions. Especially interesting in Southern agricultural tours were the descriptions of slave life and plantation economy. Remarks on phases of marketing, including transportation facilities and prices current, were common. These reports are not only an invaluable source for the agricultural historian, but contain abundant data for the general economic and social historian as well, though the latter groups have tapped these rare sources only sparingly.[57]

Certainly the most famous of the many notable traveling correspondents was Solon Robinson (1803–80) of Indiana.[58]

[56] Particularly interesting tours were made by editors R. L. Allen (*American Agriculturist*), Thomas Affleck (*Western Farmer and Gardener* and other journals), and Warren Isham (*Michigan Farmer*).

Horticultural tours were often made by the editors of such departments.

[57] There is a wide variety of data in the agricultural press of great value to the social, economic, and to a less extent, the political historian. For an analysis of these journals as an historical source, see Edwards, "Some Sources for Northwest History; Agricultural Periodicals," pp. 407–17.

Kathleen Bruce says that the "backbone of any agricultural study of Virginia in the first half of the nineteenth century lies" in the pages of the *Farmers' Register*, *Southern Planter*, and the early volumes of Skinner's *American Farmer*. Bruce, "Materials for Virginia Agricultural History," pp. 10–11. By substituting the names of local journals, a similar statement might be made regarding each of the other older states of the Union.

[58] Kellar, *Solon Robinson, Pioneer and Agriculturist; Selected Writings,*

During the period between 1841 and 1851, this popular writer made six major journeys and a number of shorter ones, visiting practically every state in the Union and portions of Canada. During this interval he corresponded for a number of journals. Most of his articles were written for the *Cultivator* and the *American Agriculturist.* He also represented the latter in a business capacity. Robinson's itinerary was carefully noted in contemporary periodicals, and farmers in all sections vied with one another for the honor of a visit from him. This shrewd and discriminating commentator devoted his attention chiefly to rural districts, and his accounts customarily furnished such details as names of individuals, places and dates, a circumstance which immeasurably enhances their value.[59] Possessed of a dry wit, pleasing personality, and the ability to tell a good story, Robinson found money a "needless commodity" on his journeys.[60] Always delighted to make himself useful wherever he went, he was called "the Agricultural Missionary of the land." [61] His articles were eagerly awaited by thousands of farmers and many of his "tours" made the rounds of the press. Probably no other man in the United States had a wider or more accurate knowledge of our agriculture in the twenty years before 1860.[62]

---

*1825–1851,* 2 vols. These volumes contain accounts of Robinson's most important agricultural tours.

See pp. 284 ff. for one of Robinson's reports on an agricultural tour in the Southwest.

[59] *Ibid.,* I, 28–29.

Robinson sometimes solicited subscriptions for journals other than the periodicals for which he wrote. *Southern Cultivator,* VII (April, 1849), 56.

[60] Kellar, *Solon Robinson,* I, 21–22, 26, 41, 322; *Union Agriculturist,* II (July, 1842), 61.

[61] *Southern Cultivator,* IV (Jan., 1846), 8.

[62] Kellar, *op. cit.,* I, 29.

In 1852 Robinson served as assistant editor of the *American Agriculturist,* and in the following year, became the editor of the *Plow,* the successor of the

SOLON ROBINSON, AT THE AGE OF NEARLY SIXTY

Frontispiece, *Facts for Farmers*, 1865.

Perhaps no American before the Civil War was more familiar with European agricultural conditions than Henry Colman (1785–1849), another popular writer for the rural press.[63] From 1843 until his death in 1849, Colman spent most of his time in Europe investigating the agricultural situation there.[64] His reports were widely published in the American farm press.[65] Patrick Barry, editor of the horticultural department of the *New Genesee Farmer,* and other horticultural writers likewise toured Europe and reported their observations.

Another type of agricultural excursion, popularized in the press, consisted of visits to the farms of distinguished American statesmen.[66] (Almost without exception, the leaders of American life were interested in farming.) These interviews, included a description of a tour about the estate with the owner,

---

*American Agriculturist.* Financial difficulties made it necessary to discontinue the latter periodical at the end of one year, at which time he became the agricultural editor of the New York *Tribune,* a position he retained until his death in 1880.

Among Robinson's publications are found: *Guano, A Treatise of Practical Information for Farmers* (1853); *Facts for Farmers: Also for the Family Circle* (1863–64); *Me-won-i-toc, A Tale of Frontier Life and Indian Character* (1867). Kellar, op. cit., I, 30–31, 34–35.

[63] Colman had previously made an agricultural survey of Massachusetts at the instigation of Governor Edward Everett and for a period was editor of the *New Genesee Farmer.*

[64] The following books were published by him: *European Agriculture and Rural Economy from Personal Observation* (1844); *Agriculture and Rural Economy in France, Belgium, Holland, and Switzerland* (1848); *European Life and Manners in Familiar Letters to Friends,* 2 vols. (1849). See *Dictionary of American Biography,* IV, 312.

[65] The popularity of agricultural tours, as a feature of the press, may be noted with a glance at the *Cultivator* for 1846. In this volume Colman reported from England; D. G. Mitchell from England, Scotland, France and Switzerland; M. Horsford from England and Germany; and John P. Norton from Scotland. In the same year, Solon Robinson reported from various sections of the United States. However, this marked concentration upon "tour" articles is not representative of the entire press.

[66] A typical report of such a visit may be found on pp. 290 ff.

an account of the farm and its management, together with interesting comments from the noted host. Especially popular were the accounts of the farms owned by Van Buren, Webster, Clay, Calhoun, Jackson, and John Randolph.[67]

These pioneer journalists faced many obstacles scarcely experienced by editors today. As late as 1841 the editor of the *Southern Planter* found himself apologizing for lack of engravings in his paper as the city of Richmond did not provide a professional engraver.[68] The unreliability of the mails and slow delivery of papers were sources of much annoyance to the subscribers and certainly of great irritation to the editors, most of whom gladly replaced lost numbers, although this sometimes necessitated sending as many as three copies.[69] While the situation was gradually improving in the East, the "Western" journals in the fifties were still complaining of delivery service. The editor of the *Wisconsin Farmer* felt that the uncertainty of the mail was "insufferable"; a subscriber to the *Prairie Farmer* complained that his paper came "provokingly irregular"; and other comments from readers indicate that delays of a month or two were not uncommon. Frequently, when the editor constituted the entire office force, as was often the case in the early years, his illness or that of a member of

[67] For examples see, *New-York Farmer*, VIII (May, 1835), 140; *Cultivator*, X (Jan., 1843), 15; n.s., VI (Jan., 1849), 9; *Agriculturist*, IV (June, 1843), 94–95; *Farmer's Monthly Visitor*, IV (Nov. 30, 1842), 175; VII (Sept. 30, 1845), 132; VII (Oct. 31, 1845), 147–48; XI (Aug. 31, 1849), 119; *New England Farmer*, XXIV (Aug. 13, 1845), 54; XXIV (Aug. 20, 1845), 62; XXIV (Sept. 17, 1845), 89–90; *Southern Cultivator*, V (Dec., 1847), 182; *American Agriculturist*, XV (Oct., 1855), 7.

[68] *Southern Planter*, I (Dec., 1841), 260.

[69] *Farmers' Register*, I (Oct., 1833), 320; I (Feb., 1834), 568–69; *Farmers' Cabinet*, I (July 1, 1837), 4; *Cultivator*, VI (June, 1839), 65; *Prairie Farmer*, XV (Jan., 1855), 37; *American Agriculturist*, XVI (July, 1857), 144; *Ohio Cultivator*, XVI (Dec. 15, 1860), 374.

his immediate family, would retard publication for weeks. Additional irregularities arose through unexpected delays while the editor was on agricultural tours; lack of suitable paper; or a breakdown of machinery.[70] However, improvement with regard to punctuality was evident as the period wore on.

Another annoyance, especially in the early years, was the persistent use of the *nom de plume*. While the editors were usually willing to publish articles when they knew the writer's identity, they begged permission to disclose the real name, especially in the case of factual material, so as to give "character" and increase public confidence in the contents of the articles.[71] The editor's comment in the *South-Western Farmer* is characteristic of the criticism made by readers: "Scarcely a subscriber drops into our office without making complaint that there is so much anonymous in our columns, and we are assured that many are of the opinion that most of what appears as communications in our paper is really composed by the Editor." [72] The usual defense offered by contributors was based upon "delicacy in appearing before the public" under any other than an assumed name, but this "modesty" was very often an unwillingness to subject themselves to the "scourge" of competent critics, or, as a particularly caustic writer in the *Genesee*

[70] *Cultivator*, V (Feb., 1839), 222; *Franklin Farmer*, III (Jan. 11, 1840), 158; *Genesee Farmer*, VII (April, 1846), 79; *Michigan Farmer*, V (April, 1847), 15; *Farmer's Monthly Visitor*, XI (Jan. 31, 1849), 6; *Prairie Farmer*, IX (March, 1849), 103; XII (Aug., 1852), 388; XIII (Feb., 1853), 84; XV (Feb., 1855), 69; XV (April, 1855), 135; XVII (Jan. 15, 1857), 18; *Ohio Cultivator*, V (Sept. 15, 1849), 280; *Wisconsin Farmer*, IV (Sept., 1852), 238; *Valley Farmer*, X (Sept., 1858), 291.

[71] *Farmers' Register*, I (June, 1833), 63; *Cultivator*, I (June, 1834), 49; *Southern Planter*, I (May, 1841), 67; *Agriculturist*, III (July, 1842), 166; *South-Western Farmer*, I (Oct. 14, 1842), 44; *Prairie Farmer*, XVI (Feb. 21, 1856), 30; *Wisconsin Farmer*, XII (July 1, 1860), 221.

[72] *South-Western Farmer*, I (June 24, 1842), 121.

*Farmer* said, "Certainly I should have been loathe to have said in my *own name* half what I said under a fictitious one." [73] It is interesting to note the frequent use of such signatures as "Cincinnatus," "Agricola," "Silvanus," "Baltimoriensis," and "Veritas"—reflections of the same classical influence noticeable in the architecture of the period. Most of the writers resorted to the use of pseudonyms at one time or another.[74]

Vast quantities of letters, carelessly written, rambling, without dates and signatures were received. Many writers, unwilling to indicate their post offices, were delighted to date letters from "such perfumed regions" as "Rosewood Glen," "Violet Lawn," "Muskmelon Creek" and "Pollywog Pond." The editors constantly admonished the farmers to write constructive communications, to deal concretely with their subjects, and to avoid "magniloquent descriptions of the beauties of farming." [75] Henry Ward Beecher of the *Indiana Farmer and Gardener*, warned his contributors against the use of so-called

[73] *Genesee Farmer*, II (April 28, 1832), 133; *Farmers' Register*, V (June 1, 1837), 87; *Southern Cultivator*, I (May 24, 1843), 87; *Southern Planter*, XIX (March, 1859), 174.

[74] Edmund Ruffin probably published more anonymous contributions from his own pen than any other editor. He did this, in part, to avoid appearing "unnecessarily conspicuous" in his paper. *Farmers' Register*, V (Aug. 1, 1837), 249. He had other reasons, however. On one occasion he called the "editor's" attention to the unfair practice of copying articles from the *Farmers' Register* without rendering proper credit, and signed his communication, "South." *Farmers' Register*, II (Feb., 1835), 581. At another time, after Ruffin had been criticized by his readers for his too frequent remarks on banks, he hid behind "Taylor of Caroline" in his next article on this subject. *Farmers' Register*, IX (July 31, 1841), 439 ff. For a list of the incredible number of anonymous articles in the *Farmers' Register* written by Ruffin himself, see Swem, *op. cit.*

[75] *Kentucky Farmer*, V (Dec. 4, 1841), 82; *Maine Farmer*, XIII (Nov. 27, 1845), 189; *New England Farmer*, XXIV (Dec. 3, 1845), 183; *American Agriculturist*, XII (Aug., 1854), 328; *Prairie Farmer*, XVIII (Aug. 26, 1858), 136; *Ohio Cultivator*, XVI (April, 1860), 120; *Cultivator*, ser. 3, VIII (May, 1860), 163.

"fine writing," which he labeled "painted emptiness." "In short [he wrote] geoponical cant, and pastoral cant, and rural cant in their length and breadth are, like the whole long catalogue of cants (not excepting the German Kant), intolerable." [76] Characteristic of the comments of his colleagues was the statement made by the editor of the *Prairie Farmer* when he wrote, "Be poetic when you have nothing else to do; but when you are dealing with facts be prosaic as dirt." [77]

One of the most difficult tasks faced by the editors was the breaking down of the farmers' aversion to agricultural journals and books. The old saying that, "Books and learning never made farmers," was often voiced by agriculturists who sneered at the idea of regarding agriculture as a science.[78] Then too, it was generally conceded that the most famous farmers, men who had rendered the most important service to agricultural science, were almost without exception poor practical farmers.[79]

There was justification for the farmer's hostility toward agricultural journals and his suspicion of everything that savored of "scientific farming," for theoretical agriculture was just beginning to sever its connections with speculative philosophy. Agricultural experimentation was often conducted unscientifically and "authorities" sometimes glibly announced theories

[76] *Indiana Farmer and Gardener*, I (Oct. 4, 1845), 294.

[77] *Prairie Farmer*, XIV (Jan., 1854), 41.

[78] *Southern Cultivator*, I (Aug. 2, 1843), 127; II (Nov. 13, 1844), 183; *Farmers' Cabinet*, VIII (May 15, 1844), 323; *Western Farmer and Gardener*, II (May 15, 1846), 150; *Farmers' Cabinet*, I (May 1, 1837), 320.

[79] Edmund Ruffin admitted that there was foundation for this general opinion and pointed out such English agricultural leaders as Lord Kames, William Marshall, and Arthur Young as examples. Indeed it was said that if an inquisitive traveler had asked a practical husbandman of Suffolk as to who was the worst farmer in the country, he would immediately have selected the most distinguished, as well as the most voluminous writer on agriculture, Arthur Young. *Farmers' Register*, II (June, 1834), 17 ff.

that appear ridiculous today.[80] Indeed, it is not surprising in those days of early beginnings that some of the teachings were utterly impractical, scientifically and economically unsound, and from the modern point of view, sheer nonsense.

The dirt farmer frequently expressed dissatisfaction with the contents of the farm press. "There is so little matter in it of use to the small, poor, middle-interest farmer . . . and so much about flowers, tulips, geraniums, etc. etc.; and so much about this and that great farm, managed by the rich and opulent, all of which is beyond the reach and calculated to discourage the great mass." [81] Another writer complained, "For instance, it seems to me that every philosopher, lawyer, doctor, merchant and loaf—gentleman of leisure, I mean,—when he has nothing else to employ him, sits down and amuses himself by writing for the agricultural papers." [82] Referring to agricultural journals, a former subscriber said, "I did take one several years ago and that had so much to tell about a new kind of potatoe, that they sold for 25 cents a pound, and after all, it wasn't no better than the long reds; and about tree corn and mulberry trees; and a good many farmers got *bit*, by believing their great stories, that I got sick and stopped it, and would not now take the gift of one." [83] "Half-baked" ideas, on occasion, found their way into the press. At one time the papers suggested placing a saucerful of chloroform under beehives so that the honey might be extracted from the hive much like "teeth are extracted by a fashionable dentist." However, reports from the farmers indicated that the bees "instead of being put to sleep for the operation only," never woke up.[84] In

[80] Gray, *op. cit.*, II, 789.

[81] *New England Farmer*, XVII (June 26, 1839), 406.

[82] *Southern Planter*, VIII (Jan., 1848), 22.

[83] *Farmers' Cabinet*, IX (Sept. 16, 1844), 52.

[84] *American Agriculturist*, XI (Nov. 9, 1853), 129.

justification of the criticism directed to the agricultural papers, A. B. Allen, editor of the *American Agriculturist*, called attention in 1845 to the "rank humbugs and ill-judged matter with which their columns" were too often overloaded.[85]

The editors in general readily admitted that fallacious material occasionally appeared in their pages. However, the statement that "probably seven-eighths of the agricultural reading of the present day is humbug, and will not stand the test of experiment," which appeared in the *Country Gentleman* of 1857, referring primarily to the agricultural press, was pronounced unfair and inaccurate.[86] Furthermore, they stoutly denied a predominance of theoretical ideas. The *Southern Cultivator* disclaimed any intention of making the journal a scientific paper filled with "speculations of philosophy, and extracts from Chaptal, Davy, Liebig, Johnston, Boussingault" and added, "What we want now is a plain account of the experiments of men of plain common sense." [87] The editor of the *Tennessee Farmer* also took this point of view, stating that it was a great fallacy to suppose that an editor of an agricultural journal constituted himself a dictator of opinion and practice for his readers.[88] The *Farmers' Cabinet* expressed a similar attitude:

> We have greatly misconceived the design of the agricultural periodicals of the day, if one of their leading objects is not to afford the practical farmers of the land an opportunity of communicating and comparing their several methods of tillage—thus embodying the opinions and experience of the whole reading community, for the mutual benefit of all.[89]

[85] *Ibid.*, IV (Nov., 1845), 335.

[86] *Country Gentleman*, IX (May 14, 1857), 313.

[87] *Southern Cultivator*, III (May 1, 1845), 72. See also *Southern Planter*, I (Jan., 1841), 1; *Rural American*, II (Sept. 1, 1857), 129.

[88] Quoted in *Pennsylvania Farm Journal*, I (April, 1851), 1.

[89] *Farmers' Cabinet*, IV (April 15, 1840), 286.

The *Valley Farmer* estimated that nine out of ten articles found in farm journals were written by practical farmers.[90] Even so, the farmers' aversion to agricultural reading was so great that the editor of the Albany *Cultivator* at the middle of the century estimated that not one in two thousand subscribed for an agricultural paper.[91]

Reimbursement for editorial services varied greatly throughout the agricultural press. Many of the editors accepted no compensation for their efforts; others, who received salaries, normally supplemented these incomes with additional earnings. A third group of editors, who were at the same time proprietors, were fortunate indeed to receive any return. Those who found it necessary to supplement their earnings frequently issued annual almanacs, engaged in book and job printing, conducted seed stores or agricultural warehouses, sold real estate, served as school teachers, agricultural chemists, or farmers. The statement by the editor of the *American Agriculturist* in 1845 characterizes the attitude of many of his colleagues upon this problem of finance. "If it were not for our extensive business and other things connected with this journal, we would not continue it another month—we could not afford to do so." [92]

Of the hundreds of journals inaugurated during the period, scarcely more than a score survived beyond five years and the vast majority less than three years. The apparent success that attended a half dozen of these papers caused many ambitious individuals to conclude that such a venture was an easy road to wealth.[93] Consequently journals sprang up like mushrooms,

[90] *Valley Farmer*, XII (Jan., 1860), 11.

[91] *Cultivator*, n.s., VII (April, 1850), 152.

[92] *American Agriculturist*, IV (Nov., 1845), 334.

[93] *Prairie Farmer*, VII (Feb., 1847), 69; *Country Gentleman*, VI (Dec. 6, 1855), 364.

the proprietors too often oblivious of the requirements essential to a successful paper; that is, a real need, sizable capital, and outstanding ability on the editor's part. In 1847 the *Prairie Farmer*, commenting on new periodicals, said that a half dozen a year were "sallying out, like bees in a January noon, to fall suddenly into the snow, benumbed, to rise no more." [94] A notice similar to the following became a familiar item in the press, "THE FARMER, by Prof. Nash of Amherst, Mass., has been discontinued, having run its publishers badly in debt." [95]

Many reasons have been proffered for the high mortality of the journals.[96] Poor business methods, subscription payments in "bad" money (for which the editors sacrificed a portion of each dollar), the loss of subscription money in the mail (the risk of which the editors normally assumed), and dishonest agents all played their part.[97] Very often subscribers were alien-

---

Particularly successful were the *American Farmer* (under Skinner), the *Cultivator*, the *American Agriculturist*, the *New Genesee Farmer*, and the *Ohio Cultivator*.

[94] *Prairie Farmer*, VII (Feb., 1847), 69.

The editor of the *New England Farmer* in 1859 claimed that agricultural journals were being inaugurated almost every week. *New England Farmer*, n.s., XI (Nov., 1859), 501.

[95] *Ohio Cultivator*, XI (Dec. 15, 1855), 375.

[96] In 1841 the *Kentucky Farmer* did not believe more than one-fifth of the journals made a reasonable profit; in 1855 it was the opinion of the *Country Gentleman* that not one in ten earned expenses; while in 1860 the *Wisconsin Farmer* declared that less than one-third netted the publishers one penny for their labors. *Kentucky Farmer*, IV (Jan. 2, 1841), 148; *Country Gentleman*, VI (Dec. 6, 1855), 364; *Wisconsin Farmer*, XII (Aug. 1, 1860), 238.

[97] *Plough Boy*, III (April 20, 1822), 374 ff.; *Farmer's Monthly Visitor*, VII (Nov. 30, 1845), 176; *Kentucky Farmer*, IV (July 31, 1841), 357; *New Genesee Farmer*, III (Feb., 1842), 17; *Prairie Farmer*, XV (Jan., 1855), 37; *Cultivator*, VI (June, 1839), 65; n.s., II (March, 1854), 98; *Country Gentleman*, III (Feb. 16, 1854), 104.

The editor of the *Farmers' Register* stated in 1841 that in spite of the largest circulation in its history, profits reached their lowest ebb because of

ated by ill-chosen material as well as by the occasional editor who unwisely wandered off into the dangerous field of politics.[98] Excessive competition was without doubt the main reason for the failure of a vast majority of periodicals, and the fact that a large body of readers defaulted in subscription payments accounts for the collapse of many. The editor of the *Farmers' Register*, in the tenth volume, claimed that the subscription arrears amounted to more than all the clear profit ever derived from the venture.[99] The *Southern Planter* in 1848 estimated $2,500 unpaid on its books, while the *Michigan Farmer* and *Prairie Farmer* in 1857 claimed $5,000 and $10,000, respectively, was due them.[100] While it was not unusual for a journal to have delinquent accounts of six or eight years, an announcement in the *New England Farmer* stating it had received forty-two dollars from a distant patron in payment of dues for the previous twenty-one years, elicited numerous editorial comments on the "beauties of the credit system." [101] The publication of "black-lists" and application of other measures proved ineffective. Gradually the editors were compelled to adopt the cash system. Notwithstanding, incredibly high unpaid balances continued to be a problem with most of the journals and proved a major cause for many failures.

Many of the articles and letters written for the agricultural

---

the "disordered state of the currency." *Farmers' Register*, IX (Aug. 31, 1841), 507.

[98] *American Agriculturist*, I (June, 1842), 73–74; IV (Nov., 1845), 335; *Farmers' Register*, IX (March 31, 1841), 163 ff.; IX (Oct. 31, 1841), 617 ff.

[99] *Farmers' Register*, X (April 30, 1842), 155.

In 1841 the editor reported between $4,000 and $5,000 in subscription arrears long overdue. *Farmers' Register*, IX (Aug. 31, 1841), 507.

[100] *Southern Planter*, VIII (Dec., 1848), 353; *Michigan Farmer*, XV (April, 1857), 120; *Prairie Farmer*, XVII (March 5, 1857), 78.

[101] *New England Farmer*, XXIV (Oct. 15, 1845), 126.

press were prompted solely by "patriotic" motives, the satisfaction of advancing a good cause; no material compensation was received by the writer. However, the editors from time to time offered financial incentives to encourage contributors. The *Farmer's Monthly Visitor* paid one dollar for every thousand words deemed "worthy of insertion." [102] Many of the papers offered a year's subscription to anyone who would write "a good article," the equivalent of a page in length.[103] Solon Robinson, one of the best-known "professional" correspondents, received one hundred and fifty dollars for ten installments of "Notes of Travel" published in the *Cultivator* during 1845.[104]

Another method, which editors quite generally adopted, consisted in offering prizes for the best essays on selected topics.[105] The *American Farmer* at one time offered a perpetual free subscription for the most creditable contribution on any strictly agricultural subject, and later more than six hundred dollars in book premiums for specified essays.[106] The *Genesee Farmer* offered a dollar book for the best short composition on one of hundreds of subjects announced in the paper from

[102] *Farmer's Monthly Visitor*, I (Jan. 15, 1839), 1.

[103] *Western Farmer and Gardener*, II (Oct., 1840), 2; *Southern Planter*, V (Dec., 1845), 282; *American Agriculturist*, IX (Jan., 1850), 37; *Northern Farmer*, I (Jan., 1854), 1–2.

[104] Robinson, "Account Book, 1840–1853," p. 27. (This interesting document is in the possession of the McCormick Historical Association, Chicago.) During 1847 Robinson received eight dollars for a review of each past number of the *American Agriculturist*—each review occupied between one and a half and two and a half pages of the journal.

[105] *American Farmer*, XIII (March 18, 1831), 7; *Cultivator*, V (March, 1838), 5; *Franklin Farmer*, II (June 1, 1839), 325; *Agriculturist*, I (April, 1841), 93; III (April, 1842), 73; *Farmers' Cabinet*, VIII (Jan. 15, 1844), 198; *Valley Farmer*, IX (Dec., 1857), 387; *New Genesee Farmer*, XIX (Jan., 1858), 36; *Prairie Farmer*, XXII (Nov. 29, 1860), 344.

[106] *American Farmer*, XIII (March 18, 1831), 7; n.s., I (April 1, 1840), 353.

time to time. This journal was eager to receive among others, articles "On Butter Making," "On the Cultivation of Peas," "On the Best Method of Fencing a Farm," and "On the Benefits of Agricultural Fairs." The same premium was offered to the women for the "Best Dozen Domestic Receipts" and for the most satisfactory answer to such questions as, "What are the Proper Duties of a Farmer's Wife?" and "What can mothers and daughters do to make farm life attractive to their sons and brothers, and prevent them from leaving the farm to engage in mercantile or professional pursuits?" [107] Cash prizes were also frequently given by the journals. For example, the *American Agriculturist* offered two one hundred dollar prizes for the best essays on fencing and dairying.[108] In judging the merits of the contributions, practical articles were given preference over theoretical treatises. As the editors rectified errors of style and composition, such defects did not detract from articles received from inexperienced writers. Thousands of these prize essays were published in the journals and a particularly fine one often made the rounds of the press.

The agricultural societies also offered awards and, in searching for methods of extirpating insect pests, gave large prizes for essays on such discoveries. They granted additional premiums in some states running as high as five hundred dollars for the best farms and the press eagerly gave wide publicity to these activities.[109]

[107] *Genesee Farmer*, ser. 2, XVII (Nov., 1856), 355; ser. 2, XVIII (Jan., 1857), 35; ser. 2, XVIII (Nov., 1857), 355.

An example of a prize essay from the *Genesee Farmer* may be found on pp. 293–94.

[108] *American Agriculturist*, XVII (Dec., 1858), 378.

[109] *New England Farmer*, V (Oct. 6, 1826), 81; IX (Oct. 15, 1830), 89–90; *Cultivator*, II (April, 1835), 30; IV (May, 1837), 48; V (June, 1838), 71; *New Genesee Farmer*, V (Nov., 1844), 90–91; VI (June, 1845), 81; *Wisconsin Farmer*, XI (Sept., 1859), 331; *Country Gentleman*, XIV (Sept. 8, 1859), 160.

The journals resorted to numerous methods for expanding their subscription lists. Often sample copies were mailed to farmers; occasionally short trial subscriptions at reduced rates were granted, and in some cases, produce of all kinds was accepted in payment of subscription dues.[110] The most characteristic device, however, was the authorization of postmasters to serve as agents for which they normally retained ten percent of sums collected.[111] Most of the papers eventually permitted anyone who volunteered, to act as agent. Frequently the lists of agents were published in the journals; the *Cultivator*, for example, in 1849 recorded approximately two thousand names from all portions of the United States, Nova Scotia, New Brunswick, and Canada.[112] A number of traveling agents applied themselves exclusively to this work, while a greater number combined it with other employment.

Salesmen received compensation in various forms. For instance, a farmer might secure his own copy free by obtaining several subscriptions from his neighbors. Very often, however, a small commission served as reimbursement, but the most effective means for increasing the circulation was the inducement offered through attractive awards. These consisted of premiums given for obtaining a specified number of subscrip-

[110] *American Farmer*, I (April 2, 1819), 5; *Genesee Farmer*, III (Jan. 5, 1833), 1; *Farmers' Register*, I (June 1, 1833), 64; *Michigan Farmer*, III (May, 1845), 32; *American Agriculturist*, XII (March 22, 1854), 16; *Prairie Farmer*, XVII (Nov. 19, 1857), 372; *Ohio Farmer*, VIII (Jan. 8, 1859), 13.

Agricultural societies and clubs were entitled to cheaper rates.

[111] Postmasters were permitted to frank letters containing remittances for periodicals. This privilege was modified in the forties and the practice continued with restrictions. *The Post-Office Law with Instructions and Forms*, Act of March 2, 1799, Section 17, p. 15; *Kentucky Farmer*, IV (Oct. 17, 1840), 61; *Farmers' Cabinet*, IX (July 15, 1845), 380; *Michigan Farmer*, VII (Nov. 15, 1849), 337; XII (Nov., 1854), 347; *Ohio Cultivator*, VI (Feb. 1, 1850), 32.

[112] *Cultivator*, n.s., VI (Nov., 1849), 356 ff.

tions and prizes offered in competitive contests for gaining the greatest number of patrons for the paper. The latter awards included a variety of items such as flower seeds, agricultural books, pieces of the "genuine Atlantic Cable," large cash prizes, Webster's dictionary, "handsome full-bred Berkshire hogs," sewing machines, and reapers.[113]

While the editors maintained that they desired to appeal primarily to the patronage of the farmer, actually, available data indicates that men in all walks of life subscribed to the journals. From the first, the so-called gentleman farmers and the Southern planters were enthusiastic supporters.[114] The *American Farmer* was always strongly supported by professional men.[115] In 1833 the list of subscribers to the *Farmers' Register*, aside from farmers, included preachers, professors, lawyers, physicians, officers of the army and navy, and a number of prominent political figures. Approximately ten percent of this total were medical men.[116] In the same year the editor of the *Genesee Farmer*, lamenting the condition of rural apathy toward agricultural advancement, stated that a majority of his readers were other than farmers.[117] The *American Agricul-*

[113] *New England Farmer*, IV (Jan. 6, 1826), 192; *Farmers' Register*, I (Dec., 1833), 448; *Southern Cultivator*, II (Nov. 27, 1844), 191; *Ohio Cultivator*, II (Feb., 1846), 17; *Michigan Farmer*, XI (Dec., 1853), 367; *Country Gentleman*, IV (Dec. 7, 1854), 360; *American Agriculturist*, XVII (Oct., 1858), 320; *Valley Farmer*, XII (Dec., 1860), 389.

[114] Eli J. Capell, a fairly well-to-do Mississippi planter, seems to have been representative of the latter group. His library indicates that he subscribed at various times in the forties and fifties for the *American Agriculturist*, the *American Cotton Planter*, the Albany *Cultivator*, the *Soil of the South*, the *Southern Agriculturist*, the *Southern Cultivator*, the *Horticulturist*, and the *Horticultural Review and Botanical Magazine*. See Wendell Holmes Stephenson, "A Quarter-Century of a Mississippi Plantation: Eli J. Capell of 'Pleasant Hill,'" p. 359.

[115] *American Farmer* IV (Aug. 9, 1822), 161.

[116] *Farmers' Register*, I (1833–34) Supplement 769 ff.

[117] *Genesee Farmer*, III (Nov. 2, 1833), 345.

*turist* in 1845 asserted that not one farmer in a hundred subscribed to an agricultural paper and added: "If it were not for the gardeners, mechanics, merchants, and professional men, who mainly support it, our paper *could not live a single year!*" [118] However, a greater interest in agricultural reading on the part of the dirt farmer became evident as the period progressed.

As farmers' activities became more specialized, editors inaugurated new departments to meet the rising needs. About the middle of the century, sections devoted to horticulture, stock raising, veterinary science, poultry, bees, and farm machinery became very popular. In addition to meeting these demands through special departments, the editors increasingly attempted to convert their journals into family papers by appealing to all members of the household. Thus they added special columns on rural architecture and current events, and instituted the "Fireside Department," "Ladies' Department," and "Children's Department," which helped to satisfy the broadening interests of rural life. These new features were often placed in the hands of specialists. Colored and special editions, illustrated sections and foreign language editions, which occasionally appeared, pointed to our "modern" trend.

The journals, in keeping abreast of the times, installed the steam press during the forties. The general format and physical make-up of the papers grew increasingly attractive and the use of illustrations became more popular. Upon occasion, the proud announcement was made that the news and market quotations had arrived by "magnetic telegraph." In harmony with this "lightning spirit of the age," the monthlies of the fifties were frequently converted into weeklies, for as Solon Robinson said, "Monthly reports of markets tell of things past

[118] *American Agriculturist*, IV (Nov., 1845), 335.

and unprofitable to a telegraphing age." [119] At the same time, the public wanted a less expensive paper. There was a gradual recognition of the wisdom in sponsoring fewer periodicals, those few to be more national in scope and of wider circulation. This would permit greater concentration of outstanding talent and also make possible cheaper rates.[120]

[119] *Plow*, I (Nov., 1852), 357.

[120] *Ibid.*, I (Nov., 1852), 357; *Prairie Farmer*, XV (Aug., 1855), 250; *Genesee Farmer*, XIX (Dec., 1858), 375–76; *Wisconsin Farmer*, XI (Dec., 1859), 455; *Rural New-Yorker*, X (Nov. 26, 1859), 382.

# Chapter V

## SPECIAL FEATURES

*You see, Mr. Editor, that what I wish is, for you to give us a more diversified* fare, *and not always the ploughing, sowing, and reaping and mowing.*[1]

FROM THE earliest issues, articles not strictly agricultural in nature frequently appeared in the journals. With the recognition of the widening interests of the readers, the editors were forced to establish special departments for particularly popular subjects. Talented contributors, in many cases, helped to make them attractive. Four of these special features have been selected for consideration in this chapter.

Articles devoted especially to children appeared occasionally in the periodicals during the first half of the century, but rarely did a journal set aside a column or devote a particular section to the especial interests of young readers. A few pioneers in this field may be noted as early as the thirties. The *Cultivator* had a "Young Men's Department" from its beginning. The *Maine Farmer* inaugurated a short-lived "Youths' Department" in 1838, while the editor of the *American Agriculturist*, though fearful of his success in interesting the young people,

[1] Letter to Editor Skinner, *American Farmer*, VIII (Aug. 11, 1826), 168.

launched a "Boys' Department" in 1845. The keynote was instruction, not entertainment.

Perhaps the most popular series of articles for children in this earlier period was written by James Pedder [2] for the *Farmers' Cabinet.* The series entitled, "Frank; or, Dialogues between a Father and Son, on the subject of Agriculture, Husbandry, and Rural Affairs," started in 1839 and continued three years.[3] These articles, "replete with the soundest truths of morality and agriculture," were intended especially for farmers' children. They were widely reprinted in other journals throughout the entire United States, and eventually published in book form, passing through several editions.

About the middle of the century, greater attention to young people may be observed and most of the journals inaugurated departments with titles such as "Girls' and Boys' Corner," "For the Children," "Young Folk's Page," and "Children's Page." Girls were given almost as much attention as boys. The new departments contained a miscellaneous collection of material, some of which was intended to instruct its young readers; much, however, was merely to entertain.

Educational articles included essays on chemistry, physics, electricity, travel, nature, and various phases of history. To foster an intellectual interest, the *Prairie Farmer* offered a prize for the best juvenile composition and many of these competitive essays were printed in the journals. A committee finally decided upon the winner and awarded him the "palm of su-

[2] James Pedder, editor of the *Farmers' Cabinet* from 1840 to 1843, had earlier shown ability in writing children's stories. In 1814 he had published a little book for children entitled, *The Yellow Shoestrings, or The Good Effects of Obedience to Parents,* which is said to have gone through 17 London editions and at least two in the United States. *Dictionary of American Biography,* XIV, 387.

[3] For a representative article of this series, see pp. 295 ff.

periority" in the form of *Dr. Kane's Arctic Expedition.*[4] The practical side was not neglected. The same journal offered a premium of $25.00 to the boy under sixteen who raised the greatest number of bushels of rutabagas on an acre of ground, the cultivation to be entirely by himself.[5] Nor did the moral training of the children suffer from lack of guidance. Frequent articles warned the youth of the evils of tobacco and spirituous liquors, the dangers of bad company, and the ill effects of bad habits. The following characteristic caution appeared in the *Rural American:* "You, young man, on the way to the ball-alley, or the billiard-room, with a cigar in your mouth, and with an appetite for mint julep—stop a moment. Are you not in a dangerous way?"[6] Articles on etiquette, behavior, and care in dress as well as suggestions on health, exercise, and the care of the body appeared in quantities.

Likewise, the recreational side was emphasized and most of the journals in the fifties amused and puzzled their young readers with an occasional enigma, conundrum, charade, anagram, rebus, or labyrinth. Frequently, diversions suitable for the long winter evenings were discussed by special writers. The popular games during this pre-Civil War period were hunt the slipper, the huntsman, blind man's buff, the rule of contrary, and cross purposes, all familiar to modern readers, although the names in many instances have changed. Indeed, the modern reader of these old agricultural journals is often reminded of the proverbial saying that there is nothing new under the sun. He wonders how many generations prior to the fifties followed instructions similar to those of a writer in the *Ohio Cultivator:* "Let a person stand with his back and

[4] *Prairie Farmer,* XVIII (April 8, 1858), 116.
[5] *Ibid.,* XIX (June 16, 1859), 376.
[6] *Rural American,* I (Jan. 26, 1856), 32.

heels close to a wall, then place a dime on the floor at a little distance in front of him, and tell him he shall have the money if he can take it up without advancing his heels from the wall."[7] Certainly it has been new to each generation of young people since that time. Other means of entertainment included poems and short stories which, while written to amuse, were intended also to elevate and inform.

The outstanding young people's department was the "Boys' and Girls' Column" of the *American Agriculturist*. The editor, Orange Judd, had a great interest in children and claimed this particular department for his own, writing many of the articles himself. In 1858 he secured the aid of one of the most famous writers of children's stories in the United States, "Uncle Frank"[8] who contributed a monthly column called "Uncle Frank's Chat with the Boys and Girls." Another column in the same journal entitled "Grandmother and the Little Girls" proved to be very popular and the letters from "Aunt Sue" which appeared from time to time were always engaging.

Each month the editor printed a number of problems and puzzles of all descriptions in the "Boys' and Girls' Column."

[7] *Ohio Cultivator*, XIII (Jan. 15, 1857), 29.

[8] Uncle Frank's real name was Francis C. Woodworth (1812–59). He was the nephew of Samuel Woodworth, the author of "The Bucket" ("The Old Oaken Bucket"). Soon after his graduation from Union Theological Seminary in 1839, he was compelled to leave the ministry because of failing health. The remaining years of his life were devoted to literature. He possessed a remarkable aptitude for rousing the interest of children, and most of his fifty-odd volumes are addressed to juvenile comprehension. Among the best known of his numerous publications are: *Uncle Frank's Home Stories* (6 vols., 1851); *Uncle Frank's Picture Gallery* (2 vols., 1852); *Theodore Tinker's Stories for Little Folks* (12 vols., 1854–58); and *Uncle Frank's Pleasant Pages for the Fireside* (1857). See *American Agriculturist*, XVII (Sept., 1858), 281; XVIII (July, 1859), 217; *Lady's Home Magazine*, XIII (Sept., 1859), 146; *National Cyclopaedia of American Biography*, V, 509. A number of his beautiful little volumes may be examined in the Rare Book Collection of the Congressional Library.

Thousands of letters from children attempting to solve them, came from all parts of the United States, swamping the editor at times.[9] The names of the young people who offered correct solutions subsequently appeared in the journals.[10] Items of an interesting and instructional nature were also found in the *American Agriculturist.* Among these were articles on the microscope, the alphabet used by the deaf and dumb, a description of the process for making engravings, illustrations of kinds of type used in printing magazines, and a simple explanation of telegraphy. The great number of articles copied from this department was a definite indication of the high regard placed upon it.

In the sports field, John S. Skinner, as editor of the first agricultural journal, set a precedent for the farm press. He encouraged sports of all kinds and included miscellaneous material of this nature in his journal. In the second number he wrote: "We have long been of [the] opinion that children, instead of being restrained too much, as they often are, *especially in Town*—ought to be indulged, and, as far as convenient, furnished with the means of exercising in all manly amusements—such as skating, hunting, shooting, racing, etc." [11] His reasons are interesting and indicate again the fear of leisure so prevalent at this time. "Frisking abroad in the open fields, enlivens the imagination, gives vigour to the body, and strength

[9] *American Agriculturist,* XVI (Aug., 1857), 186; XVII (March, 1858), 90.

[10] Only the county and state were given in the address, as the editor desired to protect his readers. He said there were a "lot of dealers in lottery tickets and gift enterprises and various humbugs," who were constantly on the lookout for victims. *American Agriculturist,* XVII (April, 1858), 123.

[11] *American Farmer,* II (April 7, 1820), 15. In a later number, he reiterates the same idea and also shows that prominent men such as Judge Marshall, John Calhoun, Thomas Jefferson, and John Adams sanctioned horseracing. *American Farmer,* VI (Nov. 12, 1824), 270.

to the mind, and above all, it leaves no time or inclination for drinking, gaming, and other low vulgar and degrading associations and amusements, so called." [12]

The experiment of including such material in his periodical seemed to have proven successful, for in 1826 the editor said: "It gives us pleasure to intimate to the lovers of field sports, that we have reason to feel assured that the appropriation of a small portion of the Farmer to *their* use and amusement, though condemned by a few, is, upon the whole, approved and encouraged. There is, in a word, a decided and growing taste for such amusements, and for discussions calculated to enhance the pleasures of those healthful diversions that necessarily conduct gentlemen from the bar-room and the gaming table into the open air and over the fields." [13] A regular column, headed "Sporting Olio," had been inaugurated in 1825 in the *American Farmer* and it was continued for a number of years. In general the *Farmer* devoted more space to sports than other members of the agricultural press during the period before the Civil War.[14]

[12] *American Farmer,* II (April 7, 1820), 15. Another writer in the same number preferred sports of the field because he felt they had "no tendency to lead young men into vicious habits" and then, too, the scenes of rural sports were "necessarily at *a distance from cities and towns.*" It is "where men *congregate,* that the vices haunt." *American Farmer,* II (April 7, 1820), 11.

[13] *American Farmer,* VII (Feb. 3, 1826), 368.

[14] John S. Skinner's interest in horses and sports is witnessed by his publication of the *American Turf Register and Sporting Magazine* (monthly) in September, 1829. This magazine was to serve as an authentic record of the performances and pedigrees of the bred horse and was designed also as a magazine of information on veterinary subjects. It included rural sports, such as racing, trotting matches, shooting, hunting, fishing, and the like. *American Turf Register and Sporting Magazine,* I (Sept., 1829), 1.

Turfmen have not forgotten Skinner's contribution to the development of the sport in Maryland; to this day an annual "John Stuart Skinner" race is run at Pimlico. Helfrich, "A Baltimore Pioneer of Farm and Turf."

Americans were an out-of-door folk and their chief sports during the period were hunting, fishing, and horse racing.[15] Therefore it is not surprising that many of the journals carried occasional articles such as "Fox Hunting in the District of Columbia," "Shooting Match on Long Island," "The Road to Health or, A Physician's Opinion of Hunting," "Trout Fishing," "Care of Horses," and frequently the pedigree and performance of famous horses as well as complete records of past races. There was a great interest in feats of pedestrianism, especially before 1830. "A young gentleman of the city of Boston has walked to Charleston, S. C. 1150 miles, in 36 days," said one paper.[16] This sort of "remarkable exploit" was recorded many times. Now and then an article on skating, a report on a "Swimming Match" or a "Leaping Match" appeared. In 1829, the attempt to banish ninepins from the city of Albany proper to the suburbs—for moral reasons as well as because the game was played at night and people were disturbed by it—brought down Skinner's editorial wrath.[17] It is surprising to note the large percentage of articles on sports copied from the English sporting magazines of the period,

[15] Krout, *Annals of American Sport*, Foreword, p. 3.

According to Editor Skinner, cockfighting was pursued in all parts of the United States in the late twenties to a much greater extent than was generally believed. He considered it a cruel sport. *American Farmer*, XI (May 22, 1829), 79.

[16] *American Farmer*, XI (March 27, 1829), 15.

[17] After a grave debate the city council of Albany finally resolved that no minor should be permitted to roll ninepins in the city, and that the amusement should be prohibited after eight o'clock in the evening. Skinner, commenting editorially, said: "What shall we have next in the way of excessive regulation? as bad in morals as in trade; smuggling is apt to ensue in both cases.—The suppression of one open gentlemanly and manly amusement is sure to give rise to two gross vices—the more vulgar and pernicious for being the more concealed." *American Farmer*, XI (June 26, 1829), 118.

especially the *English Sporting Magazine.* However, farmers were prompt to remind the editors when too much space was devoted to nonagricultural subjects, as did a correspondent in the *New-York Farmer* in 1831. He felt that the press had bestowed too much attention on the "breed, pedigree, bottom, speed and performance, of the most distinguished" race horses, and that speed was not considered agriculturally important.[18]

While the journals were opposing horse racing in the fairs in the fifties, there seems to be no evidence of any general objection to horse racing *per se.* They did not hesitate to record the pedigree and records of famous horses all over the country, but it was their feeling that horse racing did absolutely nothing to improve the horse and therefore was out of place on the fair ground. However, at an exhibition in 1860, prizes were offered for fast walking horses, since such a skill had practical utility. This suggestion was enthusiastically received by the agricultural press and was successfully tried at Elmira, New York.[19] Of course the journals carried accounts of the plowing matches (which had their sporting side), also an occasional foot race, and at times, a baseball game, especially popular in the New England fairs.

Numerous articles devoted to women's sports also appeared. While the journals were opposed to female horseback riders at the fairs, they were enthusiastic about riding for exercise elsewhere, and so voiced their opinion under the heading of "Advice to Lady Equestrians," "Lessons in Horseback Riding," and so on. Other sports such as skating, swimming, and archery were encouraged through editorials which emphasized

[18] *New-York Farmer,* IV (June, 1831), 158 ff.

[19] *American Agriculturist,* XIX (Aug., 1860), 230 ff.; XIX (Oct., 1860), 303; *Cultivator,* ser. 3, VIII (Feb., 1860), 69; ser. 3, VIII (Sept., 1860), 292; *Country Gentleman,* XVI (Aug. 23, 1860), 128.

the importance of exercise to bring about a race of healthy women.

From what has just been said, we should not conclude that a profound and widespread interest in sports existed in the United States before the Civil War. Few Americans had leisure for sports and games.[20] Consequently, there was no particular emphasis upon this subject in the journals. However, after 1850, a surprisingly large number of articles appeared dealing with sports for both men and women, indicating a wider interest.

Columns of news under the headings "Weekly Summary," "Record of the Times," "Items of Intelligence," and "News of the Week" were carried by a vast majority of the periodicals. Fearful, no doubt, that this might be considered a drifting from the avowed purpose of the journals, editors often said, in effect, that they had no interest in the transient occurrences of the day, and due to the permanent character of the journals would give no attention to matters of temporary interest. In spite of these declarations, news items continued to appear. Furthermore, it was difficult to maintain a consistent policy, since editorship often changed and ownership was frequently transferred.

The news sections often contained summaries of foreign events,[21] happenings in the state and national capitals, and almost invariably a table of current prices of grain, provisions, and cattle. The journals, predominantly sectional in character, also

[20] Fish, *The Rise of the Common Man,* p. 148.

[21] The *American Agriculturist* for a time, had the best foreign news of any of the journals. The editor subscribed to a number of European periodicals from which he gleaned his material. He complained that the make-up of this department cost him ten times the labor required to write an equal amount of editorial matter. *American Agriculturist,* II (Dec. 15, 1843), 374.

carried a record of births and deaths. Political news, as such, was almost entirely avoided. All things considered, the *Country Gentleman* contained the best news section of all the journals, as it was well organized, competently written and no expense was spared to keep its readers abreast of the times.

In general it was the unusual, the violent and catastrophic that seemed of greatest interest. Steamboat and railroad accidents were numerous and received much comment. In 1839 one journal reported that of the 1,300 steamboats built in the United States, 200 had been lost through various accidents.[22] It was claimed that many of these were the result of "brag trips" which meant that steamboat captains were attempting new records.[23] During the five years from 1848 to 1852 inclusive, there were 50 steamboat explosions, causing a loss of 1,155 lives and 475 persons "wounded," a record which dropped considerably after the passage of the Steamboat Law in 1853.[24] From the first, the dangers of railroad travel were apparent—dangers to be apprehended from the breaking of wheels or axles, from obstructions on the road, or from passing the "turnouts." Increased speed often added to the hazards. One writer in 1834 asserted that all the useful purposes of the railroads could be as fully attained by a speed of twelve to fourteen miles per hour, as by the more dangerous one of twenty or thirty.[25] News items covering the epidemic of railroad and steamboat accidents carried headlines such as "Fearful Disaster," "Horrible Destruction of Human Life," "Wholesale Murder," and so on. These accounts were often

[22] *Farmer's Monthly Visitor*, I (April 15, 1839), 63.

[23] *Maine Farmer*, VIII (Feb. 8, 1840), 40. The "abominable" practice of racing was also given as a cause for many accidents. *Yankee Farmer*, IV (July 14, 1838), 223.

[24] *Ohio Farmer*, VII (Jan. 16, 1858), 24.

[25] *Farmers' Register*, I (Feb., 1834), 528 ff.

accompanied or followed by the editors' comment on the "unwarrantable carelessness," "shameless negligence," "criminal recklessness" of the owners, companies, and agents. In order to end these "wholesale slaughterings" the *Country Gentleman* inaugurated a safety-first drive. Editorials demanded the passage of stricter laws to govern these carriers and prompt legal justice to punish officials guilty of carelessness or disobedience.[26] Even so, in 1858, the *Ohio Farmer* quoted figures showing that one person was killed or "wounded" to every 188,450 passengers carried, a record one-half as good as that of England and France. According to this editor, the redeeming feature of railway travel, as compared with the old stage coach system, lay in fewer casualties proportionate to the number carried.[27] Additional misfortunes such as fires, murders, and "melancholy" suicides were featured as entertaining news material.

Warnings against counterfeits repeatedly appeared in the journals, as the circulation of spurious notes was so common as to be a real menace to the farmer.[28]

Another subject receiving frequent reference was perpetual motion. Its "discovery," made many times during this period, was always received enthusiastically by the readers. The *Plough Boy* in 1819 announced: "This grand desideratum, the attain-

[26] *Country Gentleman*, II (Sept. 1, 1853), 144; II (Dec. 15, 1853), 384; IV (July 13, 1854), 32; IV (Aug. 17, 1854), 112; VI (Sept. 6, 1855), 164.

The editor of *Moore's Rural New-Yorker* felt that the "excursion trains, especially on single track railroads," were the "most perfect death-traps ever invented by an emissary of Satan." *Moore's Rural New-Yorker*, V (July 15, 1854), 226.

[27] *Ohio Farmer*, VII (Aug. 28, 1858), 276.

[28] Because of the frequent and gross impositions practiced, especially upon farmers, the *Farmers' Cabinet* decided to do its part by issuing, free for subscribers, a work of 16 pages containing a full description of all counterfeit notes, fraudulent issues, altered notes, broken banks, and so on. *Farmers' Cabinet*, I (July 1, 1837), announced at end of Volume I.

ment of which the votaries of science have uniformly pronounced incompatible with the laws of motion, is at length discovered by a self-taught mechanic, a Mr. Foster, of Owego in this state."[29] Although such announcements inevitably brought embarrassment to the editor, they continued to appear. Nor did the statement by the *Wisconsin Farmer* in 1856 that "such a discovery will never be made" as it is a "mechanical impossibility" dampen the interest in this subject.[30]

Outstanding rural miracles and freaks of nature were usually recounted in the periodicals. "There is now in the possession of Mr. Hayes, a butcher of Southhampton, a pig, with a *wooden leg*, on the off side before, and it appears to walk with little lameness or inconvenience," stated one paper.[31] The *Farmer's Monthly Visitor* told of an ox with a wooden leg that managed to get around satisfactorily.[32] A report of the largest hog in the world standing 7 feet, 6 inches high and weighing 1,200 pounds was made in the *Wisconsin Farmer*.[33] The *Agriculturist* described a lamb with but one eye, and that in the middle of the forehead, while the feat of a cow giving 24 quarts of milk within five hours was recorded in the *New England Farmer*.[34] Such items frequently made the rounds of the press.

The news policies of the journals were in gradual formation as the period wore on. By 1855 the *American Agriculturist* decided not to attempt to "catch popular favor, by lumbering up its pages with . . . miscellaneous matter of the

[29] *Plough Boy*, I (Sept. 4, 1819), 110.
[30] *Wisconsin Farmer*, VIII (Dec., 1856), 562.
[31] *American Farmer*, I (July 2, 1819), 112.
[32] *Farmer's Monthly Visitor*, XIII (May, 1853), 156.
[33] *Wisconsin Farmer*, V (Jan., 1853), 16.
[34] *Agriculturist*, II (Jan., 1841), 1; *New England Farmer*, VI (July 27, 1827), 7.

day," and the *Prairie Farmer* three years later, promised to avoid murder trials and police reports.[35] In general, the trend during the latter part of the period showed less emphasis upon news, especially political news—due, probably, to the great expansion of the cheap newspaper.

As was generally true of news, so rural anecdotes and jokes appeared in the agricultural press.[36] By the middle of the century many of the journals regularly set aside a column for humorous material under titles such as "The Leisure Hour," "Humorous and Witty," or "Everybody's Corner." The majority of the papers, however, used items of this nature as "filler" and treated it in a very unsystematic manner.

Of course the old stand-by conundrums such as "When is a door not a door?" and "What makes more noise than a pig under a gate?" found their way into the journals. The jokes which occasioned a laugh from the farm population a hundred years ago would scarcely be considered amusing today. They were naturally rural in character, sometimes crude, but rarely risque. Puns were extremely popular. These samples are drawn from the earlier journals.

A gentleman complimented a lady on her improved appearance. "You are guilty of flattery," said the lady. "Not so," replied the gentleman, "for I vow you are as *plump as a partridge*." "At first," rejoined the lady, "I thought you guilty of flattery only, but now I find you are actually making *game* of me." [37]

A gentleman riding through the town of —— one day met an awkward fellow, leading a hog, whom he accosted in the following manner: "How odd it looks to see one hog lead another." "Yes,"

[35] *American Agriculturist*, XIV (March 14, 1855), 16; *Prairie Farmer*, XVIII (Dec. 9, 1858), 382.

[36] A few journals, including Ruffin's *Farmers' Register*, avoided such trivialities.

[37] *New England Farmer*, I (March 1, 1823), 248.

replied the chap, "but not so odd as it does to see a hog ride on horseback." [38]

Later, the jokes often centered around the town drunk, the Irishman, or the politician. Not infrequently the city dandy, the stock broker, or the rich banker were ridiculed. The following random selections are representative of the quips that amused the rural population in the forties and fifties.

An exchange paper, announcing the death of a gentleman out west, says that the deceased, though a bank director, it is generally believed died a Christian, and universally respected.[39]

"How do you and your friends feel now?" said an exultant politician in one of our Western States to a rather irritable member of the defeated party. "I suppose," said the latter, "we feel just as Lazarus did when he was *licked by dogs*." [40]

*New Drink.*—"Mr. Guzzlefunction, I have discovered a new drink for you. Suppose you try a little."

"Well, I don't care if I do; (drinks). It hasn't got a very bad taste to it; and if my memory serves me right, it is what they call water. I recollect drinking some of the stuff when I was a lad." [41]

A young stock broker having married a fat widow with £100,000, says it wasn't his wife's face that attracted him, so much as the figure.[42]

Passenger—"Well, Mr. Conductor, what news in the political world?"

Conductor—"Don't know, sir; I haven't been to church for the last two weeks." [43]

"Pat," said a Yankee to an Irishman, as they passed a tree near

---

[38] *Ibid.*, I (Jan. 4, 1823), 184.

[39] *Prairie Farmer*, XVI (March 20, 1856), 48.

[40] *Country Gentleman*, XI (Jan. 21, 1858), 40.

[41] *New England Farmer*, XIX (Feb. 3, 1841), 248.

[42] *Ohio Farmer*, IV (May 5, 1855), 72.

[43] *Prairie Farmer*, XVI (March 20, 1856), 48.

Harlem with a rope hanging from one of its branches, "Where do you suppose you would be now if that rope had its desert?" "Faith, and I'd be a walking here all alone to New York." [44]

The following conundrums are representative.

Why are the ladies like stage-drivers?—Because they generally secure the *mails*.[45]

Why is an overworked horse like an umbrella?—Because he is used up.[46]

Humorous selections with special appeal, like other popular material, regularly made the rounds of the periodicals. On occasion, when copy was scarce, some of the editors recopied the sallies of earlier issues to obtain "filler."

[44] *Boston Cultivator*, X (Sept. 9, 1848), 292.
[45] *American Farmer*, X (Oct. 3, 1828), 230.
[46] *Ohio Farmer*, VIII (Jan. 8, 1859), 15.

# Chapter VI

## ADVERTISING

---

*We want a paper published by a person, who is not a* Proprietor *or* Copartner *in any "Agricultural warehouse" "Seed Store" or "Deposite of Farmer's Seeds and tools" and edited by a gentleman, who will not consent to lend the sanction of his name, to puff any article whatever, any further than its real merit and intrinsic value will warrant.*[1]

---

ADVERTISING, as found in the agricultural press, experienced remarkable development and expansion during the forty years before the Civil War. The early years of the journals show comparatively little emphasis upon this subject, but 1860 found high-powered, modern methods in evidence. There were, however, numerous restrictions with regard to advertising content and the space allotment throughout the entire period.

As in so many other phases of agricultural journalism, the *American Farmer* set the standard. In its first year, Skinner announced: "ADVERTISEMENTS, which are, in their nature and objects suited to a paper of this sort, such as, the sale of land, seed, live stock, implements of husbandry, new inventions, &c. &c., will be inserted *once only*, at the rate of $1.00 per square,

[1] Asa Barton in *Maine Farmer*, VIII (April 18, 1840), 114.

to be paid in advance." He also suggested that one insertion might be "as good as forty," except in cases where the law prescribed a greater number of times.[2] The *Plough Boy* took a similar stand explaining that as copies were generally preserved in files for family reference, it was evident that one insertion would answer "as well as twenty." [3]

Gradually the insertion rule of "once only" broke down under the persuasion of advertisers who knew the value of repetition. As other restrictions became more generously interpreted, advertisements increased, in spite of objections from subscribers who declared that this "foreign material" was infringing upon the reading matter.[4] Even so, certain limitations remained. The *Boston Cultivator* reserved the privilege of discontinuing any advertisements after three insertions and excluded all except those "directly connected with agriculture." [5] The *Farmers' Cabinet* accepted advertisements the subject matter of which might "correspond with the agricultural character" of the journal, and the *Country Gentleman* printed a "limited number of advertisements." [6] The *Genesee Farmer* was willing to insert "a few short advertisements of interest to farmers"; the *Michigan Farmer* limited its advertising to one page.[7] The *Rural New-Yorker* welcomed "brief and appropriate advertisements." [8]

[2] *American Farmer*, I (Sept. 17, 1819), 199.

[3] *Plough Boy*, I (Oct. 23, 1819), 166.

[4] *Cultivator*, VII (Jan., 1840), 5; *South-Western Farmer*, II (Jan. 5, 1844), 346, 353.

[5] *Boston Cultivator*, IX (Nov. 13, 1847), 368.

[6] *Farmers' Cabinet*, XII (Sept. 15, 1847), 71; *Country Gentleman*, I (Jan. 6, 1853), 16.

[7] *Genesee Farmer*, ser. 2, XXI (Jan., 1860), 37; *Michigan Farmer*, I (April 1, 1843), 29.

[8] *Rural New-Yorker*, V (Jan. 14, 1854), 20.

The Southern journals whose circulation was normally small, depended to a larger degree upon advertising income. Consequently, fewer restric-

When the editors saw the informational benefits of advertising to the readers, as well as the new possibilities for their own financial gain, they offered fewer apologies and eventually "pointed with pride," to the "helpful," "valuable," "interesting," "educational," "reliable," and "attractive" advertisements. The subscribers were urged to read them since the advertisers were considered "men of strictest integrity," a "class of high-minded, honorable business men." One editor expressed this changed attitude, when he wrote that the advertising department was "by no means the least interesting and useful department" which the *Country Gentleman* contained.[9] However "valuable" these advertisements, the editors were ever cautious about encroachments upon the reading space and this often resulted in relegating advertising to the periodical covers as well as the addition of "advertising sections."

Since the journals were always considered clearing houses for agricultural information, farmers wrote to the editors from all parts of the country seeking advice on seeds, plants, animals, fertilizers, and agricultural implements. Thus, these men became mediums whereby the farmers contacted both dealers and manufacturers. It was a part of the service of an agricultural journal to keep readers advised upon new and valuable agricultural discoveries and to aid in the popularization and adoption of these. The system of recommending products with fulsome or interested praise was called puffing, an early form of

---

tions were made. *South-Western Farmer,* XI (Jan. 5, 1844), 346; *Southern Planter,* XV (Jan., 1855), 18.

[9] *American Agriculturist,* XVII (Sept., 1858), 288; *Genesee Farmer,* ser. 2, XIX (Sept., 1858), 288; ser. 2, XIX (Oct., 1858), 322; ser. 2, XIX (May, 1859), 164. *Wisconsin Farmer,* IX (Aug., 1847), 297; *Prairie Farmer,* XX (Sept. 29, 1859), 201; *Country Gentleman,* XV (March 29, 1860), 208.

advertising. With the vast introduction of plants and animals, and the great number of new inventions placed upon the market, it is not surprising that the journals frequently puffed unworthy objects. The subject was further complicated by the fact that the officials of the journals were gradually becoming financially interested in the sale of certain of these products.[10]

Since most of the editors and proprietors were forced to seek additional sources of income, they frequently established agencies for the sale of horses, cattle, sheep, hogs, seeds, agricultural machinery, implements, and books. Thus, the seed store or agricultural warehouse was a common adjunct to an agricultural journal office. For example, the proprietors of the *New England Farmer* not only owned a large seed store, but were the manufacturers and owners of the patent right of the Howard's Plow; the editors of the *American Agriculturist* were proprietors of the New York Agricultural Warehouse and Seed Store, sole agents for Premium Plows, and manufacturers and holders of patents on agricultural machinery. The editor of the *Farmers' Cabinet* sold a patented fertilizer called Poudrette, and for a time J. S. Wright of the *Prairie Farmer* manufactured the Hussey Reaper and Atkins Automaton; while the editors of the *Southern Planter* and *Ohio Cultivator* were agents for the McCormick Reaper.[11]

[10] The *Western Farmer and Gardener* voiced the problem faced by many editors: "We do not consider it *puffing*, when we commend anything which we honestly think farmers are in want of, and which is worth their attention.—And if it should in such a case even benefit ourselves at the same time, we really cannot see where is the mischief done to anyone. The term *puffing* is often sadly abused. It is certainly the duty of an Editor to keep the public informed of all good and useful things." *Western Farmer and Gardener*, V (July, 1845), 265.

[11] *New England Farmer*, XVII (Jan. 16, 1839), 222; *American Agriculturist*, IV (May, 1845), 167; *Farmers' Cabinet*, VI (April 15, 1843), 296; *Prairie Farmer*, XII (May, 1852), 259; *Southern Planter*, VII (March, 1847), 95; *Ohio Cultivator*, VI (April 15, 1850), 120.

There were occasions when less obvious reasons prompted an editorial puff. For instance, a mill builder confidentially offered to share profits with the editor of the *American Agriculturist* in return for such assistance, but this offer was promptly exposed in the journal.[12] When the *Farmers' Register* received a glowing "editorial" all ready for the press calling attention to the shipment of certain "excellent" animals from England, Ruffin stated indignantly that "however much the practice of puffing by means of supposititious *editorial* articles" might be sanctioned by general usage, it never had obtained, and never would, in the *Farmers' Register*.[13]

While it is impossible to estimate the extent of this abuse, the public felt that many of the journals, for selfish reasons, lent their support to unworthy causes and encouraged speculative agricultural crazes.[14] The *New England Farmer*, owner of a famous seed store, was conscious that "unfavorable inferences" had been made as to its "impartiality and disinterestedness in the great cause" to which it had been "for sixteen years steadily devoted." [15] A contributor to the *Maine Farmer* wrote: "You are aware sir, that there are many journals published . . . in order to disseminate the advertisements of the proprietors for the sale of these articles, or improvements, kept in their 'Agricultural establishments, Seed Stores' or 'warehouses,' and their great usefulness, is thus, trumpeted forth every week to the community." [16] Edmund Ruffin, whose integrity was never imputed, explained the difficulties involved.

[12] *American Agriculturist*, XVI (Feb., 1857), 36.

[13] *Farmers' Register*, V (June, 1837), 125.

[14] For the part played by the journals in the agricultural crazes, manias, and fevers of the period, see pp. 59 ff.

[15] *New England Farmer*, XVII (Jan. 16, 1839), 222.

[16] *Maine Farmer*, VIII (April 18, 1840), 114.

"*We*, who have no such connexion with seedsmen or others, and have no such private interest to forward—and who have treated with distrust, scorn and contempt, all efforts to buy of us editorial puffs—yet even *we* can scarcely avoid giving some help to the progress of this widely-spread puffing system." [17] James J. Mapes, editor of the *Working Farmer* and manufacturer of Mapes' Superphosphate of Lime fertilizer, was often cited in the press as an horrible example of this practice—"a persistent puffer of humbugs, and a turner of the grindstone for the sharpening of individual axes." [18]

After the middle of the century the journals gradually divorced themselves from business connections that might tend to warp their judgment. An editorial in the *Prairie Farmer* pointed out that in order for an agricultural journal to be successful it "must have no connection whatever with any branch of business which comes between it and its duty." [19] The *Genesee Farmer* boasted that no employee of the paper had the remotest financial interest in any outside business, while the *Cultivator* and the *Country Gentleman* stated with pride, that the publisher had "no connection with any breeders of stock, or with any agricultural warehouse, or patent machines, or manures—no business connection to sway him in favor of this or that establishment." [20] The *American Cotton Planter* declared itself equally free and said, "When we speak to *praise* or *puff* anything of the kind as valuable, it is because we have found it so from actual practice or experience." [21] In an edi-

[17] *Farmers' Register*, VI (April 1, 1838), 47.

[18] *Prairie Farmer*, XIII (Dec., 1853), 458.

[19] *Ibid.*, XVII (Nov. 19, 1857), 376.

[20] *New Genesee Farmer*, XVIII (Jan., 1857), 34; *Cultivator*, n.s., IX (Dec., 1852), 393.

[21] *American Cotton Planter*, o.s., XIII (Oct., 1859), 327.

torial entitled, "A Thoroughly Independent Journal," the *American Agriculturist* stated its disinterestedness in all outside business and added, "The position we aim at, is like that of an attorney or councillor, especially employed to promote the individual and general interests of our readers, and to defend them to the utmost of our ability." [22]

The great wave of patent medicine and other fake advertisements that swept the agricultural press and newspapers soon brought disrepute to this form of publicity.[23] However, the tremendous profits realized by these advertisers established the potentialities of selling through this medium. Today the trade names of leading medicines of the pre-Civil War period may be found on druggists' shelves, although the formulae and claims have been somewhat modified. Widely advertised were Ayer's Sarsaparilla, Ayer's Cathartic Pills, and Ayer's Cherry Pectoral—the latter a "cure for cough, colds, hoarseness, bronchitis, asthma and consumption." Mothers were assured that Mrs. Winslow's Soothing Syrup would give rest to themselves and relief and health to their infants. Vaughn's Lithontriptic Mixture claimed a "*greater healing power,* in all diseases, than any other preparation" before the world, while Dr. Townsend's Sarsaparilla, "the most extraordinary medicine in the world," not only purified the whole system and strengthened the body, but it created new, pure, and rich blood, "a power possessed by no other medicine." Mexican Mustang Liniment, a volcanic oil from Mexico, brought "glad tidings" to the "halt, the lame, the sore, and the stiffjointed," for it cured "confirmed cancer," and was good for "man or beast." It was claimed that more

[22] *American Agriculturist,* XIX (Oct., 1860), 320. The position taken by the *Indiana Farmer* was almost identical. *Indiana Farmer,* VII (Sept., 1858), 174.

[23] Presbrey, *The History and Development of Advertising,* p. 300.

than four million bottles of this "one of the most perfect remedies ever offered to the afflicted," were sold during its first two years before the American public.[24]

Hair restorers, such as Lyon's Kathairon and Baker's Cheveustonique were extensively advertised with extravagant claims for cures. The former brand gave "tone and elasticity to the whole system," and the latter was famous for "its infallibility in cases of headaches." "Balm of a Thousand Flowers," used as a dentifrice, was advertised to remove the "curse of a disagreeable breath." The reader was reminded that many persons did not know their breath was bad, and since the subject was so delicate, their friends would never mention it.[25] Dr. Urban's Inebriate's Hope, "a cure for Intemperance, Delerium Tremens and Neuralga" was very effective and it was said that one package would "cure an ordinary case of intemperance," while two packages would "cure the most inveterate drunkard of all desire for alcoholic drinks." [26]

The advertisements inserted by the physicians of the period displayed the same boastfulness and extravagant assertions. The *American Farmer* carried that of Doctor Baakee of Baltimore which will serve as an illustration. After listing many of the worst ills to which the flesh is heir, the doctor announced that he was a specialist in all of them. The notice modestly stated that Doctor Baakee was one of the "most skillful and celebrated Physicians and Surgeons now living." His fame was known personally "in every principal city in the World." He added in large letters that he treated all diseases free of charge and a personal interview with the patient was unnecessary. All letters directed to Dr. Baakee (enclosing one dollar) from

[24] *Prairie Farmer,* XIII (July, 1853), Advertising section.
[25] *American Farmer,* XII (1856–57), Advertiser.
[26] *Valley Farmer,* IX (Oct., 1857), 179.

any distance, correctly stating the nature of the disease, were promptly answered, and "a package of medicines sent by mail with full instructions for treatment, free of charge." All patients with chronic diseases were "successfully treated by correspondence." Should any sufferer still doubt the Doctor's abilities, he assured them further, that he could produce one thousand certificates attesting his perfect success in curing cancer and tumors of every description without the use of the knife.[27]

Along with medicinal advertisements, occasionally there appeared opportunities to buy good land or "sure win" lottery tickets for little or nothing, offers of employment at high wages for a three-cent stamp, or the secret of acquiring quick riches for the same fee. An interesting advertisement that probably appealed to numerous farmers was Signor D. Alvear's Goldometer. The advertisement explained that the "first discovery of gold in California was made by DON JOSE D'ALVEAR, "an eminent Spanish Geologist, Chemist, and Natural Philosopher, by means of a newly invented Magnetic instrument, called The Goldometer, or Gold Seeker's Guide!" Now for the small sum of three dollars farmers were given the opportunity to purchase such an instrument with full instructions for discovering silver, platinum, coal, iron, copper, and other valuable minerals on their farms.[28]

Shortly after the middle of the century a decided reaction against the admission of patent medicine, fake, and humbug advertisements was evident in the journals. *Moore's Rural New-Yorker,* in its first issue in 1850, excluded "patent medicines and other quackery—including deceptive advertisements

[27] *American Farmer,* n.s., XIV (1858–59), Advertiser.
[28] *Farmer's Monthly Visitor,* X (Dec. 31, 1848), 188.

of all classes."[29] By 1860 many of the better journals took a similar stand.[30] Perhaps the *American Agriculturist* exercised the greatest care in the advertisements admitted to its columns, but they were in no sense guaranteed. The editor wrote: "We wish it distinctly understood, however, that we do not endorse, or take any responsibility for anything in the advertising pages, unless it be specially referred to in the reading columns."[31] The *Ohio Farmer* hints at the difficulties involved in censorship. "When we have doubts of the genuineness of an advertisement we always refuse to insert it, whether the money is sent with it or not. Still we are not infallible and may be deceived."[32] As the journals began to exclude advertisements of a doubtful character, companies sprang up whose sole business consisted in compiling and selling "sucker lists." These names and addresses were sold at a certain price per thousand and the mails then became the medium for reaching the rural "easy marks."[33]

Together with this watchfulness upon their own advertising, many journals took pride in exposing fakes and humbugs wherever these appeared. The *Ohio Cultivator*, which delighted in its accounts of irresponsible "tree peddlers," traveling "swindlers," and "sharpers," together with certain patented fertilizers, claimed that a subscriber would save ten times the cost

[29] *Moore's Rural New-Yorker*, V (Jan. 28, 1854), 34. The *Pennsylvania Farm Journal* took the same stand in its second number. *Pennsylvania Farm Journal*, I (May, 1851), 35.

[30] *American Agriculturist*, XVII (Sept., 1858), 288; *Genesee Farmer*, ser. 2, XIX (Sept., 1858), 288; *Wisconsin Farmer*, VIII (Sept., 1856), 424; *Ohio Farmer*, VII (Oct., 1858), 348; *Ohio Cultivator*, XV (April 1, 1859), 112.

[31] *American Agriculturist*, XIX (Jan., 1860), 27.

[32] *Ohio Farmer*, VII (Oct. 30, 1858), 348.

[33] *American Agriculturist*, XVIII (Dec., 1859), 355; XX (Jan., 1861), 6.

of the paper by observing its warnings against impositions.[34] Many editors advised their patrons not to purchase such humbugs as the "Celestial Rose," "New Hampshire Pine Apples," and "Chilian Guano." [35] The *Cultivator,* under the heading "FARMERS, REMEMBER THE HUMBUGS, AND DON'T BITE," gave warning that Santa Fe wheat, at that time quite popular, had made its debut at least twenty times in twenty years under as many different names. The editor of the *Ohio Cultivator* explained that German Millet had acquired the name of "Hungarian Grass," and, finally, "Honey Blade Grass," a change in name which increased its price six hundred percent.[36]

As in many other activities, the *American Agriculturist* took the lead in this reform. For a few years it carried a column headed "What the Humbugs are Doing." In order to expose these impostors, the editor answered attractive advertisements wherever found, enclosing stamps or money if requested, signing a different name to each letter, and mailing these letters from post offices throughout the country. The results were finally published in his journal and widely copied in other farm periodicals. He found, however, that as a general rule if money or stamps were sent the letters were never answered.[37] Many of the advertisements were similar to "Rev. Wilson's cure for consumption" which was available for a postage stamp. The "minister" forwarded a recipe, together with the assurance

[34] *Ohio Cultivator,* XV (Jan. 15, 1859), 24.

*The Farmers' Register,* which carried a few advertisements on the cover was always a diligent guardian of the readers' interests. *Farmers' Register,* I (June, 1833), 446; V (June 1, 1837), 125–26.

[35] *American Agriculturist,* XVII (Aug., 1858), 251; *Genesee Farmer,* ser. 2, XVI (Aug., 1858), 241–42.

[36] *Cultivator,* VIII (Jan., 1841), 13; *Ohio Cultivator,* XV (Feb., 1859), 41.

[37] *American Agriculturist,* XVIII (Aug., 1859), 231; XVIII (Oct., 1859), 296.

that he had "no mercenary motives," and a suggestion to the effect that he would sell a bottle of the "cure," which was difficult to make, for $2.00—a price much cheaper than a druggist could compound it.[38] The editor of a leading medical journal commended the *American Agriculturist* for its nostrum exposures and added, "the 'laity' do not believe *us* when we tell them such homely truths, because they think (that we think) our craft is in danger. *You* they will hear—*perhaps*." [39] Referring to the *Agriculturist*, Horace Greeley observed in the *Tribune:* "The Editor don't mean to be humbugged himself nor let any body else be, if he can help it." [40]

It is interesting and illuminating to note the reason advanced for this solicitude in exposing humbugs and constantly warning the farmer against attempts to get something for nothing. The *American Agriculturist* explains: "The *Agriculturist* circulates largely among rural people, who are not familiar with the arts of swindlers, and as we have before stated, they are, as a class, more honest themselves, and therefore less likely to be on the look-out for deception from others. Swindlers themselves understand this, and hence, nine-tenths of all their efforts at imposition are directed at rural people." [41] This comforting explanation of rather popular acceptance was also voiced by a farmer in the *Southern Planter:* "We are a confiding, trusting class, because we are honest and guileless ourselves; we judge others as we are proud to be judged—by that fair justice which springs from the purity of motive and honesty of intention." [42]

[38] *Ibid.*, XVIII (July, 1859), 198. A druggist, when shown the recipe, offered to compound it for less than a dollar, although he warned the editor against the use of it.

[39] *Ibid.*, XVIII (Nov., 1859), 345. [40] *Ibid.*, XVII (Nov., 1858), 348.

[41] *Ibid.*, XX (Jan., 1861), 6.

[42] *Southern Planter*, XX (March, 1860), 143.

Aside from the strictly agricultural advertisements (which predominated) and the so-called humbugs, there were others familiar to the readers of the journals. A frequent patron was the Illinois Central Railroad Company which advertised the sale of millions of acres of farm land. Others were Wells, Fargo and Company's California Express, Fairbank's Scales, Webster's Dictionary, and sewing-machine manufacturers—Grover and Baker, Wheeler and Wilson, and Singer. Advertisements of the New England Mutual Life and the Aetna insurance companies appeared occasionally. The *Saturday Evening Post*, a "moral paper" which any parent might "allow to go freely before his innocent sons and daughters" was widely advertised as established August 4, 1821 (with no mention of Benjamin Franklin). It is interesting to note that even the Southern journals almost totally excluded slave notices. From time to time a matrimonial advertisement was inserted, such as the following in the *Michigan Farmer:* "A Bachelor . . . desires to say to the Lady readers of the Farmer, that he is of good character, healthy, intelligent, enterprising, warm hearted and moderately good looking, abjures tea, coffee, tobacco and spirituous liquors, and now desires to be united but not *sold* to a lady of like qualifications and habits." [43]

Perhaps the most characteristic feature of the advertising of this period was the use of testimonials. It was always considered important that the author of a testimony have a national reputation, but that he be qualified to evaluate the product, then as now, was of secondary interest. At one time Grover and Baker's sewing machine met the hearty approval of Governor J. G. Harris of Tennessee, Senator J. H. Hammond of South Carolina, and Cassius M. Clay of Kentucky. At a later date, fifty-five clergymen, whose names were affixed, took "pleasure in

[43] *Michigan Farmer*, XV (Sept., 1857), 284.

recommending" the same machine.[44] On the other hand, the advertisements of the Wheeler and Wilson machine carried a list of 18 publications which favored it.[45] Webster's Dictionary, as we might expect, carried testimonials from President James K. Polk, Vice President George M. Dallas, and Washington Irving.[46] Enthusiastic letters from Professor Hitchcock, president of Amherst College, Senator Edward Everett, and Secretary of State W. L. Marcy endorsed Ayer's Cherry Pectoral. The Mexican Mustang Liniment was recommended by the American Express Company, Harnden's Express Company, and Wells, Fargo and Company.[47] If the product was awarded a premium at an agricultural fair, that recommendation was promptly noted and emphasized in the advertisements that followed. As a substitute for testimonials, the reaper manufacturers, toward the end of the period, listed names of satisfied customers with whom prospective buyers might communicate.

The famous field trials of agricultural machinery during the fifties may logically be considered in a chapter on advertising. It was the publicity gained by the manufacturers of reapers and mowers, which accounted for their enthusiasm in these contests.[48] On the other hand the agricultural societies, as well as the journals, were vitally interested in the relative merits of farm inventions, especially harvesting machinery.

[44] *Indiana Farmer*, VII (March, 1859), 437–38; *Genesee Farmer*, ser. 2, XXI (Oct., 1860), 327.

[45] *American Agriculturist*, XVII (March, 1858), 96.

[46] *Farmer's Monthly Visitor*, XI (Feb. 28, 1849), 32; *Cultivator*, n.s., VIII (Aug., 1851), 286.

[47] *Prairie Farmer*, XI (June, 1851), 273; XIII (June, 1853), Advertising section; *New Genesee Farmer*, XV (June, 1854), 199.

[48] Sometimes contests resulted from a rival challenge in the agricultural press and were fought for large money prizes. Suspicion was warranted, on occasion, that the contestants agreed beforehand that neither should win, the whole affair serving for mere publicity. Hutchinson, *Cyrus Hall McCormick*, I, 340.

A competitive grain-cutting race provided the most logical method for establishing superiority. Because life in rural communities seemed unexciting and uneventful, farmers responded enthusiastically, and lovers of sports were treated to a contest with features not unlike the plowing match and the horse race. Since these contests were of necessity held during the harvest season, it was normally impossible to combine them with the agricultural fair in the Fall.

The reaper trials, which were held in many parts of the United States, were made as entertaining as possible by shrewd competitors. Occasionally each contestant was accompanied by a band to enliven the event, and refreshments were frequently served to sustain the audience as it followed the machines in the midsummer heat. Sometimes the winner was determined by the relative volume of cheers greeting the name of each reaper as announced at the end of the contest; prizes, similar to those at state fairs, were thereupon awarded the victors.[49]

The great advantage of actual reaper trials in the field as compared with reaper displays at fairs was well illustrated at the World's Fair held in London in 1851. The McCormick reaper, while on exhibition there, afforded abundant sport for the London *Times,* which ridiculed it "as a cross between Astley's chariot and a flying machine." However, the field trial altered this picture. So successful was the demonstration that the crowd burst into three "*hearty English cheers,*" the reaper was awarded the Council Medal, the London *Times* apologized, and the machine when again displayed had more visitors than the "Ko-i-noor diamond itself."[50] McCormick "suddenly waked upon a throne higher than the Emperor

[49] Hutchinson, *op. cit.*, I, 338.

[50] *Cultivator*, n.s., VIII (Oct., 1851), 327; n.s., IX (Sept., 1852), 312; *Country Gentleman*, II (Dec., 1853), 357.

Nicholas." This award received great publicity throughout the agricultural press of the United States, and did more to advertise the McCormick and other reapers than any other single incident.[51]

Prior to the Civil War, the most ambitious reaper trial in the United States was held at Syracuse, New York, in 1857, under the auspices of the United States Agricultural Society.[52] The trials, which lasted one week, were opened with speeches by Marshall P. Wilder (president of the society), Governor King of New York, and Governor Morehead of Kentucky. Twenty-five hundred people from all parts of the country, including many national figures and representatives of the agricultural press, witnessed the trial. Thirty-seven reapers and mowers were entered for competition.[53]

While fields trials were ideal for advertising the machines and provided an infallible method for exposing out-and-out humbugs, these contests could not determine which machines were best adapted to the farmers' wants. Normal conditions did not exist at the time of these tests, which at best were inconclusive. A good team or a skillful driver might win for a second-rate machine, while a first-rate machine might lose for

[51] *Farmer's Monthly Visitor,* XII (March, 1852), 80; *Prairie Farmer,* XIII (May, 1853), 173; *Michigan Farmer,* XI (Oct., 1853), 296.

[52] The duties outlined for the judges showed the care exercised to obtain an accurate evaluation of each machine. One committee was directed "to report the precise *weight* and *cost* of each machine; another the *dimensions* of the several parts; another the *form* of the cutters; another on the *mechanical principles* involved; another the *material* used in the manufacture; another the results given by the *dynamometer;* and others variously on the quality of work performed; on the side draft; on the condition of the *stubble,* as indicating the character of the cutting, &c., &c." *Country Gentleman,* X (July 16, 1857), 48.

The judges were selected with great care from the best mechanics and practical farmers of the country. *Country Gentleman,* X (July 23, 1857), 65; XI (Feb. 4, 1858), 73.

[53] *Country Gentleman,* X (July 23, 1857), 65; XI (Feb. 11, 1858), 89.

want of good management. The machine entered was frequently not representative of the product offered on the market. Judges were sometimes prejudiced or incompetent. All these factors played a part in making the contests highly speculative. The trials of machines were necessarily so brief, that the question of durability—and other points of practical importance—could not be adequately tested.[54] "I desire to enter my protest [said one writer] against any more petty trials of reapers. They cost a great deal and amount to nothing. The decision at one trial is reversed the next week at another, perhaps with the same machines, and often the competitors can show their defeat was owing to some extraneous circumstances, as not having a suitable team, bad driving, or unfortunate management in some way." [55]

The actual field test was only a preliminary in the "battle of the reapers," for regardless of who received the prize, a vigorous war of letters ensued in the agricultural press lasting for months. If a manufacturer won the decision, he boasted; if he lost, he accused and explained. When McCormick failed to win the award at Geneva, New York, in 1852, he immediately flooded the journals with "proof" of the injustice done his reaper, asserting that the superiority of his machine "over all others included in said trial, was abundantly *proved*—the *awards* of the committee to the contrary notwithstanding." [56] This brought a retort from Manny, the victor: "The flutterings of a wounded bird are so mingled with expressions of pain, that it is perfectly natural that sympathy should be excited." He also called attention to the "many erroneous assertions"

[54] Hutchinson, *op. cit.*, I, 338 ff.; *Indiana Farmer*, VII (July, 1858), 97; *American Agriculturist*, XVI (Aug., 1857), 174; *Prairie Farmer*, XVI (July 17, 1856), 114; *Ohio Cultivator*, XIV (Aug. 1, 1858), 232.

[55] *Country Gentleman*, III (March 2, 1854), 135.

[56] *Cultivator*, n.s., IX (Oct., 1852), 355; n.s., IX (Dec., 1852), 412.

*View of the Grounds and Trial of Reapers in the Barley Field, July 22.*

**REAPER TRIALS, GENEVA, NEW YORK, 1852**

From *Cultivator*, N. S., IX (September, 1852), 312.

in McCormick's correspondence, attempting to "expose some of its errors and absurdities," and assured his readers that the committee had not "acted under the corrupting influence of bribery." [57] And so for months, to the delight of the readers, the war of letters ran its course—"the inevitable aftermath of field trials." These "wars," carried on with vulgar boasts and bravados, were quite in keeping with the blustering fifties.[58] Incidentally, the farmer became "reaper conscious."

Many practices of modern advertising were in the process of development during this pre-Civil War period. Present day big-business methods are apparent especially in advertisements of agricultural machinery and the sewing machine. By 1860, the use of the guarantee, installment buying, the goods-on-approval plan, the trade-in and other modern devices for facilitating sales were well established. As early as the middle forties, Hussey "warranted" his reaper to cut twenty acres of wheat in a day, whereas McCormick quoted fifteen to twenty acres.[59] Deferred payments were also introduced at this time, McCormick's terms on his reaper being "four or six months, from harvest with interest." Mann's reaper sold for $50.00 on delivery, and the buyer was permitted eighteen months in which to complete payment.[60] McCormick was willing to send his reaper to responsible farmers who might "desire to make a trial of it alongside, and on the same terms of *any other*, to be purchased or refused, as decided on making such trial." [61] The Singer company, on the other hand, advertised "old machines taken in exchange for new and improved ones." [62]

[57] *Ibid.*, ser. 3, I (Feb., 1853), 69.

[58] Hutchinson, *op. cit.*, I, 344, 415.

[59] *Cultivator*, n.s., I (May, 1844), 168; n.s., II (June, 1845), 181.

[60] *Indiana Farmer*, VII (Jan., 1859), 348.

[61] *Southern Planter*, n.s., XI (Jan., 1846), 6; *Cultivator*, n.s., III (May, 1846), 165.

[62] *American Farmer*, XIV (1858–59), Advertiser.

It is evident that advertisers of the period understood sales psychology, as is demonstrated by the use of high pressure methods. During the depression of 1857, readers were thus advised, "Now is the time to buy a Singer Sewing Machine. If you have Bank bills that are current to-day, they may be discredited to-morrow. Buy a Singer Sewing Machine, it is a good investment under all circumstances—if you fail entirely, it will afford you a handsome income, and if you survive the panic it is worth quadruple its present price." [63] Another appeal urged the husband to relieve his wife from the never-ending drudgery of the needle, and preserve "her from the long train of diseases concomitant thereon" for, with a Singer Sewing Machine, her cheek would "retain the roseate hue of health and her heart glow with gratitude for her deliverance." [64] The people were told that they could not afford to be without a Grover and Baker Sewing Machine for it would do better and cheaper sewing than a seamstress even if she worked "for *one cent an hour*." [65] Subtle suggestions that the price would probably be higher the following year were frequently made by the reaper manufacturers. Annually, they assured their readers that the machine was "perfected." Name calling and thrusts at competitors appeared in some advertisements. McCormick went so far as to announce that he would sue anyone who purchased a Seymour and Morgan machine because that company had infringed upon his patent.[66]

A good portion of the agricultural press was devoted to what might be termed free advertising. For instance, the editors inserted gratis engravings of new agricultural inventions, with

[63] *Valley Farmer*, IX (Dec., 1857), 231.
[64] *Ibid.*, IX (Feb., 1857), 38.
[65] *Indiana Farmer*, VII (Feb., 1859), 437.
[66] *New Genesee Farmer*, XIII (June, 1852), 196.

descriptions and remarks. The introduction of new plants and animals had real news value and was featured with puffs always reacting in some merchant's favor. The plowing matches and reaper trials featured as sporting events provided publicity for the manufacturers. However, the editors attempted to eliminate from the news columns communications containing surreptitious advertising. In the earlier years paid advertisements probably filled considerably less than five percent of the paper and by 1860 it was customary to devote from ten to twenty percent to this purpose. The *American Agriculturist* asserted that it did "not feel at liberty to occupy more than one-eighth of the paper" with advertisements; while this limitation was generally accepted in theory, it was difficult to carry out in practice as the pressure for advertising space increased.[67]

Advertising fees lacked uniformity and varied from time to time in any particular journal. The circulation of the paper determined the rate, which ranged from four to thirty-five cents a line for each insertion. Testimonials from satisfied merchants and farmers bore ample evidence that the agricultural press was an ideal medium for advertising.

[67] *American Agriculturist*, XVI (Feb., 1857), 45.

# Chapter VII

## "LADIES' DEPARTMENT"

*What! woman the same rights as man?*
*'Tis folly to suppose it—*
*She's been a long time under ban,*
*And every body knows it!* [1]

WHILE this was essentially a man's world during the pre-Civil War period, nevertheless, the journals, at an early date, recognized the need for features devoted to the interests of women and girls. It became customary for the editor, upon launching a journal, to announce that the farmers' wives and farmers' daughters would not be neglected, and that everything relating to the improvement of the domestic circle would be discussed. Until the advent of highly organized "ladies' departments" which appeared about the middle of the century, the editors inserted articles of interest to the gentler sex as good copy came to hand and as space from time to time became available.[2]

Included in the wide range of domestic topics were recipes for making bread, baked apples, chicken pie, muffins, egg bis-

[1] "Woman's Rights," *Boston Cultivator,* XV (July 16, 1853), 227.

[2] In 1824 the *American Farmer* began a column headed "Ladies' Department." The articles were selected from all manner of sources; however, very little original material found its way into these columns.

cuits, and the like taken from standard recipe books or sent in by farmers' wives. Under the heading "Hints to Mothers," subjects such as the dressing of children, care of infants, and family management were discussed at great length. Miscellaneous articles on "How to Choose a Good Husband," "Important Requisites in a Wife," "Female Education," "Woman's Fashions," "Extravagance and Its Consequences," and "Home as the Seat of Virtue" (the sanctity-of-the-home theme was so often repeated that it became oppressive) abounded in the early journals. Flower gardens received much comment. Wide publicity was given Dr. Mussey's statement, that "greater numbers annually die among the female sex by the use of the corset, than are destroyed among the other sex by the use of spirituous liquors in the same time!" [3]

These articles constituted a highly specialized field and the editors' lack of training was obvious—especially to the editors themselves. This feeling was well expressed by the editor of the *American Agriculturist,* when he wrote in 1844, "We wish some one would volunteer to make us a Ladies' Department . . . , for we feel totally inadequate to attempt anything of the kind—and yet the farmers' wives and daughters ought not be neglected." [4]

After the middle of the century, the editors enlisted their wives and, very often, outstanding feminine contributors in an editorial role; thus regularly organized departments gradually came into being.[5] This was especially true of the "western"

[3] *New England Farmer,* XI March 6, 1833, 270.

[4] *American Agriculturist,* III (Dec., 1844), 353.

[5] These editors and their columns included Mrs. Josephine C. Bateham, "Ladies' Department," *Ohio Cultivator;* Mrs. E. O. Sampson Hoyt, "Home Circle," *Wisconsin Farmer;* Mrs. L. B. Adams, "Household Department," *Michigan Farmer;* Mrs. Abby A. Miller, "Ladies' Department," *Northwestern Farmer and Horticultural Journal;* Mrs. Mary Abbott, "Family

journals, probably due to the greater need of entertainment for the farmers' wives in less settled regions. These new departments under various titles followed their own policies. Communications from many talented contributors enriched these columns whose popularity is attested by the large correspondence received from many sections of the country.

The most famous and certainly the most interesting "Ladies' Department" was that of the *Ohio Cultivator.* A detailed consideration of it may serve to illustrate the practice of the best journals. In the first issue M. B. Bateham, the editor, indicated his great interest in this field, and urged readers to contribute worthwhile material. Being a bachelor, he said he would depend entirely upon the ladies for suitable articles. However, he promised that as soon as the profits of the *Cultivator* were sufficient, he would "endeavor to find an assistant," who was qualified, and would "*consent,*" to take charge of the "Ladies' Department." [6] For the next three years, Bateham continued the work with occasional contributions from his fair correspondents—although many of these were "rather too sentimental," displaying "too much of an attempt at fine writing," and their material was often "too deficient of interest" to be used.[7] Consequently, *Miss Beecher's Domestic Receipt Book* [8] became an ever present source of comfort to the editor.

In the Fall of 1847, Bateham married Louisa Jane Lovell, a "Farmer's Daughter," and soon proudly announced that his

---

Circle," *Valley Farmer;* Mrs. Margaret J. Newton, "Woman's Department," *Indiana Farmer.*

[6] *Ohio Cultivator,* I (Jan. 1, 1845), 5.

[7] *Ibid.,* II (Jan. 15, 1846), 13; III (Sept. 1, 1847), 126.

[8] Published in 1846 by Catharine Esther Beecher, sister of Henry Ward Beecher and Harriet Beecher Stowe. At a later date, Mrs. Horace Mann's *Christianity in the Kitchen: Physiological Cook Book* (Boston, 1857) served in the same fashion when copy was scarce.

wife would take charge of the "Ladies' Department" thus fulfilling he said, "the promise made by us three years ago."[9] Although Mrs. Bateham lived but a year, her department from this time assumed the leadership in this field.[10] In 1850 Bateham married Josephine A. P. Cushman of Oberlin, widow of the late missionary to Haiti.[11] Within six weeks, she assumed the conduct of the column; under her direction the "Ladies' Department" reached the height of its popularity.[12] The second Mrs. Bateham wrote editorials on education, peace, family training, health, importance of simple and wholesome diet, regular habits, wages and management of domestics, woman's rights, temperance and so on.[13] The following advice to young ladies characteristically portrays her fine common sense:

> We do not wish you to become masculine in looks or manners, nor to aid your brothers in their laborious occupations, but we do earnestly desire that you should draw back the curtains, ventilate your rooms thoroughly, engage actively in household labor, avoid injurious habits of dress and of diet, pay strict attention to personal cleanliness, and above all, *take abundant exercise in the open air.*[14]

[9] *Ohio Cultivator,* IV (Jan. 1, 1848), 4.

[10] Mrs. Bateham, a "fine scholar" and a "vigorous writer" taught school for several years before she entered the female department of the Oberlin Collegiate Institute, and after the regular course of four years' study graduated with honors in 1846. She continued her studies one year longer, as a resident graduate. It was on the occasion of reading her graduating essay that Bateham saw Miss Lovell for the first time and she "so favorably impressed his mind and heart as subsequently to induce him to seek her acquaintance and her love." *Ohio Cultivator,* IV (Nov. 1, 1848), 167.

[11] *Ibid.,* VI (Oct. 1, 1850), 302.

[12] Josephine C. Bateham was graduated from Oberlin Collegiate Institute in 1847 at the age of 18. She taught school for a short time and soon married the Rev. Mr. Cushman. It was while they were in Haiti that he died. *Ibid.,* X (July 15, 1854), 221.

[13] Editor and Mrs. Bateham attended the World's Fair and Peace Congress in London in 1851. Numerous editorials dealt with this experience in Europe.

[14] *Ohio Cultivator,* VI (Dec. 1, 1850), 364.

One of the most popular and capable of the many active contributors to this department of the *Ohio Cultivator* was Hannah M. Tracy.[15] In her famous "Letters to Housekeepers" (signed "Maria") she dealt with a variety of subjects—games and entertainments for social winter evenings, the advisability of admitting hired help to the family fireside, and good reading for the long winter months. Under the pen name of "Aunt Patience," Mrs. Tracy in 1849 inaugurated a kind of "Dorothy Dix" column in which she endeavored to answer letters from girls who would soon be filling the "honorable and distinguished places of Farmers' Wives of Ohio." Each week the "Country Nieces" presented their problems to "Aunt Patience" to be answered in the *Cultivator*. In 1851 Mrs. Tracy went to London to attend the Peace Congress and World's Fair, and the *Ohio Cultivator* received frequent reports of her experiences and observations in Europe. While in London, she created a great sensation by appearing on the public platform in the new Bloomer costume. In this outfit, she delivered lectures on "Bloomerism" urging the abolition of the old forms of dress, not as a whim, but as a step for better health.[16] It was her belief that fashions should be held

[15] Hannah M. Tracy ("Maria," "Aunt Patience") had been a frequent contributor to the *Cleveland Herald* and the *Western Magazine* before her association with the *Ohio Cultivator*. Having been bereaved of her husband, she was now mainly dependent upon her own efforts of mind and pen to support herself and her three children. For nearly three years, she served as matron in the Ohio Deaf and Dumb Asylum, but resigned to become principal of the Female Department of the High School in Columbus, Ohio. *Ohio Cultivator*, VI (Nov. 15, 1850), 349.

[16] Amelia Jenks Bloomer (1818–94) was active as a temperance reformer and was a pioneer in the woman's rights movement. In 1849, she began publication of the *Lily*. For six years, she continued its publication, writing vigorous articles on education, unjust marriage laws, and later on woman's suffrage. Her name, through newspaper publicity and ridicule, is associated in the public mind with dress reform, however. Her costume consisted of loose Turkish trousers gathered at the ankle with an elastic

to scientific accountability and rendered subservient to the truths of physiological laws.[17] She said her speeches were kindly received and the crowds were "inordinately *curious,*" but she doubted whether her efforts did much good as "the mold of ages" rested upon the hearts of the British people.[18] Upon returning to the United States, this reformer continued to write articles on a great variety of subjects.

Frances D. Gage ("Aunt Fanny"), another outstanding contributor, was not only a poet of note, a writer of many beautiful stories for children, and author of innumerable sketches on social life, but was known, perhaps as well, for her activity in behalf of reforms.[19] She wrote and lectured against slavery, intemperance, capital punishment, and was a faithful crusader for woman's rights.

These three women who saw eye to eye on woman's rights, temperance, slavery, and female education did not permit lecture tours, family duties, and other interests to interfere with

---

band, a short skirt, short jacket, and a straw hat. *Dictionary of American Biography*, II, 385; Faulkner, *American Political and Social History*, p. 277.

[17] *Ohio Cultivator,* VII (Nov. 15, 1851), 349.

[18] *Ibid.,* VII (Dec. 1, 1851), 366.

[19] Frances D. Gage (1808–84), was born in Marietta, Ohio. On January 1, 1829, when not yet twenty-one, she married James L. Gage, a lawyer of McConnelsville, Ohio. In spite of the demands made ultimately by a family of eight children, she found time for reading, writing, and speaking on many "reform" subjects of the day. In 1853, the family moved to St. Louis, where she was soon in difficulties because of her antislavery proclivities. *Dictionary of American Biography*, VII, 84–85; see also "Frances D. Gage," by Elizabeth Cady Stanton in Parton, *Eminent Women of the Age*, pp. 382 ff. After her husband failed in business and in health in 1857, Mrs. Gage wrote more and more for the agricultural papers—especially, the *Ohio Cultivator*, *Ohio Farmer*, and *Prairie Farmer*. In 1861 she again moved to Ohio and took charge of the "Home Department" of the *Cultivator*, a position she retained until October 1, 1862. *Ohio Cultivator*, XVIII (Oct. 1, 1862), 309. She served in the Civil War in various capacities.

For Mrs. Gage's poetical contributions see pp. 192–93.

their literary activities.[20] Mrs. Tracy was president of the Ohio Woman's Rights Association; Mrs. Bateham was president of the State Temperance Society of the Women of Ohio; and Mrs. Gage, an excellent public speaker, frequently made lecture tours to discuss the controversial subjects of the day.[21]

It would be unfair, however, to infer that the "Ladies' Department" of the *Ohio Cultivator* was completely filled with propaganda for reform. Poems were extremely popular and were given a prominent place.[22] Many of "Aunt Fanny's" letters to "My Dear Cultivator Girls" dealt with commonplace subjects such as the proper dress for young ladies at the fair. "But go to the Fair," she urged, "don't trouble your minds about your dress. The plainest and simplest is always most beautiful upon young folks. The less gewgaws and fancy trimmings you have, the more like good, substantial women you will look." [23] The question of proper education for farmers' daughters was often presented to "Aunt Patience" by her "Country Nieces." She tried to break down the idea that girls who attended female seminaries necessarily would come home "proud, and think it disgraceful to perform the ordinary household duties." Hints on how to keep preserves, how to make dresses, and how to entertain young men, all found their way into this corner of the journal. The fact that in one year alone, fifty different correspondents contributed to its columns is evidence of the wide interest which existed in this department. The editor was justly proud of this record, as well as the

[20] Miss Sarah Coates, an ardent supporter of woman's rights and temperance, wrote frequently under the *nom de plume* "Chamomile."

[21] *Ohio Cultivator*, VIII (June 15, 1852), 189; *Ibid.*, IX (Feb. 1, 1853), 45.

[22] See pp. 192–93.

[23] *Ohio Cultivator*, VII (Sept. 15, 1851), 285.

F. D. GAGE

Frontispiece, *Poems* by Mrs. Frances Dana Gage, Philadelphia, J. B. Lippincott & Co., 1867.

manner in which the great variety of subjects had been treated. She was proud also, that over one hundred valuable recipes had been published for the housewives.[24]

Very few of these departments took the "advanced stand" of the *Ohio Cultivator* in regard to the so-called reforms of the period. Certainly this department was not representative in this respect. Mary Abbott, who conducted the "Family Circle" of the *Valley Farmer*, took the ultraconservative attitude. Her position on the wider activities of women may be seen in her comment on horsewomanship at fairs: "If it is right for women to make a public display of themselves at fairs, they have a right to vote at the polls, give temperance lectures, preach, or in fact do anything that the sterner sex may do, even drinking and smoking." She continued, "None but Bloomers and Woman's Rights, and females of the baser sort would thus display themselves." [25] The majority ranged between these extremes.

Through these departments the journals provided a veritable clearing house for domestic and social ideas. All manner of subjects were discussed and debated at length. For instance, a remark in the *Michigan Farmer*, that the poverty of farmers in general was due to the idleness and extravagance of their wives and daughters brought such an avalanche of letters that the editor found it impossible to print them.[26] The *Prairie Farmer* favored dancing in an editorial and this resulted in a bitter discussion lasting months. "I am sorry [wrote a correspondent] to be again found arrayed in opposition to the avowed sentiments of yourselves and many patrons, but reli-

[24] *Ibid.*, XI (Jan. 1, 1855), 12.
[25] *Valley Farmer*, V (Dec., 1853), 437.
[26] *Michigan Farmer*, XII (May, 1854), 152; XII (July, 1854), 211.

gious duty compels me to enter my earnest protest against your position as to dancing." [27] Other controversial subjects included spiritualism, watercure, and women as physicians.

Occasionally, the unfavorable conduct of the sterner sex became the center of discussion. Men drank too much and the women voiced their displeasure. Moreover, the presence of Women's State Temperance Societies all over the country was concrete evidence that they intended to do something about it. Smoking and swearing were likewise in disfavor. A correspondent complained that even in the street women were "nearly suffocated by ungentlemanly men" smoking in their faces.[28] Tobacco chewing was another target of wrath. "I do not know a more disgusting practice than that of defiling cars, and public rooms, and private parlors even, in this way," wrote one woman.[29] "Men look solemn, talk grave and spit," and always "finish a sentence in conversation by a spit," complained another writer. She added hopelessly, "Will the time ever come, when the spittoon, that disgusting reminder that people spit, will be removed from our parlors, steamers and cars?" [30]

A matrimonial bureau inaugurated in 1857 provided a unique and interesting feature of the *Rural American*.[31] Editor Miner announced that he stood ready to help all modest, industrious young ladies to secure husbands and the only recommendation he required was, that they be "up with the sun," and not "ashamed to be seen at a washtub, at the oven,

[27] *Prairie Farmer*, XIX (April 14, 1859), 234.

[28] *Valley Farmer*, V (Nov., 1853), 406.

[29] *American Agriculturist*, XVIII (June, 1859), 183.

[30] *Ohio Farmer*, IV (Aug. 18, 1855), 132.

[31] A similar feature had been very successful in the *Northern Farmer* in 1852. It was necessary, however, after a few years to "shut down the gate" on the applications as the editor was getting too much business "to afford the time that their cases demanded." *Rural American*, II (July 15, 1857), 112.

nor with a broom" in their hands.[32] He insisted that there were thousands of both sexes who would never get married because they were too modest or bashful to let anyone know they were in want of partners. In some localities, there was an actual dearth of young men while young ladies were "as plenty as blackberries in harvest time." On the other hand, there were towns in the West in which perhaps a hundred young men resided who really wanted wives and not a dozen young ladies could be found.[33] Young ladies responded promptly to his suggestion as the subject was one that "fired their souls." [34] In those days of early marriages, a girl was "available" at sixteen, was "getting along" at eighteen, and was ever mindful of the possibilities of being an old maid in her twenties. Letters from the girls were printed in the journal; sometimes accompanied with editorial comment and recommendation. For six cents in stamps, a bachelor could obtain from the editor the address of a young lady whose letter especially appealed to him. Sometimes the same information was sent to a dozen different applicants; the editor then stepped aside to allow Cupid free range.

The young men were not timid and frequently outlined their marital requirements in full. The following is typical:

> MR. MINER:—If there is any young lady reader of your paper possessing in her own right a farm, and wishing a husband to assist her in carrying it on, and who is well educated in both domestic and literary matters, I would inform her in a private and confidential correspondence of myself and references, and receive the same from her.
>
> Biddeford, Maine, Oct. 9, 1857. HORATIO BAKER [35]

---

[32] *Rural American*, II (July 15, 1857), 112.

[33] *Ibid.*, II (Oct. 15, 1857), 160. [34] *Ibid.*, II (Sept. 15, 1857), 144.

[35] *Ibid.*, II (Nov. 16, 1857), 176.

According to the editor, thousands of young ladies were enthusiastic subscribers to this popular journal. We have only his word to substantiate the boast of numerous romances initiated in this department, since grateful married couples neglected to send in their testimonials.

A large portion of the space devoted to the gentler sex, especially in the later years, was filled with warning and advice to young ladies on every phase of living. Novel-reading in particular was in bad repute. One writer, in describing the effect on young women who became engrossed in this pastime said, "their business is neglected, sick friends are forgotten, the Sabbath is profaned, and the worship of God in His sanctuary, exchanged for a lounge o'er a sick'ning dream." [36] Girls were warned against the use of slang at all times. Surely, stated one journal, no lady would speak of taking a "snooze," instead of a nap; she would not call pantaloons "pants," or gentlemen "gents"; a poorly dressed man never looked "seedy"; and an amusing incident was never "rich." All these "vile" expressions were "detestable from the lips of ladies." [37] Contemporary fashions always provided good material. It was one editor's settled conviction in 1852 that the fashion of that age was "carrying more to early graves than all the diseases and contagions in the world." [38] Dancing was considered immoral

[36] *Michigan Farmer,* II (May 1, 1844), 47.

[37] *Rural American,* I (Aug. 1, 1856), 224.

[38] *Northern Farmer,* I (Oct., 1852), 156. The same thought was expressed in *Moore's Rural New-Yorker,* IX (Nov. 13, 1858), 368; X (March 19, 1859), 96.

Nevertheless, every style had its defenders. One young lady, in favor of the hoop-petticoat claimed it was "very airy"—and also kept men at a distance. She added: "It is most certain that a woman's honor cannot be better entrenched than after this manner, in circle within circle, amidst such a variety of outworks and lines of circumvallation." *Rural American,* I (Feb. 2, 1856), 36.

from the point of view of many. Even the piano, recognized as a "fashionable extravagance," did not escape suspicion, for who ever heard that the ability to play such an instrument helped any wife to cook a good dinner? [39] Many felt that the whole scope of female education was to make young ladies into "mere parlor ornaments." [40] The editor of the *Agriculturist* declared there was little time for the training of young ladies, because the marrying age came "so soon after quitting the bottle." [41] Other stand-by editorial subjects included: "Marry in haste and repent at leisure" and "Early Marriage." It was generally urged that a girl should wait until she was twenty before taking this important step. Young ladies were definitely advised to dress plainly, rise early, work hard and make themselves "worthy" of a good farmer husband. Surely no farmer would want a wife who was a "bundle of gewgaws, bound with a string of flats and quavers, sprinkled with cologne, and set in a carmine saucer." [42]

In advice to young ladies, the eternal theme "of the good old days," the idea that somehow the contemporary generation was less moral, less industrious, less healthy than past generations runs through the articles in the journals. Time was, said one correspondent, when "young ladies with blooming cheeks,

[39] *Country Gentleman*, IX (Feb. 27, 1857), 145; *Genesee Farmer*, IX (April, 1848), 115; *Farmer's Monthly Visitor*, XIII (May, 1853), 155. A writer claimed, "The very families, often, whose daughters are drumming on the piano to the neglect of domestic duties, don't know where the money is coming from to purchase a barrel of flour or a load of wood. Such is the tyranny of fashion, and the nonsensical desire of having to keep up appearances." *Moore's Rural New-Yorker*, VII (July 19, 1856), 233.

[40] Most of the articles on female education laid stress on the need for "practical" training and frowned on nonutilitarian artistic "accomplishments."

[41] *Agriculturist*, IV (Jan., 1843), 18–19.

[42] *Ibid.*, XIII (May, 1853), 155.

buoyant spirits, and healthful activity" might be seen at the wash tub at six in the morning.[43] Another writer lamented: "The present stock of girls, growing and grown, have not the vigorous health which is essential to become the mothers of men and women—such men and women as were their ancestors." [44] In speaking of the women in his grandmother's day, one editor observed: "With them accomplishments and ornament, romance and moon-gazing, were minor objects—and made secondary to domestic duty." [45] Under "changes noted," the *Prairie Farmer* expressed the same attitude:

> To spin and weave, to knit and sew,
>   Was once a girl's employment;
> But now to dress and catch a beau,
>   Makes up her whole enjoyment.[46]

During the middle of the century, a group of women writers, who in many respects resemble our present day newspaper columnists, made their appearance. Although often completely independent of the "ladies' departments," this interesting group may be conveniently considered at this point. Not restricted in subject matter, nor necessarily voicing the ideas or opinions of

[43] *Northern Farmer,* I (Oct., 1852), 156; cf. *Michigan Farmer,* II (May 15, 1844), 55.

[44] *Country Gentleman,* XIV (Nov. 17, 1859), 323.

[45] *Farmer's Monthly Visitor,* XIII (Jan., 1852), 22.

The hardship which the household industries had imposed upon women in New England was frequently discussed. The agricultural revolution which occurred in this region before the Civil War, wrought vast changes in household economy. At this time the farmers' wives and daughters seemed relatively idle, since the household textile industry had been gradually taken over in a large part by the new factories. Very soon however, the farmers' daughters in large numbers found employment in the factories of Lowell, Lawrence, and Fall River. Bidwell, "The Agricultural Revolution in New England," pp. 693 ff.; *Cultivator,* n.s., VI (Sept., 1849), 281–82; n.s., VI (Nov., 1849), 346–47.

[46] *Prairie Farmer,* XVI (Feb. 7, 1856), 24.

the editors, they wrote, as one of them said, on whatever pleased or interested them.[47] Perhaps best known of this group were Frances D. Gage, already mentioned in connection with the "Ladies' Department" of the *Ohio Cultivator,* and Anna Hope. The latter, whose early life is unrecorded, was described in 1858 as a wife, mother and sister with earnest dark eyes, beautiful to look upon, gentle and mild in temperament, and graceful in manner.[48] Both women were widely traveled and keenly alive to the vital movements of the day. Frances Gage wrote for the *Prairie Farmer,* the *Ohio Farmer,* and the *Ohio Cultivator* on the benefits of travel, spiritualism, birds of St. Domingo, breadmaking, female medical colleges, capital punishment, and a variety of other topics.[49] Anna Hope, who contributed principally to the *American Agriculturist* and the *Ohio Farmer,* wrote on varied topics including deportment at the table, dangers of fire works on the 4th of July, quack medicines, the Saratoga water cure, cheerfulness, tenement houses in New York, gifts for the holidays, and dresses for traveling.[50] These writers had a large and enthusiastic following.

Before concluding the discussion of the "ladies' departments," mention should be made of the great quantity of material found therein relating to the care of the sick—naturally within the woman's province. Since physicians were often inaccessible to country folk, readers of the farm papers, particularly mothers, were vitally interested in home remedies and "health reform." Articles on diet, hygiene, exercise, and the

[47] *Ohio Farmer,* VII (March 6, 1858), 76; VII (July 17, 1858), 232.

[48] *Ibid.,* VII (May 29, 1858), 172.

[49] An article on capital punishment by Mrs. Gage may be found on pp. 298–99.

[50] An article on dresses for traveling by Anna Hope may be found on pp. 300–1.

like, as well as "cures" for every conceivable ailment were published. Very often such material was used as "filler," and upon occasion it was found in special columns under "Health," "Health and Disease," and "Medical Facts for the People."

Among the thousands of suggested remedies furnished by subscribers, the following are characteristic. Horseradish was recommended as an excellent cure for "hoarseness, cough, sore throat, and disease of the lungs." Those who suffered from rheumatism were urged to take daily a mixture of saltpetre and brandy. Toothache sufferers were assured relief by such a simple expedient as cutting the nails on Friday.[51] By the middle of the century a number of editors had become skeptical of the efficacy of these popular panaceas, and papers such as the *American Agriculturist* and the *Prairie Farmer* refused to print "medical recipes." [52] After printing a remedy for hydrophobia, the editor of the *Wisconsin Farmer* characteristically voiced the growing distrust: "We publish the foregoing at the request of our lady friend, and hope it is as infallible as she believes; for if so it is invaluable. But we confess that we should be somewhat afraid to rely upon it, especially if a dog was *very mad.*" [53]

In spite of such editorial doubtings, the average reader had the utmost confidence in practically any concoction recommended by another layman.

There was at this time a growing lack of confidence and a prevalent cynical distrust in the medical profession.[54] The situ-

[51] *Farmer's Monthly Visitor,* X (Jan. 31, 1848), 8; *South-Western Farmer,* I (April 1, 1842), 25; *American Agriculturist,* XVI (Feb., 1857), 43.

[52] *American Agriculturist,* XVIII (June, 1859), 183; *Prairie Farmer,* XIX (March 24, 1859), 186.

For the attitude toward patent medicines, see Chapter VI.

[53] *Wisconsin Farmer,* X (March, 1858), 105.

[54] Shryock, *The Development of Modern Medicine,* pp. 241, 249.

ation was summarized editorially in the "Ladies' Department" of the *Ohio Cultivator* in 1858: "Medical science seems at the present time to be in utter confusion. The public confidence in physicians is decidedly shaken, and the medical faculty have apparently about as little faith in each other." [55] The popular protest against the old-school practice centered and developed around a number of medical sects and health cults.[56] Homeopathy, hydropathy, Thomsonism, mesmerism, and the like were discussed and evaluated as possible alternatives to the "druggers." Some correspondents expressed the opinion that the situation would improve if more thorough training were demanded of doctors or if women entered the profession, and possibly, if excessive dosing were prohibited. One editor, who felt that doctors were more interested in curing disorders than in preventing them, suggested a solution commonly offered today: "We should like to see physicians like ministers—each with a particular charge and salary." [57]

The pages of the agricultural press, during the fifties, furnished a battleground for conflicting views on woman's rights. The acrimonious debates however, were not restricted to the "ladies' departments." [58] This movement had been formally launched in the United States in 1848 at Seneca Falls, New York, where the first woman's rights convention was held.

The crusade was supported by a large group who desired justice for their pet grievances. Thus "Susan," writing in the *Northern Farmer*, advocated that children should legally belong to the mother, that women should be allowed to vote at

[55] *Ohio Cultivator*, XIV (Jan. 15, 1858), 29.

[56] Shryock, *op. cit.*, p. 250.

[57] *Ohio Cultivator*, XIV (Jan. 15, 1858), 29.

[58] As we have seen, the introduction of the Bloomer costume, as a dress reform in the early fifties, inaugurated another hotly debated subject.

elections, to sit on juries, and to enter Congress.[59] Other correspondents demanded that all professions and occupations be opened to women on an equal basis with men, and that decent compensation be granted to them. In general, their objectives included the elimination of legal, educational, and vocational limitations.[60] Reminiscent of earlier cries of injustice, "Taxation without *representation* is tyranny" became the militant slogan.[61] "Camilla," in the *Boston Cultivator,* touched a popular chord when she included in woman's rights the handling of "a little of the 'chink'" that the husband jingled in his pocket and refused to give up without an itemized account of every petty expenditure.[62]

On the other hand, many readers were shocked at the demands of their sex, and protested vehemently. One expressed it thus: "Woman's rights are, to stay at home and bring up their children, and not be running over the country preaching woman's rights."[63] Even though the vote were gained, "Almy" could not "believe any *intelligent, noble-minded woman,* true to her interests, and the welfare of her family, would be willing to descend from the holy and sacred precincts of *Home,* to sacrifice and degrade the finer sensibilities of the soul in associations not congenial to her nature." The opposition frequently used the argument that the proper sphere of woman was "an Heaven ordained one," and that it would be better for her to "fill well the station already assigned to her; than to be continually harping about rights from which every true

[59] *Northern Farmer,* II (Sept., 1855), 425.

[60] *Ohio Cultivator,* VIII (June 1, 1852), 171.

[61] *Country Gentleman,* II (Nov. 17, 1853), 320; *Moore's Rural New-Yorker,* VIII (Jan. 24, 1857), 22.

[62] *Boston Cultivator,* XIV (May 15, 1852), 157.

[63] *Ibid.,* XIV (Sept. 18, 1852), 301.

woman shrinks."[64] "Azile" did not want to be an "orator" or a "senator," and felt that a woman should "throw around her a halo of goodness, gentleness and affection." "Woman!" she admonished, "there is a higher, a more powerful claim upon you than political strife."[65]

While most of the journals opened their pages to a free discussion of this controversial subject, few of them heartily supported the movement. The *Ohio Cultivator* was an outstanding exception. Its editor was proud of the fact that he employed only "lady bookkeepers" and mailing clerks in the *Cultivator* office; that he himself had carried around a hat for the offerings at a suffrage lecture; and that he had even stood on the platform and introduced Mrs. Bloomer "in that terrible dress."[66] As previously pointed out, the "Ladies' Department" of this journal was a hotbed of "radical" thought. Other journals did all in their power to ameliorate certain of the harsh conditions under which many women lived, and enthusiastically introduced various labor-saving devices to reduce household drudgery; but upon political aspirations such as voting, the journals were extremely conservative.

In general, the periodicals were "suspicious" of the movement and in many cases openly hostile. To the *Rural American*, the advocates were either married women who wanted

[64] *Moore's Rural New-Yorker,* VIII (Feb. 14, 1857), 56; VIII (Jan. 24, 1857), 32.

[65] *Boston Cultivator,* XIV (May 22, 1852), 165.

[66] *Ohio Cultivator,* XVI (May 15, 1860), 158–59. The editor of the *Ohio Farmer* also supported woman's rights, although less aggressively. To him, this "popular agitation about the women," was the "legitimate working out of the declaration of independence." *Ohio Farmer,* V (July 5, 1856), 106.

In general, the movement evoked greatest enthusiasm from the journals of the West.

to "wear the breeches," or disappointed old maids, who, for the lack of a better object vented their indignation on the "cruelty and inhumanity of the sterner sex." [67] The editor of the *Country Gentleman* had "too much respect for the ladies to wish to see them haranguing public audiences, stumping the country as candidates for political office, crowding aside their husbands to deposit the first vote in the ballot box." [68] According to *Moore's Rural New-Yorker*, "every conceivable reason" refuted this folly, while the *Valley Farmer* deemed it "contrary to the whole spirit of Divine Revelation for women to leave their homes and all that ought to be dear to them, to travel about preaching or lecturing to gain masculine fame and notoriety." [69]

The editor of the *Northern Farmer* summarized his objections in this manner:

WOMAN'S RIGHTS—AN ACROSTIC

W–hen women breeches seek to wear,
O–f politics e'en claim a share,
M–idst rowdies gather in and out,
A–nd at the polls do loudly spout,
N–o longer then will Love be known,
S–ince discord o'er the land will roam,

[67] *Rural American*, I (March 15, 1856), 88.

[68] *Country Gentleman*, III (March 16, 1854), 172–73. The editor of the *Pennsylvania Farm Journal* expressed the same thought. *Pennsylvania Farm Journal*, III (Aug., 1853), 139.

[69] *Moore's Rural New-Yorker*, VII (Sept. 6, 1856), 288; *Valley Farmer*, V (March, 1853), 112.

Mrs. Abbott of the *Valley Farmer*, who fought tooth and nail against wider activities of women, nevertheless, urged the training of woman physicians. This apparent inconsistency was occasioned, no doubt, by the failure of medical science to make satisfactory progress against female diseases and excessive infant mortality. Moral delicacy and extreme modesty on the part of women prevented researches by male doctors that otherwise might have improved conditions. *Valley Farmer*, V (June, 1853), 228; Branch, *The Sentimental Years*, pp. 265 ff.

R—emember that the words of Paul,
I—n proper spheres place you all;
G—od never will be found to lie,
H—owever much you may deny.
T—hat NOW you're just where you should be,
S—ecure, protected, loved and *FREE.*[70]

Although this controversy frequently waxed violent, it was merely a preliminary skirmish in the great campaign that was to be waged during the following sixty years.

In addition to the various features in the "ladies' departments," which have already been considered, are found illuminating articles describing journeys made by the women editors throughout the United States and, in a few instances, Europe. These accounts portray: conditions of travel by stage, rail and boat; pictures of city and country landscapes; the habits and customs of the people.[71]

[70] *Northern Farmer,* II (Feb., 1853), 28.
[71] For a typical report of such a trip by Mrs. Mary Abbott, see pp. 302 ff.

# Chapter VIII

## RURAL POETRY

*Let monied blockheads roll in wealth,*
*Let proud fools strut in state—*
*My hands, my homestead and my health*
*Place me above the great.*[1]

RARE IS the agricultural periodical in which there fails to appear some form of poetry.[2] Poems by "standard" poets as Henry Wadsworth Longfellow, James Russell Lowell, Oliver Wendell Holmes, William Cullen Bryant, and John Greenleaf Whittier were frequently copied by the journals after making their appearance in the Metropolitan press. It was the exceptional journal of our period that did not include, for instance, "The Village Blacksmith."

The agricultural press had its poets, men and women, whose

[1] *New England Farmer*, I (March 29, 1823), 280; the opening stanza of "The Farmer" by Thomas Green Fessenden, the editor of the *New England Farmer*. Nathaniel Hawthorne, who was a lodger in Fessenden's home during the last two years of the editor's life, considered him the "most noted satirist of his day." He said that in the course of time, Fessenden's name, "once the most familiar, was forgotten in the list of American bards," since "his poems, being connected with topics of temporary interest, ceased to be read." Hawthorne, *Works*, XII, 246 ff. Many of Fessenden's poems printed in the *New England Farmer* made the rounds of the agricultural press.

[2] The word "poetry," as here used, implies no literary merit, but merely the use of a verse form.

followers were legion in the forty years before the Civil War, but whose names today are forgotten, save the few listed in the standard biographical dictionaries. By far the greater number of these rural pre-Civil War poets were of the gentler sex. An authority, writing in 1850 and referring to the United States, declared that one of the most striking characteristics of that age was the number of female writers, especially in the field of poetry.[3] Very few received any financial return for the "effusions" that flowed from their pens in countless numbers.[4] The compensation lay, no doubt, in the satisfaction of unburdening self in an age barricaded with artificial restrictions and limitations. If the quality of the verse lacked much to be desired, the editors never had reason to complain of the quantity. That everybody was not equipped to write poetry rarely occurred to this ante-bellum generation—except to the editors, whose diplomacy was often taxed to the utmost.

The "delight of melancholia," perhaps more than any other mood, pervades the poems found in the agricultural press. This is especially true after 1840. While it would be unfair to say that this generation enjoyed its sorrows, nevertheless it can be said that it made the most of them. Some of the outpouring was sincere and real, but all too much of it was sentimental mush and crass sentimentality. The *Boston Cultivator* particularly was cluttered with "sob" verse. One writer in this journal expressed the thought of many when she wrote that some termed the graveyard a place of moody reflection, while to her it afforded "serious, yet pleasurable themes for the dwelling of thought." [5] Thus we have

[3] May, *The American Female Poets*, Preface, p. v; Griswold, *The Female Poets of America*, Preface, p. 4.

[4] Certain writers, such as Lydia Howard Huntley Sigourney and Frances D. Gage, who gained their living from the pen, were exceptions to this rule.

[5] *Boston Cultivator*, XII (May 18, 1850), 158.

THOUGHTS IN A CEMETERY

In walnut grove, in pensive mood,
I sat in Nature's bower;
A marble stone in sight there stood,
That spoke of life's last hour!

The brook ran gently at my feet,
It murmured, soft and low,
"The grave will soon be your retreat—
Prepare before you go!" [6]

The melancholy-minded "Charlotte" also found an attraction in

THE BURIAL GROUND

When Nature sinks to sweet repose,
And silence reigns around,
In solitude I love to tread
The quiet burial ground.

No sorrow does it give to me,
To tread with weary feet
The green-roofed chambers of the dead,
Where I shall shortly sleep.[7]

The character of innumerable poems is expressed by these titles: "When I could Die," "The Sick Child," "Tears," "The Song of the Bereaved," "When the Baby Died," "Our Lost," and "Funereal Thoughts." [8] In attempting to explain this mor-

[6] *Ibid.*, XII (Nov. 2, 1850), 354. [7] *Ibid.*, XI (Dec. 22, 1849), 410.

[8] As this emotional age progressed, there appeared more often in the journals verse such as this from the *American Farmer*, X (June 13, 1828), 102:

MOTHER, WHAT IS DEATH?

Mother, how still the baby lies—
I cannot hear his breath;
I cannot see his laughing eyes—
They tell me this is death.

bid emphasis, E. Douglas Branch says in *The Sentimental Years*:

> The soul had been liberated, but social decorum still limited its opportunities for satisfying expression; therein, perhaps, lies the explanation of much of this inverted self-love, this pleasure in weeping at one's own bier—or the pleasure in prolonging a legitimate sorrow as long as one more tear, or one more poem, could be squeezed from it.[9]

The poems reflected somewhat the changing moods of the first of the century. In the earlier years, it was quite common to paint in verse the farmstead as the seat of virtue and the city as the home of vice, a Sodom of the times. Associated with this aversion to the city was the fear and distrust of bankers. The farmer was advised to shun the door of a bank as he "would an approach of the plague or cholera," while the farm boy and girl, especially the "innocent" maids, were warned of the trap that "vice and folly" set for them in that place of corruption and sin.[10] Time and again the warning rang out not to tempt fate by going to the city. Solomon Southwick,[11] editor of the *Plough Boy*, made a typical appeal in this poem

TO A COUNTRY GIRL,
*Who expressed a wish to lead a town life*

Sweet Mary, sigh not for the town,
Where vice and folly reign;

[9] Branch, *op. cit.*, p. 154.

[10] *New England Farmer*, XIII (May 27, 1835), 368.

[11] Solomon Southwick (1773–1839), who frequently signed his articles "H.H. Jr." or "Henry Homespun, Jr.," wrote many poems for the *Plough Boy*. He was active in politics and was twice an unsuccessful candidate for governor of New York. He was editor of numerous political newspapers. Later in life, he became a religious and moral enthusiast, delivering frequent lectures on the Bible, temperance, and self-improvement. *Dictionary of American Biography*, XVII, 413–14.

Spurn not the humble homespun gown
  That suits the rural plain.

In ev'ry street the city's glare
  Doth simple hearts betray:
And simple hearts who wander there,
  Are sure to lose their way.

The tradesman plays his wily part
  To take the stranger in:
The profligate displays his art,
  The modest maid to win:

He lures her to perdition's brink
  By ev'ry treach'rous scheme,
Then leaves the hapless wretch to sink
  In pleasure's guilty stream!

. . . .

Then, Mary, sigh no more to rove,
  Or change your native fields,
The rural walk, the verdant grove,
  For all the city yields.

And when some swain, of soul sincere,
  Shall seek your love to gain,
Trust to his faith, nor ever fear,
  That you shall trust in vain.

So shall your rustic life be spent,
  With every blessing crown'd,
Within your doors, shall sweet content,
  And faithful love be found.

And when your infant offspring rise,
  A mother's smile to greet,
The joy that sparkles in their eyes,
  Shall your own bliss complete!

Your tide of life, thus even flowing,
  Will ebb at last, 'tis true;
When calm, with Hope your bosom glowing,
  You'll bid the world adieu! [12]

On the positive side, the theme of the glorification of the farm and rural life was often voiced.[13] This motif constantly recurs also in the utterances of our great national leaders. According to Washington, agriculture was "the most healthy, the most useful, and the most noble employment of Man," while Jefferson was convinced that farmers were the "chosen people of God," in whose breasts He had made "his peculiar deposite for substantial virtue." [14] As in agricultural speeches and editorials, so in the poems, the "happy plodding farmer" lived a contented, easy, chaste, and independent life.[15] All was

[12] *Plough Boy*, II (Aug. 19, 1820), 89. This poem was copied in the *New England Farmer*, II (Oct. 13, 1823), 96, and the *American Farmer*, V (Nov. 7, 1823), 264.

[13] The theme of contrasting "the sins and temptations of the city with the innocence and purity of the country" and the philosophical implications of this point of view are developed in the agricultural writings of such famous men of antiquity as Plato, Aristotle, Cicero, Virgil, Horace, and others. For an interesting survey of this ideology, see Johnstone, "Turnips and Romanticism," pp. 224 ff.

[14] *American Agriculturist*, I (April, 1842), 1.

[15] Many of the poems run along the following strain:

THE FARMER'S SONG

I envy not the mighty king
  Upon the splendid throne—
Nor crave his glittering diadem,
  Nor wish his power mine own:
For though his power and wealth be great,
  And round him thousands bow
In reverence—in my low estate
  More solid peace I know.

*Yankee Farmer*, I (July 20, 1835), 120. We may quote also in similar mood:

beauty. The farmer was indeed a man set apart. The persistence of this theme and the fact that it was frequently accompanied by implicit admission of cultural inferiority of rural life, points to a belief that it was, in part, a defense mechanism. The following poem is a good illustration.

### THE MILK-MAID'S SONG

I wake to breathe the purest gale,
  That first comes o'er the wood,
Now hie me with my white milk-pail,
  To milk pure nature's food.
And when my easy task is done,
  I'll hasten to the grove,
The sky-lark's plaintive song to con,
  Or listlessly to rove.
    For I love the morn before its dawn,
      And stars just growing dim,
    I love the song of the wild-bird throng,
      Their grateful morning hymn.

I've heard them tell of Ladies fair,
  With robes of purple hue,
Whose cheeks ne'er felt the sultry air,
  Or garments brushed the dew.
And then, I've thought how fair I'd be
  Without this ugly tan,
And half resolved, the sun to flee,

---

THE FARMER'S LIFE

I love the farmer's quiet life—
His peaceful home, devoid of strife,
  With gay contentment bless'd.
I love the virtues of his heart,
Which peace, and joy, and love impart
  Around his tranquil rest.

*Southern Cultivator*, I (Oct. 11, 1843), 168.

Or *fight* with veil and fan.
But I love the morn the early dawn
Just blushing o'er the hill
The sunset rays of the fairest days,
And the rippling of the rill.

. . . .

I envy not their pomp and pride
With 'boasted modesty',
My humble home, by woodland side,
Has far more charms for me;
Their balls and op'ras they may love,
And hear the organ's swell;
Give me my fragrant happy grove,
The tinkling cow-worn bell.
For I love the grove, the sweet pure grove,
The tinkling of my bell;
The purest sound that earth has found,
The echo of the dell.[16]

Occasionally, a "Working Farmer" objected to this sort of idealizing, too conscious of the hardships of farm life and all too oblivious to the "skylark's plaintive song." " 'Tis very pretty [wrote a farmer] to talk and sing all about the pleasures of a farmer's life! Upon paper he has but little to do and less to care for; all is blue sky and sunshine, and after the *pleasures* of the day, he has only to retire to his cheerful hearth, where there is nothing to vex or annoy him; enjoy himself in his 'old oak chair,' and sing 'over the hills and far away.' " [17]

The increasing interest in nature, noticeable in the United

[16] *Maine Farmer,* VIII (June 6, 1840), 176.

[17] *Ibid.,* VIII (Aug. 29, 1840), 272. He continues: "But the most amusing part of it is to hear her talk—after her *easy task* of milking is done—to hasten to the grove to *con* the sky-lark's *plaintive* song—there she is unfortunate, for the sky-lark never sings until after the dawn and then his notes are proverbially *sprightly,* but that's nothing—or to rove about *listlessly,* all this before the dawn of morning too."

States after 1825, later manifested in Transcendentalism, was conspicuous in the farm journals. After 1830 numerous articles called upon the farmers for a closer study of nature and a finer appreciation of her beauties. Nature, after all, was God's Nature and the farmers' possession. The whole idea was in harmony with the rural-agrarian theme already mentioned. The rural poets turned with added zeal to "their" nature and bounteous was the flow of rhyme devoted to trees, brooks, birds, mountains, fields, and flowers. Even today, pressed flowers and leaves found between the pages of an occasional journal tell of the interest of this past generation. Mary Robbins, who belongs to this group, was a highly respected writer of the *Ohio Farmer* and wrote with "fine poetic sensibility." [18] The editor declared that these verses by her were "lines worthy of Whittier."

### A SABBATH IN THE COUNTRY

A Sabbath in the country, still, serene!—
The very woods seem lifting up their hands
In silent worship, and the brooks, unseen,
Sung songs of gladness o'er the lowly sands.

"Sweet Nature's worship is the best," I said,
"And all her days are Sabbaths—prayer and psalm,
And ceaseless benediction, here are shed—
Here may earth's weariest, saddest ones find calm.

Here, far removed from all the fevered life,
That thrills the city's pulses night and day,
The soul may know itself, not all at strife
With God—not yet a hopeless cast-away." [19]

---

[18] Mary Robbins (Whittlesey) was born in Elyria, Lorain County, Ohio, in 1832. Coggeshall, *The Poets and Poetry of the West*, p. 575.

[19] *Ohio Farmer*, VIII (Nov. 19, 1859), 376.

In the late twenties, the journals printed many poems reflecting the wave of temperance reform that was sweeping the country.[20] Indeed these poems continued to be popular to the time of the Civil War. The following tearful lament by "Clara" is representative.

THE LITTLE BOY'S APPEAL TO HIS PARENT

Oh! father, won't you sign the pledge,
And never drink again;
And let us go to that dear home,
From whence we long since came.

. . . .

Oh! we were very happy then,
And now we're very sad;
If you will only sign the pledge,
We all should be so glad.

Dear mother now is very dull,
And weeps when you are gone;
She does not seem so happy now,
As when in that dear home.

I cannot bear to see her grieve,
It makes dear Ellen weep;
And supperless we go to bed,
And cry ourselves to sleep.

. . . .

Oh! we should be so happy then,
And suffer no more pain;
Oh father won't you sign the pledge,
And be a man again.[21]

---

[20] See p.77 for the attitude of the journals on the temperance movement.
[21] *Boston Cultivator,* IX (Jan. 30, 1847), 35.

Temperance songs occasionally appeared. A fragment of a drinking song written "by a member of a temperance society" follows:

See drinkers of water, their wit's never lacking,
Direct as a railroad and smooth in their gaits;
But look at the bibbers of wine, they go tacking,
Like ships that have met a foul wind in the *straights*.
Then hey for a bucket.[22]

Since more attention was devoted to the children's departments after the middle of the century, poems directed to youngsters became common. Mrs. E. O. Sampson Hoyt,[23] who conducted the "Home Circle" of the *Wisconsin Farmer*, wrote singularly felicitous and amusing poems for children. The following contribution is typical of many that appeared above her signature.

THE LITTLE BIG MAN

Little Tommy had a mind
To be a Little Man,
Before he'd learn to spell such words
As If, and But, and Can.

He did not like to go to school,
And stand up in his class,
But rather play beside the brook,
Or lie upon the grass.

. . . .

[22] *New England Farmer*, XV (March 15, 1837), 288.

[23] Mrs. Hoyt, a native of Athens, Ohio, was extremely precocious as a child, having a volume of poems to her credit before she was fifteen years of age. Her eyes failed early in life, leaving her almost solely dependent on her friends. She married John W. Hoyt, M.D., who later became editor of the *Wisconsin Farmer*. Mrs. Hoyt assisted her husband in his editorial duties for many years. Coggeshall, *op. cit.*, p. 575.

Once, when Mama and Pa were gone
  To spend the summer day,
Our little Tommy took the freak,
  The "Little Man" to play!

He got into his father's boots;
  Put on his father's hat;
With spectacles before his eyes,
  In father's arm-chair sat.

. . . .

A great cigar he lighted next,
  With many a little puff;
'Til, lit, at last, our "Little Man"
  Had smoke and fire enough—

For, finding soon the bitter thing
  Was very bad to taste,
He threw it straight into the fire
  With such an angry haste,

That headlong after it he fell,
  Hat, spectacles and boots!
While furious cries for "help, and haste,"
  The nursery-maid salute.

"Why, little Master," said the maid,
  "Just be a Little Man,
And help yourself from out that plight,
  As quickly as you can!"

. . . .

Then Tommy thought he would not try,
  Again, to be a Man,
Until he'd learned to spell such words,
  As If, and But, and Can.

So, putting on his little sack,
  With apron, bib and blue,

He went to school with little folks,
  To see what *he* could do.

And very soon he learned to spell
  Much longer words than Can,
And while a very little boy,
  Was quite a little Man.[24]

The wider use of machinery, especially the growing popularity of the reaper, was also reflected in the rural verse of the period.[25] Frances D. Gage,[26] one of the most popular and prolific contributors to the journals, was unusually sensitive to the changing scene. "The Sounds of Industry," written by her, was probably copied more frequently than any other poem appearing in the agricultural press.

THE SOUNDS OF INDUSTRY

I love the banging hammer,
  The whirring of the plane,
The crashing of the busy saw,
  The creaking of the crane,

[24] *Wisconsin Farmer*, X (Nov., 1858), 436.

[25] Some correspondents felt that the patent-right machines were fatal to poetry. As one writer said, "Singer's Sewing Machine that *never* sings is no compensation for the loss of the blue eyed girls that sewed and sang in the old homesteads. Love's own initials were sometimes stitched along the hem of the handkerchiefs we used to carry, but now, be they ever so beautiful, there is nothing *human* about them, for we know those 'devil's-darning needles' whipped through and through their borders in a minute or two. . . . The poets and the prophets are a brotherhood, but the poets and the *profits* are strangers, forever." *Prairie Farmer*, XXII (Oct. 11, 1860), 235.

[26] Many of Mrs. Gage's poems appeared in the journals, especially the *Ohio Cultivator*, *Ohio Farmer*, and *Prairie Farmer*. She also wrote numerous articles of general interest for the "ladies' departments." See p. 165 for her contributions to that section of the *Ohio Cultivator*. In 1867 a book containing about ninety of her poems was published by J. P. Lippincott & Co., Philadelphia.

The ringing of the anvil,
  The grating of the drill,
The clattering of the turning-lathe,
  The whirling of the mill,
The buzzing of the spindle,
  The rattling of the loom
The puffing of the engine,
  And the fan's continuous boom—
The clipping of the tailor's shears,
  The driving of the awl,—
The sounds of BUSY LABOR,
  I love, I love them all.

I love the plowman's whistle,
  The reapers' cheerful song,
The drover's oft repeated shout,
  As he spurs his stock along;
The bustle of the market man,
  As he hies him to the town;
The halloo, from the tree top
  As the ripened fruit comes down.
The busy sound of threshers
  As they clean the ripened grain,
And the huskers' joke and mirth and glee
  'Neath the moonlight on the plain,
The kind voice of the dairyman,
  The shepherd's gentle call—
These sounds of active industry,
  I love, I love them all.[27]

---

[27] *Ohio Cultivator*, VII (Jan. 1, 1851), 16. Seven years after this initial appearance, the editor of the *Ohio Cultivator* said that the poem had made the rounds of the press once a year and had been "copied a thousand times"; that it had "been copied, miscopied, and plagiarized until its own mother would not recognize it." So in 1858 it was reprinted from the original stereotype plate. Other poems by Mrs. Gage were scarcely less popular and many were republished in England. *Ohio Cultivator*, XIV (Nov. 15, 1858), 351; VII (Dec. 1, 1851), 366.

Of "Reaper" poems, the following is one of the best. It was difficult at this period for a poet to avoid moralizing even in dealing with so prosaic a subject as the reaper.

PRAIRIE RONDE, AT HARVEST TIME

But hark!—the rattling "Reaper,"
Here it comes, with noisy din,
And the grain shrinks before it,
Like good intentions before sin!
One rides upon the Reaper,
Waving oft the reaper's wand,
And every pass he makes
Lays a sheaf upon the land;
Now, now, busy binders,
Bind the grain, with might and main,
For the ground must all be cleared
Ere the Reaper comes again.
Thus, in ever lessening circles,
Round and round the field they go;
Nor must the noble, panting horses,
Yield a jot to failing forces,
Nor slacken to a pace more slow!
O, band of strong cradlers, with regular sweep,
Your vocation is gone!—'tis the "Reaper" must reap! [28]

In spite of poetic outbursts on the new farm machines, the virtues of honest hand labor were never forgotten. "Hands thus bent on rural toil" continued to be extolled up to the time of the Civil War. Such is the theme of the following poem by Mrs. L. H. Sigourney, one of "the most famous of the female bards of her country." [29]

[28] *Michigan Farmer*, VIII (Oct., 1850), 297 ff. No author of this poem is indicated.

[29] Lydia Howard Huntley Sigourney (1791–1865) devoted most of her life to writing. By 1830, she was contributing prose and verse regularly to

RURAL INDUSTRY

Work, Mowers, Work!
  Sweep the swathe well;
How the rich cloverfield
  Sighs where it fell;
Like a meek Christian
  Yielding its breath,
Blessing its murderers
  Even in death.
Work, Mowers, Work!

. . . .

Work, Reapers, Work!
  Wide o'er the plain
'Neath the sharp sickle
  Heap the ripe grain,
Ready and willing,
  Its life-purpose won,
So dies the good man,
  All his work is done.
Work, Reapers, Work!
Hands thus bent on rural toil,
Take no part in crime and broil,
But on households high and low,
Strength and happiness bestow,
O'er the board sweet comforts spread,
Cheer the famished poor with bread.
Work, Reapers, Work! [30]

The mania for versifying was spreading far and wide and the output almost engulfed the editors with a deluge of in-

more than twenty periodicals. The farm journals printed innumerable poems and articles by her. A total of sixty-seven books echoing the conventional sentiments came from her pen. The theme of most of her writing is death. *Dictionary of American Biography*, XVII, 155 ff.

[30] *Prairie Farmer*, XXII (July 5, 1860), 10.

different verse.[31] Nevertheless, the agricultural journals continued to encourage these embryo poets. Whatever was new or startling—the emotion, the impulse, or the passion of the hour—elicited a response and set the rhymsters to work. The following titles suggest the wide range of subjects treated: "The Atlantic Cable," "The Sewing Machine," "Our New Baby," "Don't Go to California," "I Will not Love," [32] "The Anti-Hoop Faction," "Death of President Taylor," and "Ought the Union of these States be dissolved?" Indeed, so broad were the interests of these humbler poets of the pre-Civil War period, that relying on their myriad verses one might almost write a history of the rural life and thought of the time.

[31] Even advertisements carried an occasional ditty. Kostar's Rat Poison was advertised as follows:

As June approaches,
Rats and Roaches
  From their holes come out,
And Mice and Rats,
In spite of Cats,
  Gaily skip about.

*American Agriculturist*, XVII (June, 1858), 190.

[32] Love was always a favorite theme and sentimental verse was common. Much of the verse on this subject dealt with the suffering love brings. The following by "Mira" suggests either a desire to avoid this tragedy or perhaps a subtle fear lest she escape it.

I WILL NOT LOVE

I will not bow me at the shrine
  Where love-sick maidens kneel,
I will not mourn a broken heart,
  That only love can heal!

Sly Cupid, I defy thy power,
  I scorn thy wily art;
Ne'er shall thy shafts an entrance find
  To my young, giddy heart!

*Boston Cultivator*, XVI (Jan. 7, 1854), 7.

# Chapter IX

## THE AGRICULTURAL FAIR

*Time rolls his course along—*
*And every Fair occasion*
*With us turns out a song,*
*Or sure a good oration;*
*When both are well combin'd,*
*To help along the story,*
*The Day, with hearts in union join'd,*
*Becomes the Farmer's glory.*[1]

THE AGRICULTURAL FAIR, like the agricultural journal, has exerted a tremendous influence upon American rural life. Because of the close relationship and intimate coöperation between these agencies for improvement, the following pages have been devoted to the fair. Furthermore, a discussion of this distinguishing feature of the rural community, based almost exclusively upon the farm press, will serve to illustrate the value of these periodicals as a rich repository for the study of agrarian institutions.

During the twenty years following 1810, the fair witnessed a period of great popularity. It was not until the decade prior to 1860, however, that it really came into its own. At this time,

[1] "The Farmer's Glory," *Farmer's Monthly Visitor*, I (Nov. 20, 1839), 173.

the journals were replete with material upon the fair. Indeed, frequently more than twenty percent of the paper was devoted to this subject, particularly in the Fall of the year. Many issues dealt almost exclusively with the fair.

The agricultural fairs and the journals, in the main, sought the same objectives. They were coworkers in a cause best advanced by mutual efforts.[2] Indeed, for the successful conduct of a fair, it was generally considered essential for the press "to excite in the public mind an interest in agricultural and mechanical improvement." [3] On the other hand, the journals also benefited, as agricultural societies offered thousands of subscriptions each Fall as premiums. The *Indiana Farmer*, for example, in 1858 reported with appreciation, "Since the State Fair, subscriptions come in ten times as fast as before." [4]

A great many editors were also officials of the societies and frequently served on committees in charge of the fair. The Fall was a busy time for these men. Week after week they visited the exhibitions to transact business and obtain copy; very often they were called upon to act as judges and to deliver the agricultural addresses.[5]

Elkanah Watson may well be called the "father of the

[2] *Country Gentleman*, XII (Nov. 11, 1858), 304.

[3] *Indiana Farmer*, VII (May, 1858), 43.

[4] *Ibid.*, VII (Nov., 1858), 253.

[5] The activity of the editors during this season, especially in the fifties, can be noted from the following schedule in the *Ohio Cultivator:* "At this writing, (Sept. 28th) Mr. Bateham [editor] is absent attending Fairs at Wheeling [county fair] and other places eastward. Next week he proposes to leave again on a visit to State Fairs in Michigan, Illinois and Indiana. We [associate editor S. D. Harris] have engaged to hold a talk at Mt. Vernon [county fair] on the 4th, at St. Clairsville [county fair] on the 5th, visit the Salem Horse Show [Ohio and Penna. Horse Show] on the 10th, Bucyrus [county fair] on the 12th, and talk at Mt. Gilead [county fair] on the 13th. We also have it in contemplation to meet with the Wyandots [county fair] between the 18th and 20th." *Ohio Cultivator*, XI (Oct. 1, 1855), 296.

agricultural fair" in America.[6] It was he who in 1807, exhibited two Merino sheep in the public square at Pittsfield, Massachusetts, thus founding the American cattle show or fair.[7] Three years later he persuaded twenty-six of his neighbors to join him in an exhibition of livestock on the village green.[8] Encouraged by this successful venture and desiring to make it an annual event, he organized, in the following year, the Berkshire Agricultural Society, consisting of these exhibitors and other Pittsfield farmers.[9] This became the first permanent fair association in the United States.[10] The organization and methods of this society furnished a striking contrast to the older, or "literary" agricultural societies that flourished in the eighteenth century.[11]

Watson saw the failures of the older societies and wisely decided to "seize the bull by the horns . . . by touching a string which never fails (if properly directed) to vibrate in unison with all, viz.—self-love,—self-interest, combined with a natural love of country." [12] He enlisted the aid of the farmers' wives, as well as that of the local clergymen. In the course of time, new features were added; ultimately the exhibitions of farm products and of domestic manufactures became the central feature of an elaborate program lasting two days. This program included a street parade, plowing match, a public

[6] Bailey, *Cyclopedia of American Agriculture*, IV, 292.

[7] Bidwell and Falconer, *History of Agriculture in the Northern United States*, p. 187.

[8] Watson, *History of Agricultural Societies, on the Modern Berkshire System*, p. 118. This is generally considered the first fair in the United States. However, the Columbian Agricultural Society of Georgetown, District of Columbia, held a fair a few months earlier. U. S. Patent Office, *Annual Report: Agriculture* (1859), p. 23.

[9] Bidwell and Falconer, *op. cit.*, p. 187.

[10] Bailey, *Cyclopedia of American Agriculture*, IV, 292.

[11] Bidwell and Falconer, *op. cit.*, p. 187.

[12] Watson, *op. cit.*, p. 132.

meeting in the village church opened by prayer, an address, songs, and finally, an agricultural ball.

Societies on this plan spread rapidly, and by 1819 Watson estimated that at least one hundred existed in the United States.[13] They reached their peak in the period from 1820 to 1825, due largely to the allotment of state funds to county organizations.[14] The journals enthusiastically supported such ventures, and the editors themselves took a leading part, filling their pages with reports of the societies and the cattle shows or fairs which constituted their chief feature. The events of the jubilee were described in full. Premium awards were listed, and frequently the agricultural address was given in detail.[15] However, between 1825 and 1840, when state funds were withdrawn, societies successively disappeared. Agriculture, dur-

[13] *Plough Boy*, I (Nov. 27, 1819), 205.

[14] Neely, *The Agricultural Fair*, p. 71; Bidwell and Falconer, *op. cit.*, p. 189.

[15] General Lafayette, on his visit to the United States in 1824, attended the Western Shore Cattle Show of the Maryland Agricultural Society. He was escorted to the field of exhibition by a committee of nationally prominent and distinguished men and an "honorable and numerous body guard of *substantial sunburnt Farmers*." The award of premiums was made by "*the gallant, the disinterested* SOLDIER OF LIBERTY, *the veteran companion of* WASHINGTON, and the unvarying friend of America." Indeed, "striking was the mixture of alacrity and diffidence, of pride, and of reverence, with which every one stepped forward to receive his premium, with the smiles and the good wishes for one of the noblest champions that ever drew his sword in defence of human freedom!" Then the farmers "formed themselves into two lines, between which the General passed, most graciously shaking each one by the hand." Later in the ceremonies, Lafayette proposed the following toast: "The Seed of American Liberty, transplanted on the other Shore, oppressed, not destroyed, by every sort of European Weed—may it rise again, vigorous, and pure, and cover the soil of both Hemispheres." After attending the theater in Baltimore that evening, to witness the "School for Scandal," General Lafayette was invited to the home of John S. Skinner, editor of the *American Farmer* "to meet a large party of Ladies and of Agricultural Gentlemen." *American Farmer*, VI (Nov. 26, 1824), 287–88.

ing these fifteen years, remained on a generally low level.[16]

Beginning with 1840, American agriculture experienced a remarkable era of expansion.[17] The revival of agricultural interests resulted in the formation of hundreds of state and local agricultural societies. The fact that state aid again became available, no doubt, played its part. By 1847 an added interest was engendered by the return of better prices on farm products.[18] The Commissioner of Patents in 1858 listed 912 state and county societies, five-sixths of which were established after 1849.[19] Available data found in the periodicals indicates that this estimate is conservative.

The leading feature of these rapidly increasing associations was the fair and, by the middle of the century, we find it assuming the center of the American stage. Indeed, the years between 1850 and 1870 witnessed such remarkable expansion of this institution, that it has been designated as "the golden age of the agricultural fair." [20] In September, 1858, the *Indiana Farmer* predicted that no less than 500 fairs would be held that season.[21] Most of them were promoted in the North, although they did prosper in the three livestock states of Kentucky, Tennessee, and Missouri.[22]

Perhaps the greatest fairs of the golden age were held by the United States Agricultural Society in the fifties.[23] In 1853 the first of its great fairs was held at Springfield, Massachusetts.[24]

[16] Neely, *op. cit.*, p. 71. [17] *Ibid.*, p. 73.

[18] Bidwell and Falconer, *op. cit.*, p. 317.

[19] Bailey, *Cyclopedia of American Agriculture*, IV, 292.

[20] *Ibid.*, IV, 292.

[21] *Indiana Farmer*, VII (Sept., 1858), 164. For other estimates see *American Agriculturist*, XV (Sept., 1856), 271.

[22] Gray, *History of Agriculture in the Southern United States*, II, 785.

[23] U. S. Patent Office, *Annual Report: Agriculture* (1859), p. 24.

[24] Fairs before the Civil War were held as follows: Springfield, Ohio, 1854; Boston, Mass., 1855; Philadelphia, Pa., 1856; Louisville, Ky.,

The magnitude of these annual exhibitions is illustrated by the figures of the 1859 Chicago Fair. More than 40 acres of land were enclosed and a "well beaten" half-mile ring was provided. There were 1,000 "tight roofed" stalls and pens and six "handsome commodious" halls—a total of 150,000 square feet of roofed space for exhibition purposes. Two power presses printed daily papers and bulletins, and a telegraph office was available. Twenty thousand dollars in premiums were offered and the total number of articles entered from practically every state in the Union was 2,549, classified under 123 departments.[25] Delegations representing eighty-one societies from twenty-seven states and territories were present.[26] The peak attendance for one day was estimated at 50,000.[27]

The state fairs were of next importance. By 1855 annual exhibitions were held in nearly every state in the Union. They were conducted by state agricultural societies or, as in Ohio and Indiana, by state boards of agriculture.[28] These fairs were of a migratory nature, thus gaining a wider audience but entailing additional expenditure. The cost of fixtures for an exhibition approximated $8,000.[29] Premiums frequently amounted to $16,000; entries generally ranged between 2,000

---

1857; Richmond, Va., 1858; Chicago, Ill., 1859, and Cincinnati, Ohio, 1860. All of these exhibitions were self-sustaining; that is, the receipts met the disbursements. *Ibid.*, p. 26.

25 *Prairie Farmer*, XX (Sept. 15, 1859), 173; XX (Sept. 22, 1859), 187.

26 U. S. Patent Office, *Annual Report: Agriculture* (1859), pp. 26–27.

27 *Cultivator*, ser. 3, VII (Oct., 1859), 308.

28 Neely, *op. cit.*, p. 95.

Perhaps the most famous state fairs were those sponsored by the New York State Agricultural Society. The leading spirit in this society for many years was Benjamin Pierce Johnson. *Dictionary of American Biography*, X, 90.

29 *Valley Farmer*, XII (Nov., 1860), 327.

and 2,500; and attendance on the "big day" was likely to vary between 25,000 and 60,000.

County fairs sprang up like mushrooms after 1850. Probably well over a hundred were held in Ohio in 1859, for example.[30] Many county and local associations owned considerable tracts of land enclosed with board fences, the interior admirably fitted for display purposes. Hundreds of dollars in premiums were given annually in books, silver plate, cups, diplomas, and agricultural journals.

The state fair during the fifties was in general representative of these annual events. In order to visualize such an occasion of the golden age, the following synthetic sketch is presented. This picture is drawn from innumerable descriptions found in the farm periodicals. Let us fancy ourselves a contemporary and enter into the spirit of the times.

Fairtown is a "scene of jam, cram, bustle, crowd, and commotion." The streets are thronged with people and the dusty roads are crowded with vehicles from the country, while long trains bring curious visitors from remote localities. It is said that 2,700 people came in on a single train this morning.[31]

[30] In 1859 Ohio had 88 counties, in each of which with one or two exceptions, was a regularly organized and accredited county agricultural society, auxiliary to the State Board of Agriculture. In addition, there were in several of the counties, fully organized, independent, county societies, quite as active and efficient as the regular societies. District associations with membership from parts of several counties were equally strong and active; also, there were many societies representing several townships, and there were many single township organizations which rivaled the best county societies of the state. *Ohio Cultivator*, XV (July 15, 1859), 216.

[31] It was customary for the railroads to charge half fare for visitors attending state fairs.

Many complaints were registered in those early days against the accommodations offered by the railroads as well as the rudeness of the conductors. The editor of one journal commented on the "character of the cars," the "insolence of some of the conductors," and the "equally deficient"

The schools of the city are closed and numerous factories have closed their gates so the operatives may have a holiday. A festive spirit is in the air.

All the hotels in the city are filled, one editor reports that "fifteen gentlemen had to take lodging in *four single beds.*" [32] Many strangers who could not find rooms in hotels have had to sit up all night.[33] Numerous temporary boarding and lodging houses have sprung up, prices are soaring and much talk is heard about the "gouging landlords." [34] The problem of victualling the visitors continues to worry the authorities.

The road from the city to the fairgrounds, a distance of two or three miles, is covered with buses, buggies, carriages, wagons, sulkies, barouches, in fact, every means of transportation which can be spared from the nearby towns and cities has been put

---

transportation of stock. *Pennsylvania Farm Journal*, I (Nov., 1851), 243. Another editor appeared quite relieved, that the conductors did "nothing unnecessarily to annoy the passengers." *Wisconsin Farmer*, X (Nov., 1858), 418.

[32] *Ohio Cultivator*, VII (Dec. 15, 1851), 378.

[33] At another state fair, the editor of the local agricultural journal tried to do his part. "We have thirty engaged already for our house and barn, which, as all the world knows, is not large," said he. *Wisconsin Farmer*, X (Oct., 1858), 383.

[34] As the golden age advanced, attempts were made to remedy this situation. In some of the states, contracts were made with the hotels and boardinghouses binding them to charge only their usual rates during fair week. Temporary buildings were erected to house guests. More and more it became customary to set apart 30 or 40 acres of land near the fairgrounds for camping parties. Thousands of people lived in these temporary communities.

In 1858 the Illinois State Fair was held in the small town of Centralia. The Illinois Central Railroad furnished 2½ miles of cars on its sidetracks to provide sleeping accommodations for the people who attended the fair. A successful fair was always good publicity for the state and the railroad. Gates, *The Illinois Central Railroad and Its Colonization Work*, p. 282.

Although conditions were improving, state fairs continued to be managed amateurishly because of their migratory nature.

into requisition. And still the crowd is so great as to exhaust every means provided. Many "patriotic" individuals "foot" it out and back. All day long the city pours out a continuous stream of people. Dust, insufferable dust, suffocating dust, covers everything like a mantle on all sides.

The avenue leading to the fairgrounds is lined with booths, tents, and hastily constructed stands which have been erected by thrifty "turnpennies" as refreshment saloons, confectioneries, restaurants, and cigar shops. A number of side shows are attracting attention. Among them are "The infant drummer, the greatest living curiosity in the world," "Tom Thumb, the smallest man in the World," "The Boy with six fingers on each hand, and eight toes on each foot," and "The Learned pig." A small charge is collected from the curious.[35]

A fee of twenty-five cents is paid at the gate, although editors and clergymen are furnished tickets gratis.[36] Upon entering the grounds, the visitor is confronted with halls, pavilions, tents, stalls, and booths. Refreshment stands are numerous. Many large placards politely requesting all to "beware of pick-pockets" are conspicuous. (The light fingered gentry are now firmly established at the fairs.) The business office of the Agricultural Society, the police office, and the editors' hall are on the grounds. Near the entrance is a general intelligence booth where visitors may also report articles lost or stolen. The

[35] As time wore on, the "side shows" were gradually working their way inside the gate and becoming a part of the fair proper.

[36] By the middle fifties a system of selling badges for $1.00, which entitled a whole family to enter the grounds, had to be abandoned by many states. "One man buys a badge [explained an official] and takes his family inside the gate, then gives the badge to his boy, who goes out and hands it to a neighbor, who takes in his family, including the boy, and soon a whole neighborhood may get in for a single dollar." Farmers frequently sold their badges, after the first day, to "sharpers," who in turn sold them for full price at the gate. *Ohio Cultivator,* X (Nov. 1, 1854), 321.

agricultural journals of the state and, frequently, of distant states have erected tents with their names conspicuously displayed. Here the editor, who is anxious to greet his old friends and make new ones, may be found. He will transact such business as he would in his office and is especially ready to collect past due payments or advance reasons why one should subscribe to his periodical. It is here that the editor distributes sample copies of current and back numbers. These men are a friendly, popular group. Very often two or three of them may be found talking together or making the rounds of the exhibits.

Four or five main halls are well worth visiting. Fine Arts Hall is a "perfect jam during the continuance of the Fair." You will find on display lithographs, daguerreotypes, photographs, ambrotypes, engravings, pencil drawings, and oil paintings. There are always "several paintings claimed as the work of old masters," and a few "rare and valuable paintings." Numerous pieces of sculpture, including busts of Washington, Webster, and Jefferson, are conspicuously arranged. Hart's celebrated bust of Henry Clay rests on a pedestal near a bust of Cicero. You may purchase these if you wish. Farther down the hall are samples of penmanship, plain and ornamental; exhibits of fancy work; knitting and crocheting; and as is frequently the case, jewelry and silverware. Among many other exhibits, is a display of musical instruments. Several Chickering pianos are continually surrounded by groups of interested young ladies and throughout the day we constantly hear well-known refrains. "Old Folks at Home" and "Call Me Pet Names" are the most popular pieces this year.[37]

The favorite spot for the women is Floral Hall. Much care

[37] The first of these was written (1851) by Stephen Foster whose songs were especially popular during the fifties. The second song (1851) was representative of many introduced during the decade prior to the Civil War, with words not dissimilar in character to most of the "popular

is devoted to the design of the hall as a whole, with special emphasis on the centerpiece. Each Fall this central theme suggests a new note. This year it is a fountain and suspended over it, a gigantic inverted artificial lily; last year it was "a colossal pyramid made of flowers and fruits." A few years ago it was especially attractive in the style of a "grand artificial temple, formed in the shape of a tower rising to the roof of the tent and constructed of sheaves of wheat, oats and evergreens." Floral Hall usually contains pleasing arrangements representing groves, grottoes, waterfalls, and fountains. At one end we meet beautiful flower beds, plants, and cut flowers; along the sides, large displays of choice varieties of fruit. Here a "fine band of music" discourses "sweet sounds."

In Manufacturers Hall, we hear the constant click of sewing machines: this long array is at all times surrounded by a crowd of interested women. The most popular makes seem to be the Grover and Baker, Wheeler and Wilson, and Singer. There are innumerable other items on exhibition—carpets, quilts, muslins, silks, laces, and a countless variety of articles for personal adornment and household use. One end of the hall con-

---

musical hits" of today. Indeed popular music became articulate in the United States during the decade in which the fair assumed the center of the American stage. Spaeth, *Read 'Em and Weep*, p. 33. The type of sentimental song which appealed to this generation may be seen in the editorial comments of the *Genesee Farmer*. When the new song was introduced in 1856 called "Jeannie Marsh of Cherry Valley" (words by General Morris and music by Thomas Baker), the editor wrote that it was "light, sparkling, and graceful as is the fair theme of both poet and musician"; that it possessed all the elements of immense popularity, and would "soon be found on every piano throughout the country." The first four lines follow:

Jeannie Marsh of Cherry Valley,
At whose call the muses rally;
Of all the nine none so divine
As Jeannie Marsh of Cherry Valley.

*Genesee Farmer*, ser. 2, XVII (March, 1856), 100.

tains a display of Howe and Fairbank scales, a knitting machine, and several other kinds of machinery.

The "continual roaring, humming, rattling, hammering and puffing" attracts many men to Power Hall. The whole interior of this establishment is in motion and each particular machine is doing something to gain the attention of the crowd. The Triumph grist-mill, Hancock's lath-sawing machine, two type-making devices, portable steam engines, a window sash machine; these, among others, are in operation. Near the entrance another mechanism is rapidly turning out hundreds of smooth shingles. Miscellaneous articles such as a contrivance to prevent explosions of steam boilers, a locomotive model, and a great variety of saws are nearby.

Finally, we come to Mechanics Hall where all manner of agricultural implements are shown. Sixty different patterns of plows are on display. John Deere of Moline, who as usual is a large exhibitor, presents twenty plows in seven different patterns, as well as other farm implements. Well known reapers such as the McCormick, Kirby, Hussey, Ketchum, and Allen are on hand.[38] Besides threshers, corn planters, potato planters, and stump pullers, we are confronted with cider mills, presses, and washing machines.[39] Patent gates, fire engines, model houses, water filters, guns, pistols and accoutrements, even seeds, boots and shoes are found here.

Frequently the fairs include a vegetable or agricultural hall containing products of the field, garden, and dairy. The classifications vary. A special section is always reserved for the ex-

[38] After 1848 Cyrus McCormick began more and more to shun fairs since awards were frequently based on testimonials and editorial "puffs," rather than actual trials. Hutchinson, op. cit., I, 347 ff. This situation caused many manufacturers to refuse to compete for premiums, although they were willing to display their products for the benefit of advertising.

[39] After 1857 steam plows were great attractions at state fairs.

hibition of horses, mules, asses, cattle, sheep, swine, and poultry shipped from widely scattered points.[40]

The center of the fairground contains that seemingly inseparable adjunct of the agricultural fair, the trotting track, within which is located a spacious pagoda for the judges and reporters. A grand stand capable of accommodating many thousands of spectators has been erected. This site also provides the setting for the running and trotting matches, as well as the ladies' equestrian display. Additional spectacular attractions include Professor Rarey, the horse tamer, who "demonstrates his science and skill" by taming the "untameable"; a trial of steam fire engines; a military review; and a balloon ascension. Meanwhile, the visitor must not overlook the more serious scheduled events, such as the president's opening address, the farmers' informal evening discussion, the plowing match, the official awarding of premiums, the sale of animals at auction, and the annual agricultural address.

A quotation from the *Ohio Farmer* provides an appropriate climax to our sketch of a mid-century state agricultural fair.

> At two o'clock this afternoon, the people collected in the amphitheater and upon the opposite hillside, to witness the grand cavalcade of premium stock in the horse ring. Col. Branch led the pageant, followed by a brass band which played martial music. Serious bulls and solemn jackasses stepped measuredly, handsome horses displayed their graces, the people scanned admiringly and so the Fair ended.[41]

Three important features of the state fair, namely, the annual agricultural address, the farmers' informal evening dis-

[40] Farmers could afford to send entries long distances as it was customary for the railroads to convey stock, machinery, and other articles for exhibition at state fairs free of charge. This was an important factor in the success of state fairs. *Wisconsin Farmer*, VIII (Oct., 1856), 451.

[41] *Ohio Farmer*, VIII (Oct. 1, 1859), 313.

cussion, and the plowing match (already mentioned) were particularly emphasized in the press. These warrant and shall receive special attention.

So important were the annual addresses that prominent men were, without difficulty, prevailed upon to deliver them. The reports of the fairs, as found in the pages of the farm journals, abound with the names of distinguished visitors such as Martin Van Buren, Daniel Webster, Millard Fillmore, Henry Clay, William H. Seward, Salmon P. Chase, General Taylor, Abraham Lincoln,[42] Stephen A. Douglas and many others. This was especially true of state and national fairs.

Many of these notables were invited to deliver the annual address featured, as a rule, on the final day. Frequently the state, which sponsored the fair, thus honored its governor, and these were indeed formal occasions. When the Honorable John A. Dix read before the New York State Agricultural Fair the speech prepared by the late Governor Silas Wright, the platform was occupied by officers of the society, women comprising the committee upon household productions, the governor, state officers, and other distinguished citizens. Among the latter were ex-Presidents Van Buren and Tyler, Governor Hill of New Hampshire, and other noted gentlemen from various parts of the Union.[43] Especially popular as speakers

[42] The *Wisconsin Farmer*, reporting the Illinois State Fair in 1858, noted that there was much earnest discussion and interest manifested in political matters, pending the approaching senatorial contest in Illinois. The editor said: "Lincoln and Douglas, both, have brave and zealous friends, full of boasts and cheering. . . . As we were leaving for home, about half past eleven, the train from the south arrived with Judge Douglas on board, and as he got out, a round shout went up from the crowd for him. At the same time, also, a shout of cheers went up for Lincoln, who was getting into the cars to come north; as he had been most of the day at Centralia and the Fair." *Wisconsin Farmer*, X (Nov., 1858), 418 ff.

[43] *Cultivator*, n.s., IV (Oct., 1847), 321.

were William H. Seward, Daniel Webster,[44] Salmon P. Chase, Horace Greeley, Stephen A. Douglas, and Edward Everett. These annual addresses were printed by the local journals and frequently copied in full or in part by those of other states.

The speeches of the twenties were freighted with classical illusions, flowery rhetoric, patriotic sentiments, and what proved to be bad agricultural advice. All in all, they followed a rather regular pattern, emphasizing the importance, dignity, value, and profits of agriculture. Subtle praise for the farmer was usually sprinkled throughout the address and often, so that the dullest listeners might not feel neglected, fulsome flattery was added. The oration delivered by the historian, George Bancroft, before the New York State Agricultural Fair in 1844, "though not perhaps as practical or strictly agricultural, as the more utilitarian of his hearers desired," was nevertheless "dignified, chaste, energetic and sublime." His chief object was not to enter into the details of practical agriculture, but to "exhibit the dignity and importance of this great art, in whose pursuits some of the most eminent statesmen of our country had exerted their powerful minds." [45] Josiah Quincy, Jr., of Massachusetts, delivered the address the following year, and "without the slightest aim at lofty eloquence, its arguments in favor of the superiority of agricultural occupation, over the fretful and feverish life of speculation and ambition, were powerful, convincing, overwhelming." [46] Solon Robin-

[44] A speech was demanded of Webster practically every time he stepped into a fair ground. His address before the New York Agricultural Fair at Rochester in 1843 was punctuated with "Bravo," "Cheers," "Applause," "Great cheering," and when he finished he received "Nine cheers." *New England Farmer*, XXII (Oct. 4, 1843), 106–7.

[45] *Cultivator*, n.s., I (Oct., 1844), 316; *New Genesee Farmer*, V (Oct., 1844), 82.

[46] *Cultivator*, n.s., II (Oct., 1845), 315.

son, the most important agricultural writer of the period, pronounced "the address just what it ought to be on such occasions," and said it was received with great applause.[47] He noted "friend Greeley" some distance away, taking in every word, no doubt, for publication in the *Tribune*.[48] On the other hand, Abraham Lincoln, who felt that farmers received too much praise, avoided this sort of thing in his address before the Wisconsin Agricultural Society Fair: "I presume I am not expected to employ the time assigned me, in the mere flattery of the farmers, as a class. My opinion of them is that, in proportion to numbers, they are neither better nor worse than other people." [49] Then followed a common-sense, helpful talk to the farmers of the West, in which among other things, he made some interesting observations on "labor" and indicated a recognition of the revolution already started as a result of the introduction of power machinery.[50] Plainly, these occasions were not ideal for theoretical and scientific discourses. Professor Mapes, the editor of the *Working Farmer*, while attempting to "instruct" the farmers at the Indiana State Fair in 1858, found that not one in forty on the grounds would hear his entire speech.[51]

The county and local fairs also featured annual addresses but these were more likely to be of an intimate, practical nature. Occasionally men of national prominence spoke, more

[47] Kellar, *Solon Robinson*, I, ix.

[48] *Ibid.*, I, 516–17 (from *Daily Cincinnati Gazette*, Oct. 1, 1845).

[49] *Wisconsin Farmer*, XI (Nov., 1859), 390.

[50] For portions of this speech see pp. 305 ff.

[51] *Ohio Farmer*, VII (Oct. 30, 1858), 346.

On the other hand, P. T. Barnum, who had served as president of the Fairfield (Connecticut) Agricultural Society, occasionally delivered agricultural addresses which were always "highly entertaining." The editor of the *Southern Planter*, somewhat suspicious of the speaker's knowledge, commented on one of these speeches and said that he did the author justice when he confessed "to having read far worse garbage in a yet more diluted form." *Southern Planter*, XI (Feb., 1851), 44.

often, it was an official of the county society, a local celebrity, or an editor. These editors were popular and willing speakers, who probably sensed the real needs of the community. Their attitude is characteristically expressed by the editor of the *Ohio Cultivator*, who said he would speak whenever requested and would do so cheerfully, in his "plain off-hand way without money or price." [52] Orange Judd of the *American Agriculturist*, delivered an address before the Litchfield, Connecticut, Agricultural Society that so delighted his audience that they immediately and unanimously voted to award him the title of "Professor." [53] This new manner of conferring degrees caused great amusement among his colleagues in the editorial fraternity.

There was evidence, however, in the late fifties that the annual agricultural address, like the plowing match, was a dying institution and that it would soon be replaced.

One of the most useful innovations of the golden age fairs was the farmers' informal evening discussion. At these gatherings, especially popular at state fairs, knotty agricultural problems were threshed out by large groups of interested farmers. This interchange of experience and opinion grew into one of the most valuable attractions of the annual autumn jubilee.

The subjects for discussion, announced in advance, depended upon the geographical location of the fair, and included topics such as rust in grain, culture of wheat, plowing and drainage, the breed of sheep best adapted to the state, fruit culture, and the like.

These groups included farmers from all parts of the country, frequently agricultural authorities such as Solon Robinson, and always editors of the journals. An evening discussion held

[52] *Ohio Cultivator*, XVI (July 1, 1860), 200.

[53] *Ibid.*, XII (Oct. 15, 1856), 312; *American Agriculturist*, XV (Oct., 1856), 307.

at the Freeport, Illinois, State Fair in 1859 is representative, and receives a full length report in the *Wisconsin Farmer*. It is summarized in part as follows:[54] The subject previously announced for Wednesday evening was "Plowing and Drainage." At 8 o'clock possibly 500 farmers gathered in the great tent reserved for that purpose, to obtain the facts concerning the value and best methods of these operations. The meeting was organized by appointing K. K. Jones of Quincy, chairman; and J. W. Hoyt of the *Wisconsin Farmer*, secretary. They proceeded at once to a lively and interesting discussion.[55] In addition to a great many farmers, J. W. Hoyt and D. J. Powers of the *Wisconsin Farmer*, Colonel S. D. Harris of the *Ohio Cultivator*, and Charles D. Bragdon of the *Prairie Farmer* took an active part.[56]

Usually an editor served as secretary and recorded the proceedings of these meetings. Some journals delegated special representatives who traveled great distances to attend and to report the discussions. Although the annual aggregate attendance was relatively small, it included the leaders in the crusade for better agriculture. The ideas and experiments of these leaders, together with the dirt farmers' actual practice, were reported in the press for the benefit of readers throughout the country.[57]

[54] *Wisconsin Farmer*, XI (Oct., 1859), 374.

[55] For a partial report of the discussion at this meeting, as found in the *Prairie Farmer*, XX (Sept. 22, 1859), 177 ff., see pp. 311 ff.

[56] An especially interesting evening meeting was held in 1855 at Elmira, N.Y., during the New York State Fair. The leaders in this discussion on fertilizers were Solon Robinson, agricultural editor of the New York *Tribune*, James Vick of the *Genesee Farmer*, Sanford Howard of the *Boston Cultivator*, and Joseph Harris of the *Country Gentleman*. A report of this meeting may be found in the *Cultivator*, ser. 3, III (Dec., 1855), 364–65.

[57] To the student interested in the development of the science of agriculture, these discussions afford indispensable data on the exact contemporary attitudes toward the various questions considered.

No less interesting but certainly more spectacular than the farmers' informal evening discussion was the plowing match. Such a contest was held at the exhibition of the Columbian Society (Georgetown, D.C.) on the 20th of May, 1812; this was the first field trial of implements in America.[58] The plowing match possessed utilitarian value in that it tested and demonstrated various types of plows, but as victory depended also upon skillful manipulation of the implements and animals, it readily developed into a sporting event. It was not a trial of speed, but a test to prove the best plow and the best plowman.[59] This innovation was an important feature of the Berkshire Cattle Shows and retained its popularity until the Civil War. In the early days, it was the greatest local sporting event of the year, and a victory brought with it considerable prestige in the community.

The Honorable Timothy Pickering, after many triumphs in the political arena, became the hero of Essex County, Massachusetts, in 1820, when at the age of seventy-five he won the plowing match "in competitive trials with the sturdy yeomen of Essex County" and "bore off the first premium," because "of the superior performance and superior utility of his plough." [60] He was not only a practical and scientific farmer, but "was pre-eminently distinguished as a ploughman." [61] A number of places in New England adopted uniforms consist-

[58] U. S. Patent Office, *Annual Report: Agriculture* (1859), p. 24.

[59] Solon Robinson, editor of the *Plow*, indicated the basis upon which awards should be made. "The prize should be awarded first to the plow, second to the plowman, third to the team, and finally to all doing the best work in the easiest, not the shortest manner. It is not for a trial of speed of horses or oxen. It is to prove which is the best plow, and who does the best work. The speed should be that of an ordinary day's work." *Plow*, I (Nov., 1852), 338.

[60] Essex Agricultural Society *Transactions*, "The Trustees Account of the Cattle Show and other Exhibitions, at Topsfield, Oct. 5, 1820," p. 18.

[61] Pickering, *The Life of Timothy Pickering*, I, 334.

ing of rifle frocks with belts and wreaths of straw around the hats. The contestants, in this regalia, created quite "an animated and pastoral effect." [62] Frequently matches for boys under eighteen were added to the regular contests.[63]

By 1843 the plowing match of the New York State Agricultural Fair was so famous that "thousands of spectators" including Van Buren, Governor Bouch, Webster, ex-Governor Seward, came to witness the "honorable strife" of eighteen contestants.[64] This "exciting" event became more and more spectacular. Frequently a band accompanied the contestants to the field; the audiences grew larger and more vocal.[65] An unpopular decision brought roars of disapprobation from the spectators.[66]

The plowing match of the agricultural fair during the fifties met a mighty rival in the trials of speed on the race track. Both events were justified on the ground that they promoted agriculture, but the perfecting of the plow nullified the utilitarian aspect of the plowing match, and since it lacked the romantic appeal of the horse race, it was gradually superseded.[67] The *Prairie Farmer* in 1860 reports this turn of events, citing the Illinois State Fair: "There were very few persons out to witness the trial, and little interest manifested." [68] The people had tasted the greater speed and excitement of the race course; the plowing match now seemed tame and dull.

[62] *Plough Boy*, III (Nov. 17, 1821), 193.

[63] For a brief, official report on a plowing match of the early period, see pp. 314–15.

[64] *Cultivator*, X (Oct., 1843), 155.

[65] P. T. Barnum was considered "as great in a plowing match as in a JENNY LIND concert, a traveling menagerie, or New-York Museum," for "he was in the midst of the busy scene, full of life, energy, and glory as a Connecticut farmer." *Plow*, I (Nov., 1852), 337. He was active in the Bridgeport, Conn., Agricultural Society.

[66] *Farmers' Cabinet*, XI (Jan. 15, 1847), 178.

[67] Neely, *op. cit.*, p. 209.

[68] *Prairie Farmer*, XXII (Sept. 20, 1860), 185.

In the early fifties, the editors began calling attention to the "degeneration" taking place in the fair. The first American agricultural fairs as organized by Elkanah Watson were essentially neighborhood institutions. As long as they continued to be primarily local events, commercialized amusement did not present a serious problem, but the tremendous expansion of fairs that came after 1848, brought about a marked change.[69] To those with eyes to see, it was plain that the fairs of 1855 had drifted far from the early ideals. No group saw this "degeneration" more clearly than the editors of the journals and no group fought more vigorously to maintain the original pattern.

By the middle fifties, a flood of articles appeared indicating this opposition. One editor wrote, "We fear these agricultural festivals are rapidly losing their original character and design." [70] Another, "We hope the day is not far distant when a return to Agricultural Fairs, as they were at first constituted, will be made." [71] A third one declared, "We deeply regret this change in the character of our fairs and we are glad to say that almost the entire agricultural press of the country is speaking out against it." [72] With regard to the innovations which wrought these changes, one editor strongly felt "their effect in the demoralization" of fairs should be exposed.[73] An equally severe critic, writing on the subject of "Perversion of Fairs," maintained it would take only a few more "years such as the last three, to debauch" the whole population.[74]

[69] Neely, *op. cit.*, p. 189.

[70] *Farm Journal* as quoted in *American Agriculturist*, XV (Dec., 1855), 60.

[71] *Ohio Farmer*, VII (Nov. 13, 1858), 362; *Pennsylvania Farm Journal*, V (Nov., 1855), 321.

[72] *Country Gentleman*, VIII (Nov. 27, 1856), 347.

[73] *Prairie Farmer*, XXII (Sept. 27, 1860), 194.

[74] *Ohio Farmer*, VII (Nov. 13, 1858), 364.

Horse racing with its attendant vices was probably the most obnoxious innovation. It was introduced for the improvement of driving horses, and the development of speed and stamina in light-harness horses.[75] However, the race horse having gained admittance to the fair grounds, completely captivated the fair-goers. "Senex" in the *Country Gentleman,* told of attending fairs in 1846; he stated that a few hundred people attended; no admission charge was made and "all the spectators . . . were busily occupied in examining the fine animals, and learning where the breed could be procured, how to raise such large and fat oxen, how those fine squashes and excellent watermelons were cultivated." Ten years later he said the grounds were enclosed with a high fence, a "snug fee" was exacted from every one of the 20,000 visitors and the "race-course was the great object of attraction, throughout the day." He also noted that "nineteen twentieths, at least, of all the spectators at the fair, were intently watching the performance the day through." "Senex" felt that "a great change had come over the spirit of the dream." [76]

When horse races were introduced, most of the periodicals were "willing to wait and watch their influence, before speaking of them in terms of disapprobation." [77] However, by 1856 the agricultural press had formed almost a solid front against this sport at fairs. "In all our horse talk and writing [said the *Ohio Cultivator*], we have spoken disparagingly of that class of horses the only merit of which is, that they are merely fast. These gaunt, leggy spiders that can do nothing but run, are about as useless in the world as those fancy gentlemen in flashy vests, who generally attend them in their airings." [78]

[75] Neely, *op. cit.*, p. 191.
[76] *Country Gentleman,* VIII (Oct. 2, 1856), 219.
[77] *New England Farmer,* VIII (Dec., 1856), 563–64.
[78] *Ohio Cultivator,* XII (Dec. 1, 1856), 361.

The benefits anticipated had not materialized. On the contrary, the feeling was general throughout the press that the encouragement of mere speed in horses resulted in nothing of real value [79]—indeed, it brought about the "pernicious habit of gambling" and caused a "growing tendency to transform fairs into temporary race-courses." [80] The handwriting was on the wall. "Every year, at our principal fairs, the race-track is made larger, while the stalls for cattle are conparatively neglected." [81] Thus the *Ohio Farmer* voiced the thought of hundreds of editorials during the fifties when it demanded that "*horse-racing* be transferred from the Fair Ground to the Race Course." [82]

However, the forces behind the race track were strong. The management knew well its effect on the box office, and fairs that met expenses were rare. In addition, thousands of people had found an exciting diversion that appeared to gain respectability by its introduction into the fair.[83] Of course the gambling fraternity, the jockeys, and the sports journals were enthusias-

[79] *Ohio Farmer*, VII (Nov. 13, 1858), 364.

[80] *Cultivator*, ser. 3, IV (Dec., 1856), 379.

[81] *Ibid.*, ser. 3, IV (Dec., 1856), 379.

[82] *Ohio Farmer*, VII (Nov. 13, 1858), 362. Articles confirming this attitude may be found in *Country Gentleman*, VIII (Nov. 27, 1856), 347; XII (Nov. 4, 1858), 292; XVI (Oct. 18, 1860), 256; *Valley Farmer*, VII (Oct., 1855), 418; *Prairie Farmer*, XIX (July 21, 1859), 40 XX (Nov. 3, 1859), 273; XXII (Nov. 8, 1860), 290; XXII (Oct. 18, 1860), 242; *American Agriculturist*, XV (Dec., 1856), 342; XVI (Jan., 1857), 20; XVI (Nov., 1857), 247; XVII (Sept., 1858), 272; XVIII (June, 1859), 171; XVIII (Dec., 1859), 358; *New England Farmer*, VIII (Dec., 1856), 563 ff.; *Ohio Cultivator*, XII (Dec. 1, 1856), 361; *California Culturist*, I (Sept., 1858), 173; *Northwestern Farmer and Horticultural Journal*, IV (July, 1859), 218; *Cultivator*, ser. 3, IV (Dec., 1856), 379.

[83] This new found respectability was scouted by some editors. "We protest against people exhibiting either their hypocrisy or ignorance, by countenancing at these fairs what they choose to ignore at the legitimate place for such exhibitions—the *race course*. . . . Out upon hypocrisy!" *Prairie Farmer*, XX (Nov. 3, 1859), 273.

tic. In fact the sports magazine *Porter's Spirit of the Times* gleefully approved of it, and in 1856 when a trotting course appeared in Boston on the agricultural fair grounds, it carried an editorial entitled, "A Few Inches Nearer the Millennium."[84] However, the *New England Farmer,* sadly commented, "A blight has fallen on our cherished hopes. . . . We are surprised by the suddenness with which our puritan motions have been overwhelmed and the original intention of our associations swept aside!"[85]

Another thorn in the flesh of the agricultural press was the innovation at first innocently known as the ladies' equestrian display.[86] It comprised an exhibition of horsemanship of intricate and severe evolutions, with prizes awarded to the most talented. The following account in the *Ohio Cultivator* describes this feat:

> No less than nine ladies entered their names as competitors for the prize (a silver pitcher) which was to be awarded for the best female horsemanship. The weather was fine, the grounds spacious and well arranged, and with the 8,000 or 10,000 admiring spectators surrounding the vast circle, the scene was one of exciting interest as well as beauty. A band of music, in the centre of the circle, played lively airs, while the nine beautiful equestrians, filled with the joyous spirit of the occasion, performed their evolutions in a manner that excited the admiration of the entire multitude.[87]

---

[84] *Porter's Spirit of the Times,* I (Sept. 20, 1856), 40; I (Nov. 1, 1856), 144.

[85] *New England Farmer,* VIII (Dec., 1856), 563.

[86] Instead of a ladies' equestrian display, the Virginians had a men's tournament. The gentlemen cavaliers, who participated, assumed knightly names, wore ancient costumes, and were mounted on fleet chargers. The successful knight was rewarded with the "blissful privilege of selecting the fairest lady and of crowning her Queen." *Wisconsin Farmer,* XI (Feb., 1859), 49.

[87] *Ohio Cultivator,* VIII (Oct. 15, 1852), 314.

As time went on more handsome prizes were offered. In 1854 a horse

When this display was first introduced, no one questioned its propriety; everybody conceded that it was an attractive feature of the fair.[88] Even the journals were enthusiastic and urged the societies to offer premiums for the encouragement of this sport, which proved very popular.[89] As one observer later said, these shows attracted the multitude, and it rushed out, "with eyes staring and mouth agape, to see the ladies ride fast horses, *a la jockey style*"; another noted that the enthusiasm evinced was such that with difficulty the marshals curbed the "boisterous approval of the multitude." [90]

Gradually the crowds demanded real races by the participants—then the trouble started. This development brought down the wrath of the agricultural press. In the opinion of the *Ohio Farmer*, such activity at the fairs was a "humbug and a nuisance"; the *Prairie Farmer* denounced it as a "ridiculous, outrageous practice," conscious of the "folly and immodesty of such an exhibition"; the *Indiana Farmer* definitely declared it "out of place and conducive of no practical benefit." [91] Great objections were made to the betting and gambling going on among the "gentry" and the *American Agriculturist* felt that if there were still "any one who would be willing to have a sister, wife, or relative, take active part in these 'Exhibitions,' he ought to be present at one of them, and hear the coarse, ribald jests, and the vulgar, low-life expressions addressed to the fair riders." [92] Mrs. Abbott of the *Valley Farmer*, as usual,

---

valued at $200 was awarded at the Allegheny County, Pa., Fair. *Cultivator*, ser. 3, II (July, 1854), 227.

[88] *Cultivator*, ser. 3, I (Nov., 1853), 332.

[89] *Ohio Cultivator*, VIII (Oct. 15, 1852), 314.

[90] *Country Gentleman*, VIII (Nov. 27, 1856), 347; *Ohio Cultivator*, X (Oct. 15, 1854), 312.

[91] *Ohio Farmer*, VII (Nov. 13, 1858), 364; *Prairie Farmer*, XVIII (Nov. 25, 1858), 342; *Indiana Farmer*, VII (Oct., 1858), 215.

[92] *American Agriculturist*, XVII (Nov., 1858), 332.

took an extreme position. She said: "They [the equestriennes] go against all the rules laid down in the bible to regulate their conduct, and are getting wise above what is written." [93]

Unfortunately, the forces were too strong to be turned back. The people were amused and the gate receipts bore ample testimony of success. No doubt *Porter's Spirit of the Times* reflected the attitude of thousands in an age harassed by restrictions and traditions of society. It stated, "The fine race courses which have been established at nearly all the agricultural fairs of the season, and the beautiful ladies who have not hesitated to perform the parts of jockies over them, have sounded the knell of this obstinate bigotry." Referring to the conservative agricultural press, it wrote, "How foolish, then for the cold-back croakers of the refrigerating press to attempt to write such practice and inclination down." [94]

In the middle fifties, still another innovation causing a flurry of excitement was the baby show. Some fairs offered prizes as high as $50.00 for the "handsomest and finest" babies. Normally however, the awards were more modest. Many journals refused to take this fad seriously, feeling it would soon die out; others, such as the *Prairie Farmer,* felt a "disgust" and "repugnance" for the idea, labeling it a "feature foreign to the object for which fairs were instituted." [95] The *Valley Farmer,* considered it a "foolish, wicked, and absurd exhibition." [96] The craze soon spent itself; the barrage of editorials probably had little effect one way or the other.[97]

[93] *Valley Farmer,* VI (Nov., 1854), 433.

[94] *Porter's Spirit of the Times,* I (Nov. 22, 1856), 192.

[95] *Prairie Farmer,* XIV (Nov., 1854), 403; XV (Aug., 1855), 233–34.

[96] *Valley Farmer,* VI (Nov., 1854), 433.

[97] In 1854, a year of "Shows," there was a "National Baby Show" at Springfield, Ohio, with only 20 entries. In the following year the "National Baby Show" was held at P. T. Barnum's Museum in New York, where 143 entries made it a record event. Editor Holmes of the *Maine*

WHAT THEY DO AT FAIRS

From *Prairie Farmer*, XXII (September 6, 1860), 149.

Manifold were the added "attractions" which perverted the original aims of the fair as established by Elkanah Watson. Among them were the side shows, "the veriest humbugs," suited to the "lowest minds and most depraved taste." [98] They included the swindlers with their "erasive soaps," "tooth ache cures," "headache cures," "silver polish," gilded "jewelry," and mock "auctions," and since the press was unable to exclude them, it took measures to expose them.[99] The *American Agriculturist* was especially effective in this respect. A widely copied example of such "humbug" was a "tooth wash warranted to remove all dark color, etc., etc., from the teeth immediately, and give them a pearly whiteness." The editor examined this "wash" and found it consisted of water and muriatic acid, which in itself explains the effectiveness of the product. The wash merely dissolved the surface of the teeth and, if used over a period of time, disintegrated them completely. The agent told the editor he was going to the New York State Fair where he could sell bottles of wash "like hot cakes to the green country chaps." [100]

The desire to get something for nothing led many visitors to try the various games of chance. Many "horrible examples" were cited in the journals, but there is little evidence that these warnings had any appreciable effect.

Certainly the fairs of the fifties were unlike those of the twenties. The modest fair of the early period, combining prayer and singing in the village church, was a far cry from the "stupendous" exhibitions of the golden age. The journals

---

*Farmer* said, "the good sense of the people generally, will condemn this exhibition." *Maine Farmer*, XXIII (June 14, 1855), 98; *ibid.*, XXII (Oct. 12, 1854), 166.

[98] *Ohio Cultivator*, IX (Oct. 1, 1853), 297.

[99] *Prairie Farmer*, XXII (Nov. 8, 1860), 290.

[100] *American Agriculturist*, XI (Nov. 9, 1853), 136.

vainly attempted to retain the original emphasis and to exclude, in the vernacular of the fifties, the "monkey-shines." A farmer, writing in the *Country Gentleman*, summarized his reaction to a county fair when he wrote, "I thought of Barnum." [101] Such was the heart of the editors' criticism.

Nevertheless, considering the fair in retrospect, certain important achievements should be noted. This institution hastened the development of class consciousness among the rural population, and as it was the predominating form of associated effort among farmers it performed a great service in creating an atmosphere of social unity.[102] Without doubt, this feeling gave rise to a new community of interest and a new sense of their importance as an economic group.[103]

The social and recreational features should not be overlooked. These fairs, occurring at the close of the farmers' busy season, were regarded essentially as holiday festivals, where distant friends met either on matters of business or for social enjoyment. Upon these occasions the farmers' wives were drawn out of seclusion to enjoy competition in domestic industry. Probably the rural population had too few holidays and fewer social advantages than residents of villages and cities. At any rate, country folk concluded that if fairs brought only relaxation and a season of social intercourse, they would pay for all they cost.[104]

[101] *Country Gentleman*, VIII (Oct. 16, 1856), 250.

[102] Bailey, *Cyclopedia of American Agriculture*, IV, 292.

[103] Bidwell and Falconer, *op. cit.*, p. 193.

Indeed the intense public interest in their occupation tended to increase this feeling of importance. The extent of this is indicated by a correspondent in the *Ohio Cultivator* who said that farmers' sons and daughters formerly thought it a disgrace to be born and reared in the country. Suddenly they discovered that all eyes were turned to the rural districts, and they began to put "on aristocratic 'airs.'" *Ohio Cultivator*, VIII (Nov. 15, 1852), 343.

[104] *American Cotton Planter*, o.s., XIII (Nov., 1859), 332.

The conception of the fair as a great educational institution was advanced from the first. Indeed, the cattle show, introduced by Elkanah Watson, was the earliest agency for agricultural education which appealed to the masses. Some people were of the opinion that next to the common school, the fair accomplished more for the dissemination of knowledge and development of local and national resources, than any other facility in the country.[105]

The fair helped to expedite improved methods of farming and introduced new types of stock, seeds, grain, and fruit. Through spirited competition, the best breeds of animals were imported and bred with the object of greater improvement.[106]

Perhaps the fair's most important contribution to the farmers' education was the introduction of farm implements and machinery. One of the earliest and best illustrations was the cast-iron plow which became very popular between 1820 and 1830.[107] Thereafter, many other agricultural devices—reapers, mowers, threshers, corn planters, and cultivators—were introduced at the exhibitions, and accounts of them in the agricultural journals, no doubt, reached thousands of farmers unable to attend.

As true appraisals of the various items on display, however, the decisions of the judges were often valueless and the prevailing dissatisfaction was aired many times in the press. Charges of favoritism and incompetency on the part of the judges were frequent, as it was difficult to secure trustworthy

[105] At times extravagant and ridiculous claims were made in the agricultural press. One editor told the story of a young man recently graduated from college, who declared that he had learned more in three days at the state fair than in his entire college course. *Indiana Farmer*, VII (Sept., 1858), 164.

[106] *Valley Farmer*, XII (Nov., 1860), 327.

[107] Bidwell and Falconer, *op. cit.*, p. 192.

and experienced men who were willing to serve.[108] The trouble was due, in part, to the radical defects in the system of making awards. Unfortunately, the exhibits were poorly classified and no accepted standards existed. Very often mere "bigness" determined the awards in the animal division. Machines and implements were granted premiums without actual trial on the field. As one judge said, "Those machines which we had seen tried, we knew about; of the others we had to guess." [109] Another farmer complained that machines which he scarcely would have sheltered were "dignified by Silver Medals; and others of third or fourth rate value were assigned first premiums." [110] Many felt the awards were "little else but a farcical jumble of mistakes." [111] Toward the latter part of the period, certain exhibitors of implements and machinery refused to compete for premiums, but nevertheless displayed their products. This procedure proved to be good advertising and at the same time brought these inventions before the farmers.

The introduction of the sewing machine in the fifties illustrates the coöperation between the agricultural press and the fair in an attempt to popularize an important domestic device. Little or no attention was given to the sewing machines on

[108] It was a thankless office and the first experience as a judge was frequently sufficient. One judge, reporting his experience, said he was urged to accept the office as there was "not much to do," "honor to be acquired," "silk badge with Judge in big letters to stick in a button hole," "free dinner ticket," and so on. "All seemed to go on well for a time; exhibitors were all smiles and compliments; it was *Judge* here, and *Judge* there; 'make room for the *Judges.*' . . . But when the premiums were awarded, '*presto change!*' . . . we soon found we had waked up eleven wrong passengers and only one right one, and he happened to be the one that received the premium. We sneaked off, leaving all the honor and some of the liquor on the ground, hoping that 'out of sight out of mind,' but we were sorely mistaken." *Ohio Cultivator*, XI (Oct. 15, 1855), 314.

[109] *Ohio Farmer*, VIII (Oct. 8, 1859), 321.

[110] *Ohio Cultivator*, XI (Oct. 15, 1855), 314.

[111] *Wisconsin Farmer*, XI (Dec., 1859), 437.

exhibition at the London World's Fair in 1851, yet many factors were at work to bring this invention before the people.[112] The journals abounded with correspondence and editorials, while special writers, such as Anna Hope and Frances D. Gage, wrote articles with instructions for the use of these machines. The editor of the *American Agriculturist* stated with conviction: "In the single invention of the sewing machine, mechanical skill has probably done as much for the relief of household toil, as it has done for the labors of the field, in all its inventions combined." [113] This editor, faced with an avalanche of inquiries, purchased two different makes for experimental purposes, and later published his wife's comments on them.[114] Frequently more than twenty models of various types were entered at the larger fairs. A number of journals offered sewing machines as premiums for obtaining a required number of subscribers. By 1857 a writer reported that sewing machines were taking the country by storm, and the following year, Wheeler and Wilson Company boasted they had exhibited them in all parts of the country, winning first premiums in twelve states and "hundreds" of county fairs.[115] This device was thoroughly established before the Civil War, mainly through the efforts of the fair and the agricultural press.[116]

The fair also became the great testing ground for the steam plow, which the agricultural societies and journals hoped would revolutionize farming. They felt it would be especially valuable in bringing the middle-western prairies under rapid cultivation. During the late fifties, thousands of dollars were

[112] Lewton, "The Servant in the House," p. 559.

[113] *American Agriculturist*, XVII (April, 1858), 121.

[114] *Ibid.*

[115] *Ibid.*, XVI (July, 1857), 161; *Prairie Farmer*, XX (July 7, 1859), 13.

[116] More than 130,000 sewing machines had been sold by 1860. Humphrey, *An Economic History of the United States*, p. 217.

offered for an efficient, economical steam plow, a plow that would successfully compete with animal power. The Illinois State Agricultural Society offered $5,000 in prizes for such a plow; the Illinois Central Railroad Company offered $1,500; and the New York State Agricultural Society, $250.[117]

In competition with a dozen or more machines at various fairs, J. W. Fawkes's model was considered the most practicable.[118] Almost over night this unknown mechanic found himself successful, taking "rank with Fulton, Stephenson, and all other benefactors of our race." [119] It was said that "Lancaster County," proud as she was "of her Calhoun and Buchanan," would be "prouder still of her Fulton and her Fawkes." [120] Fawkes's popularity is registered in an account by a correspondent of the *Ohio Farmer* who wrote, "Fawkes is the great card now, and when he gives a couple of puffs of his whistle, and the great steam horse speeds over the ground, dragging the gang of plows, almost every man, woman and child leave horses and sulkies to their own fate, and follow this blower-up of antiquated notions." [121] A high point of enthusiasm was attained, facilitated by the agricultural press. Furthermore, thousands of people had actually seen the plow in operation. The statement in the *Wisconsin Farmer* that "the steam-plow is a

[117] *Prairie Farmer*, XVIII (Nov. 18, 1858), 321; *Genesee Farmer*, ser. 2, XX (July, 1859), 206; *Country Gentleman*, XI (March 25, 1858), 192. The Illinois Central Railroad company with its great landed interests, would benefit from the sale of land following the successful introduction of great labor-saving machines. More immediately, it would gain in the vast number of passengers to be carried to the fairs to witness the trials of steam plows. *Valley Farmer*, XII (Jan., 1860), 5.

[118] Fawkes, a mechanic from Lancaster County, Pa., after three years of despairing struggle and utter privation spent in advancing his idea, submitted to the world "the product of his genius." *Country Gentleman*, XIV (July 28, 1859), 65.

[119] *Prairie Farmer*, XVIII (Oct. 21, 1858), 257.

[120] *Country Gentleman*, XIV (Aug. 18, 1859), 106.

[121] *Ohio Farmer*, VIII (Sept. 24, 1859), 309.

**FAWKES'S STEAM PLOW**

From *Emery's Journal of Agriculture and Prairie Farmer*, XVIII (October 7, 1858), 225.

success and Fawkes is immortal!" was generally accepted.[122]

The agricultural authorities were not so easily overwhelmed. It is true, however, that Fawkes was awarded the Grand Gold Medal of Honor by the United States Agricultural Society.[123] On the other hand, the Illinois State Agricultural Society, while admitting that his machine added materially to the receipts of the fair and gave general satisfaction to thousands of visitors, felt it had not met the requirements of their award.[124] By way of encouragement, they donated $1,500 for the advancement of his enterprise.[125] After all, the real utility of the steam plow could not be estimated by the enthusiasm of the crowds. The committees' official reports were only mildly enthusiastic; their records were studded with comments such as "the trial was not conclusive," the machine "developed certain defects," or the test was "not completely successful."

By 1860 the editors came to the conclusion that the steam plow could never be used economically.[126] M. L. Dunlap, editor of the *Illinois Farmer*, and an authority on the subject, summarized his observations saying, "we have given up this pleasing hope as an illusion now completely dispelled." [127] The trials, however, turned many eyes to the rural districts and their needs; increased the attendance at the fairs tremendously; and amused thousands of people. And the people wanted to be

[122] *Wisconsin Farmer*, XI (Oct., 1859), 376.

[123] *Prairie Farmer*, XX (Sept. 29, 1859), 199.

[124] *Prairie Farmer*, XXI (Jan. 12, 1860), 24. Noting his ability to draw crowds, the Virginia Agricultural Society offered Fawkes $1,000 and expenses merely to exhibit his machine at their state fair in 1859. *Ohio Farmer*, VIII (Sept. 24, 1859), 309.

[125] *Prairie Farmer*, XXI (Jan. 12, 1860), 24.

[126] *Genesee Farmer*, ser. 2, XXI (Jan., 1860), 17.

[127] Quoted in *Country Gentleman*, XVII (June 20, 1861), 394.

Fawkes, himself, by 1862, had lost faith in the traction principle and claimed that plowing could best be done by a windlass. *Ohio Cultivator*, XVIII (Sept. 1, 1862), 270.

amused. The editor of the *American Cotton Planter* sagely observed, "the thirst for knowledge is often wanting on the part of the visitors." [128]

[128] *American Cotton Planter*, o.s., XIII (Nov., 1859), 332.

# Chapter X

## SIGNIFICANCE OF THE AGRICULTURAL PRESS

*There is probably no other country in which so large a number of Agricultural Journals, in proportion to its population, is circulated, as in the United States—and we believe that it may with truth be said, that to these Journals we owe much of the progress which has attended our Agriculture during the present century.*[1]

AMERICAN AGRICULTURE experienced remarkable progress from 1819 to the days of the Civil War. This was due, in part, to the educational impetus of the farm press together with other agencies such as agricultural societies, clubs, and fairs.

Scientific farm practices quite generally accepted today, but new to many of the early nineteenth-century cultivators, were advocated in season and out of season by the agricultural press. To the doubting and conservative farmer, the periodicals, with remarkable unanimity, pointed out the advantages of deep and horizontal plowing; ditching and draining of wet lands; variety in cultures; the use of fertilizers such as guano, marl, and

[1] *Journal of the United States Agricultural Society*, I, No. 2 (1852–53), 144.

animal manure; crop rotation; selective breeding of livestock; the proper feeding of animals and their adequate shelter in winter. These among other practices were advocated by the press and there were indeed few farmers in 1860 who had not in some measure followed the advanced ideas.

One of the most remarkable achievements during the four decades before the Civil War was the improvement of farm implements and machinery. When markets were gradually opened and real profits could be realized through farming, a vigorous demand for labor-saving devices naturally developed. As we have seen, outstanding among the advances on the mechanical side of farming was the development of the plow. Beginning in the thirties, other important inventions, including the reaper, mower, thresher, corn planter, and cultivator, appeared in rapid succession. The press gave enthusiastic reception to these mechanical devices and coöperated with the societies and fairs in popularizing them.

Agricultural education long advocated by the journals, after a slow and halting beginning, gained wide support by the middle of the century. This resulted in the establishment of several state supported agricultural colleges and finally led to the passage of the Morrill Act of 1862.

The advances just considered, among others pointed out in previous chapters, in a large measure, were directed toward the liberation of rural folk from physical drudgery and mental apathy. Taken in the aggregate, these changes relating to the occupation of farming, constitute the agricultural revolution in the United States at this time.[2]

[2] For a survey of agricultural achievements in specific sections, see works such as Bidwell and Falconer, *History of Agriculture in the Northern United States, 1620–1860*; Gray, *History of Agriculture in the South-*

Too much, however, should not be claimed for agricultural progress generally in the United States during the forty years prior to the Civil War. The conservatism attached to the practice of farming is traditional and unique. It is therefore not strange that "agricultural revolutions" proceed with glacial dignity and the benefits, at first, are embraced by comparatively few people. At the close of this period, aversion to "book farming" was still prevalent; old methods with out-moded implements were still conspicuous; and "moon farming" retained many enthusiastic advocates. Nevertheless, if the primitive state of agriculture at the beginning of the nineteenth century is taken into consideration, it is evident that substantial progress had been made—progress sufficient, perhaps, to have justified the reasoned judgment of Edmund Ruffin, a contemporary agricultural authority:

> Notwithstanding all the existing obstacles and difficulties, American agriculture has made greater progress in the last thirty years [1821–1851], than in all previous time. This greater progress is mainly due to the diffusion of agricultural papers. In the actual absence of all other means, these publications, almost alone, have rendered good service in making known discoveries in the science, and spreading knowledge of improvements in the art of agriculture.[3]

In historical retrospect, these neglected farm periodicals constitute an invaluable repository of data for economic and social historians who are eager to investigate almost any phase of American life during the pre-Civil War period. Even the

---

*ern United States to 1860;* Craven, *Soil Exhaustion as a Factor in the Agricultural History of Virginia and Maryland, 1606–1860;* and Bidwell, "The Agricultural Revolution in New England."

[3] *Cultivator,* n.s., VIII (Feb., 1851), 91.

political historian will occasionally find an item of interest.[4]

Since these journals provided the principal clearinghouse for agricultural information, they are naturally an unrivaled reservoir of such material. The historian in this field will find abundant data, although undigested, on subjects such as crop rotation, irrigation, chemical fertilizers, agricultural education, theories of soil fertility, and soil chemistry. Much attention is devoted to horticulture, stock raising, dairying, and kindred subjects. Agricultural statistics are scattered throughout these periodicals and, while they are numerous, they are, in the main, unorganized and incomplete. The steps in the introduction and popularization of new farm implements and machinery, as well as new plants and animals are here described. No other contemporary papers reveal so vividly the resultant agricultural crazes then sweeping the country.

Of particular value to the social and economic historian are the accounts of agricultural tours, as well as the correspondence from farmers in all parts of the United States. These records state the exact farming conditions over a wide area at given times, and afford an excellent opportunity for critical analyses and comparison of farming practices. They eloquently disclose the problems faced by the pioneer agriculturists. The reports often include comments upon climate, transportation facilities, marketing, and prices current. Phases of rural life, including manners, morals, amusements, social customs, religious ideas and practices, together with a share of the intellectual diet of the country folk, are revealed. While these accounts portray general living conditions, the reader will find the descriptions of slave life and plantation economy in the South of especial

[4] The indexes of the journals, while frequently faulty, are always helpful. Those of the *Cultivator*, *Farmers' Register*, and *Country Gentleman*, for example, are of particular value.

interest. In these pages, the slaves live again; we see them in their huts, at work, and note how they were cared for, disciplined, educated, fed, and clothed.

A host of topics relating to rural life such as the development of agricultural societies, fairs, and farmers' clubs is probably not elsewhere so strikingly portrayed or so minutely related as in the farm press. The plowing match and the reaper trials receive detailed attention. The drift of sons and daughters from the farms cityward and the abandonment of the older regions for the West—these movements with their many ramifications, may be followed in the periodicals.

For the social historian, sections devoted to the interests of women are of inestimable value. Here may be found a treatment of the fashions of the period, touches of intimate home life, and discussions of social, economic, and moral problems. (In the leading journals this material was usually assembled in a special department with a competent woman in charge. Occasionally, the lady feature writers enriched the pages with editorials on pressing problems of the period.) Not the least interesting of the many discussions are the debates on timely topics such as the woman's rights question.

The interests of the younger generation may also be traced in the sections set aside for them. Sports, games, riddles, puzzles, and other forms of contemporary entertainment for the field and fireside are treated. Educational articles abound. Moral instruction including warnings against gambling, the use of tobacco and alcohol, appear repeatedly. Frequently, are found children's stories always of a wholesome type.

Many fields are here opened to the investigator. For instance, he may trace the general development of the architecture (with the aid of woodcuts and other illustrations) and study in detail the farmhouse and its setting. Material on the expansion of the common school and the development of internal improvements is found in abundance. The discussions often drift into varied subjects such as spiritualism, Puseyism, Millerism, mesmerism, Fourierism, and phrenology. In this crusading era, the cries of the reformers, who attacked the use of tobacco, pled for temperance, and agitated belligerently for world peace, are recorded. An occasional outburst against the theater, dancing, and even swearing is found.

The trends, interests, and thought of the period are vividly reflected in the "poetry," short stories, advertising, news items, and jokes. Much space is devoted to rural health, patent medicines, and home remedies. Movements, including Grahamism, hydropathy, and homoeopathy, all in protest against the regular medical practices, are dealt with in detail.

While the discussion of politics was avoided, the subject could not be completely ignored. Occasionally articles and correspondence appeared on slavery, presidential elections, platforms and candidates, as political issues. It should be borne in mind that the heroes who appear in these pages are not the great political figures of the time, but the pioneers in the crusade for better agriculture. A wealth of biographical data on the lives of these neglected agrarian leaders await the historian.

Since the United States was predominantly agricultural during these years, it is clear that the farm periodicals afford a fairly accurate cross-section of the life of the times. It is unfortunate that the files of these sources are so incomplete and so widely scattered throughout the libraries of the United States. This probably accounts for their general neglect by historians in the past.

# Part II

## SELECTED ARTICLES FROM THE JOURNALS

# IMPROVEMENT BY DRAINING

[This article by John Johnston, the father of tile drainage in the United States, appeared in the *Cultivator*, n.s., VI (March, 1849), 89–90.]

EDS. CULTIVATOR—Since Mr. Howard [editor of *Cultivator*] was here, in June, 1847, I have dug over 1,200 rods of drains, and the greater part is laid with tiles. I should have had 800 rods more done, if I could have got tiles enough in the fall. But I have now made arrangements by which I expect not to be disappointed in the future.

There is certainly, no way that a farmer can expend money that will yield so much interest as in draining wet, moist, or even *damp* land—especially if he intends to grow winter wheat. If the land is what is called wet, and of easy drainage, (I mean such as has a porous subsoil to the depth of two to two and a-half ft.,) the additional crop of wheat or Indian corn, the first season, will do more than pay the outlay in draining, and on moist or damp soils, where more wheat will grow than on wet soils, it may take the additional crop of two seasons to pay the cost of the drainage; but on *hard-pan* land, (that is, where the hard-pan comes within eight or ten inches of the surface,) I think the time has not yet arrived that we can drain such land profitably, in this country; but on all lands that can be worked with a spade two and a-half feet deep, it will pay well to drain, if the soil is injured by surface water.

My drains have cost me about forty cents per rod, all finished; but I have now made a contract for tiles at a cheaper rate. I am also, going to lay *pipes* two inches in diameter and fourteen inches long, which will answer all purposes, except where there is considerable run of water. The cost of the pipes will be less than that of the tiles, and by filling the drains with the plow, (I have always filled with the shovel till last fall,) I think I shall be able to finish my drains for 30 cents a rod.

Last fall, I exchanged ten acres of land with a neighbor. I got ten acres of wet land for ten acres partially drained. I made the exchange in order to get an outlet for my drains, through the ten acres I obtained, part of which was so wet, that when I made the exchange, cattle and horses were liable to mire on it; and it produced only coarse wild grass like the wet prairies of the West. The day I made the exchange, I set six men to ditching the new piece, and now a pair of horses will draw a wagon loaded with two tons, over any part of it that is drained, and by far the worst of it is done. The eyes of my neighbors are on that piece of land, and it would grieve me were I not to succeed with it; but I have no doubt I shall make it as good land for grass or summer crops as there is in the state of New York.

I am sanguine that the crop of wheat could in many instances, be doubled or more than doubled by thorough draining. I am aware that damp or moist land is by some held to be best for oats. I do not admit the correctness of the idea. In the first place, such land is frequently situated in hollows, where it has the benefit of the wash of uplands, and of course has become rich at their expense, and another thing, the dry land, in this part of the country, is almost always hard run with wheat, and generally brings a fair crop, which damp or moist land seldom or never does. For these reasons it has every possible chance to bring a bulky crop of oats, when the season suits. But let the dry land have the same rest with the damp land, and in nine cases out of ten, I have no doubt that the former will produce the heaviest crop of oats. The damp soil may produce the largest bulk, but put them on the scales, and the latter will be found wanting. Such is my experience. And with grass, also, wet or damp land generally produces the largest bulk,—but every man that has pitched as many loads as I have done, knows that the hay from dry land is double the weight of that grown on wet land, when the produce of both is in the same state of dryness; and the feed, as well as in the hay from dry land, is far more nutritious for any kind of stock.

I would say, therefore, drain, for the good of all crops, and for pasturage. It often grieves me, (or I might say *vexes* me,) to hear gentlemen whom I highly respect, and who profess to take science as their only guide in farming, say—"Is there no danger of making your land too dry?" Or—"are you not afraid that the most valuable salts will be drained from your farm?" Others say,—"We always look upon your farm as rather too dry." Thus, if I had not entertained strong faith in my own practice, I should have been deterred from draining by the opinions of others.

STALL-FEEDING CATTLE.—Since Mr. H. was here, I have built large barns, sheds, and cattle-stalls, and have enough to fill them all; and when I get 2,000 rods more of drains made, I have no doubt I shall require more barn room. I have fifty excellent steers, 3 years old last spring, feeding for some of the eastern markets. They were all, with one exception, fit for the butcher, when I commenced feeding hay and meal. I intend to feed them till March or April.

*Near Geneva,* N.Y.
Dec. 21, 1848.

# AGRICULTURAL CLUBS

[Report on a Delaware club, *Cultivator*, X (May, 1843), 77.]

A FARMER'S CLUB has been formed by our friends near Wilmington, Del., on a somewhat novel plan. It consists of twelve members only, who meet on the first Tuesday of each month, at the house of one of the members in rotation, at 10 o'clock, A. M., when "an examination," says the Delaware Gazette, "is made by the club of all that pertains to the farm, stock and cultivation of their host—his fields, his fences, farming utensils, mode of applying manure, rotation of crops, &tc., &tc. The conveniences and accomodations of his farm house, barn, piggery and poultry yard, are all matters of observation and discussion. At an early hour a plain farmer's dinner tests the thrift and cookery of his *better half*—her bread and butter, her savory meats and pies, well fatted poultry, her cheese, milk and cream, rich, fresh and cool from the just admired dairy, all afford practical themes at the dinner for discussion of their merits, and of woman's worth; as far as practicable, the products of the farm are required to be used for this part of the entertainment. Politics and political matters are at no time alluded to or admitted. After dinner, agricultural subjects are discussed and experiments reported; agricultural works and journals exchanged, noxious weeds noticed, and all the agricultural improvements and publications since the last meeting are passed upon and reviewed—seeds, plants, new grains, &tc., distributed—the entertaining member for the next month is agreed upon, and the club adjourns, *always early* to attend to the *feeding and foddering at home*, before dark. The gentlemen who compose this club, consist of Messrs. Bryan Jackson, C. P. Holcombe, John W. Andrews, Jesse Gregg, Samuel Canby, Henry Dupont, J. Boies, J. W. Thomson, Francis Sawden, William Boulden, George Lodge, and Major Joseph Carr.

# AN ADDRESS

*Delivered before the Agricultural Society of Cayuga County, at their anniversary meeting, held on the 7th day of February, 1820*

By David Thomas, *President*

[*Plough Boy*, I (March 4, 1820), 316.]

. . . THE VULGAR OPINION respecting the origin of this plant chess is too well known to need a particular recital; but perhaps all of you do not know that some are indifferent about its mixing with seed-wheat or seed-rye, alleging that it is never produced by its own seed. When error of opinion results in a practice so preposterous it is time to enter our protest. Perhaps we have all been told of the appearance of this plant in fields of grain where it was never sown; but this seed is so small as to render its detection by a careless observer improbable. It is true that botanists have given us long lists of mule or hybrid plants, but the chess has never had a place assigned in this catalogue. It is not even pretended by the advocates of this belief, that the seed-wheat from which this monster is said to rise, was the offspring of vegetable adultery; they admit that the wheat plant may rise perfect from the ground, but after being injured by cattle, or in unfavourable situations, its nature becomes changed; and the stalk, instead of being crowned with the golden grain, is only loaded with the shrivelled chess. Now, it would be safe to assert that nothing analogous to such transformation can be produced from the vegetable kingdom. It may not be irrelevant, however, to remark, that chess, though a weaker plant than wheat, is yet more hardy; and accordingly where that grain is thick and flourishing, the chess drops among the stubble; but where cattle, and excess of moisture have injured the wheat, the other springs with renewed vigour and fills the vacancy.

But I am well aware of the inefficacy of reason in combatting inveterate prejudices which have been cherished from infancy; and to convince the believers of that doctrine that it is founded in mistake, and unworthy of enlightened minds, I shall refer to facts that admit not of contradiction. The chess is a perfect plant, as different from wheat as the latter is from rye or barley, with seed completely capable of vegetating, and known in science by the name of *Bromus secalinus.*—The Botanist, whose observations are incomparably more close and accurate than the assertors of this doctrine, would no sooner admit this plant to be a degeneracy of nature, because it grows in our wheat fields, than the Zoologist would admit the sheep to be the degenerate offspring of the cow because they feed in the same pasture. . . .

# THE NECESSITY FOR A PROPER SYSTEM OF INSTRUCTION IN AGRICULTURAL SCIENCE

[Professor Norton, in the *Cultivator*, n.s., IX (March, 1852), 107–8.]

Analytical Laboratory, Yale College, New-Haven, Conn., Jan. 31, 1851.

EDS. CULTIVATOR—I do not propose to take up the above subject in its broadest sense, but to confine myself to a comparatively limited field. I shall say little at present as to the want, felt more and more every day by an increasing majority among our farmers, of educational institutions especially adapted [to] their wants, but would call attention to a point which has been overlooked by many in their zealous advocacy of the general cause. It seems to me, that a chief reason for the annual failure of so many plans bearing upon the educational interests of the farmer, may be found in a real scarcity which exists, of men competent to take charge of the proposed institutions. To those who have never reflected upon this subject, my assertion may seem a strong one, when I say that if any six states of the Union were within the present year to make provision for the establishment of state agricultural schools, or colleges, within their respective borders—were to endow them largely in every department, to furnish them with libraries, implements, museums, apparatus, buildings, and lands, they could not find on this continent the proper corps of professors and teachers to fill them. I will even go farther than this, and say that if in your own state of New-York, a large institution were planned out, and all proper departments of instruction pecuniarily provided for, it would be a difficult matter to fill them satisfactorily with thoroughly competent men. Enough of those who would gladly accept such appointments as might be offered, could doubtless be found, but the question is, would they be just such instructors as the farmer requires?

There are certain points relative to which he demands information from various branches of science, and this information to be of value must be *correct*. Mistakes, blunders, misconceptions, from the heads of a great state school, sent forth under authority, and promulgated rapidly, would cause infinitely greater mischief than our going without a school altogether for some years to come. For such reasons, extreme caution should be used in the selection of instructors for any large or influential school, and for such reasons among many others which might be adduced, I have ventured to say as above, that we really have not in the country the men that are needed.

If the farmers of any state were to select persons to impart instruction, or to serve as examples, in any practical department of their business, would they be contented with mere professions, or mere hearsay reports, of their success or skill? Above all would they not be disposed to question the expertness of one who professed to have made himself familiar with every department of mere mechanical labor, in the lapse of a very short time? If teaching the use of the plow in the best possible manner, and under every circumstance, were for instance the object, would they be content with a man who could only show the experience of one or two years in the use of that implement? By no means; they would say—we can do as well as he can ourselves, and do not need such instruction as this; we want a master of the subject, one who has studied it thoroughly in every department of practice, and has brought an intelligent mind to bear upon all the variations of use and construction in different districts. With a man of less acquirements than these in any practical matter, no community of farmers would be satisfied; they would not receive his advice with respect, and would not consider his opinion as worth much more than that of any other intelligent individual.

I think all will agree with me, that these views are correct with regard to subjects of pure practice, and that most farmers would act in accordance with them. Now I ask, why do not the same views obtain with regard to the teaching of science? We see men who are in all ordinary circumstances, shrewd and sagacious, swallowing every fable that comes to them in a scientific guise.

The merest charlatan may take up his books and mysterious looking apparatus, and having familiarized himself with a few hard names, is able to persuade the mass of those who meet him that he knows everything within, upon, and above the earth, that explains the action of nature's laws. Allow me to say a few words in direct reference to the falsity and even absurdity of such pretensions.

In speaking of the mechanical operations of husbandry, such as plowing, I have said that as a general fact, entire proficiency could not be attained within one or even two seasons; a long course of experience was necessary. Is it then so much easier to read the laws of nature, or rather of God, which bear upon those wonderful structures of plants and animals that we see about us! In the growth of the humblest weed that flourishes by the wayside, a series of changes, transformations, and metamorphoses, goes on, which as yet the highest effort of the human intellect has failed to fully explain and elucidate.

To produce the feeble stem which we crush under our feet in passing, the powers of earth, air and water, have joined with those of the far distant sun, and during its short life, it has been an example of a complication of most wonderful laws, imposed by the Almighty Maker of all. He has seen fit in his wisdom to ordain, that every step in knowledge must be won by toil and exertion, and thus it is in the present case; we are only able to slowly unfold the wonders that are occurring on every side, during the every-day experience of life. The field, too, widens as we advance, until we find that every step has its consequence, every breath of air its appointed mission, every drop of dew its office to perform; we discover that we are in the midst of causes and results, of which our knowledge is quite limited; that the threads we have seized only guide us to new and more difficult labyrinths of investigation. What we know dwindles away, when we compare it with the sum of that which we desire to know.

The true student of natural science, then, the true follower of patient, earnest, truth-seeking research, grows not bolder, but more modest, as he wins his way; he knows that his highest reach of knowledge is, and ever must be, limited; he feels each day so many wants yet unsatisfied, sees so many problems yet partially solved, or totally inexplicable, that he leans constantly towards caution, rather than rashness, and is disposed to qualify his strongest convictions on all theoretical points.

Of those who are not thus impressed by the advance of years, and the increase of experience, it must be said that their opinions cannot be entitled to great confidence. One who can promptly and confidently settle every question proposed, who has no doubts as to his own ability of decision on the most intricate and complicated problems, must be either a man who has advanced very far beyond the range of the other votaries of science in his own day, or one who is not able to appreciate the difficulties which surround him, and who is not, therefore, a safe guide. There is a third supposition in the above case, which is to consider such a man designing and unscrupulous, but this is, let us hope, the rarer alternative.

I might go on at great length, but these hints will, I think, be sufficient to show that farmers must not only have instruction, but that they must have it of the right character. It is obvious that every person who comes along, claiming to be highly scientific, should not be taken upon trust, but should be tested in some way, as to the soundness of his pretensions. Let the evidence of other scientific men be brought in, and let satisfactory proofs be required of his ability to do what he professes. This is not said with a view of recommending any particular person or persons, as to be followed implicitly, but with the desire of arousing more caution than has hitherto been exercised in these matters. "All is not gold that glitters," and all is not true science, that is high sounding.

It is for such reasons that I have said—we have not at present a sufficient number of the proper men to found and continue our agricultural schools, in a manner that will satisfy the expectations of the community. The training of such men, then, is a work of great importance, and even urgency. It is a work that cannot be accomplished at short notice; one or two years will not do it; we want those who have had extensive experience, who have availed themselves of every advantage for the acquirement of reliable knowledge, and who have learned to know what the necessities of the farmer are.

Among the wants of the farmer I consider this lack of first rate instructors, one of the most pressing and urgent; it is useless for him to establish schools, unless he can find proper teachers, and he ought not to be driven, by their premature establishment, into any dependence on those who can only mislead and disappoint him.

Here is a most promising field for enterprise and energy; here are many openings that within a few years must be filled. Those who now enter upon the study of science as applied to agriculture, will find their acquisitions in immediate demand. If but fifty or one hundred intelligent young men, would for the coming few years, devote their efforts to the acquisition of the various branches of science connected with agriculture, they would control the whole field, and be able to sweep away the glaring errors which are now so prevalent. We could then commence with schools in all directions; quackery and ignorance would decrease, and a great and rapid advance would be visible in every quarter.

Let us, then, while we are agitating the subject of instruction, not forget to urge upon our young men of ability, the advantages of fitting themselves as instructors; there cannot be too many of them for years to come, and they, therefore, need not fear that the profession will be overstocked.

# AGRICULTURAL COLLEGES

[*New England Farmer*, n.s., IV (June, 1852), 267–68.]

MESSRS. EDITORS:—For a few years past, and up to the present time, there has been much discussion in our Legislature, and also in many of the public prints, in regard to the utility and expediency of establishing an institution of the above character in this State. And if I am right your paper, the *New England Farmer* advocates the utility of the scheme. Now as I take the *Farmer*, I have an opportunity of seeing, and examining the arguments which the different advocates put forth in favor of the project. Yet I have utterly failed to see its feasibility. This may arise from a limited knowledge and a want of a better understanding of the subject. It is true, I profess to be nothing more than a humble human individual in the world, and make no pretensions to the superior wisdom and judgment of those more exalted in life. But as I just observed, I have never been enabled to see, from all the arguments which I have seen advanced upon the subject that the agricultural interest of the State would be benefited by the establishment of such an institution. At least, to that extent which it had ought to, considering the *outlay and expense which would accrue to the State in establishing such an institution and maintaining it.* And I believe if the *truth* were known, that the opinions of *nine-tenths* of the practical farmers of the State would be found to correspond with mine in considering such an institution entirely unnecessary, and would prove an utter failure, so far as it would tend to benefit their interest.

As regards the means and modes for obtaining an increased and necessary knowledge for the improvement and advancement in agricultural pursuits they already exist, and are afforded to a very great extent in the numerous books and periodicals devoted to that interest, and also in the numerous agricultural and horticultural

Exhibitions of the day. These afford a better and cheaper means of instruction than a College would. The farmers, so far as I am acquainted, do not desire any institution of the kind. They would not try to avail themselves of any of its advantages. They consider that it would not be managed upon a system calculated to improve their economical modes of husbandry. They consider that it would amount to little else than a grand and expensive experiment, whereby the State would subject itself to a great and enormous expense with no corresponding return, excepting, conferring upon a few men a rich and lucrative office, which they never were by experience calculated to fill, and an attempt to educate a few of the sons of certain families in an art which they are entirely unfitted by nature and habit to follow, and one which they *rarely* would adopt as an occupation.

I believe it to be a grand idea with the advocates of the scheme, that it will furnish an opportunity for the sons of some of our rich men, and also for some of the sons of those engaged in some of the more genteel pursuits in life, to obtain a knowledge of the art of farming.

Now *the art cannot be taught to any advantage, except by practice.* He who undertakes to teach it, ought to have acquired his qualifications by absolute practice: whether such an one would be selected to fill the Executive chair *is very* doubtful. He ought to have a familiar acquaintance with most kinds of tools and implements, appertaining to the business, and also of the various kinds of farm work, which I fear, on a personal acquaintance, would have no very *fascinating attractions* to *some* of the young gentlemen who are now so eager to acquire their knowledge. My opinion in [on] the inadequacy of the means proposed, rests upon their aversion to labor. Their previous lives unfit them for the severe tasks of the farmer. I judge partly from experience, but more from observation. The writer of these lines, has been engaged a part of his life in a pursuit much more favorable to his ease than that of his present occupation, with this consolation to comfort him, that it was *a little* more profitable. He also has been acquainted with many instances of young men from the country, who left their houses in early life to seek a fortune sufficient to enable them to return and purchase a farm. But after being engaged a few years in persuits less toilsome, and more profitable, they lost all inclination to return to their original occupation. The above are some of my reasons for believing that it would not be a *wise policy* for the State to try the experiment of an agricultural college.

Yours respectfully,

W.A.

*Newton Centre*, 1852.

# THE AMERICAN NAVAL OFFICERS' CONTRIBUTION TO AMERICAN AGRICULTURE

[Editorial by John S. Skinner in the *American Farmer*, IV (Aug. 16, 1822), 161.]

WE HAVE on other occasions expressed, and yet more sensibly felt, that the agricultural interest is already, and is likely to be still more indebted to the patriotic disposition and liberality of the officers of our Navy. Though the field of enterprize, and of *nautical* research is comparatively restricted by a state of peace; nevertheless our officers are not idle, at home or abroad.—The letters that we have often and recently seen from beardless youths in that service, really do credit to the corps—they give us unaffected descriptions of foreign towns, scenery, habits, productions, government, &c. which, if not profound, are yet sprightly and amusing, and clearly indicative of that eager curiosity which is the pioneer of knowledge and parent of attentive observation; which promises the soundest fruit, when they shall have attained the vigour of maturity. Hence we find in our officers of higher grade and more experience, the variety of solid attainments; the facility of communication, and the polish of manners which fall only to the enviable lot of those who are blessed with active minds, the opportunity of observation by travel, and the sacred ambition to employ these advantages for their own honour and their country's good. But the idea we would more particularly intimate is, that to us, there appears to be something in the nature of their avocation which leads them to *agricultural* pursuits; whenever their feelings are no longer engrossed by the all-absorbing anxieties, and thirst for professional distinction, which naturally belong to a state of war. Proof on proof might be given in support of this suggestion; hence is it, that researches are now made in every foreign country, by officers of the navy, for whatever may add variety and value to the stock of agricultural materials in their own: and in further corroboration, let us add with pride, that of all classes, having regard to their numbers, from which support could be expected to a journal almost altogether *agricultural,* the greatest *proportion* of subscribers to the Farmer, singular as it may appear, consists of *officers of the Navy.*—Again—It is to them we owe it, limited as their pay is known to be, that, within a few years past we have received through our publick ships—the fine horses of Arabia, of Barbary, and Peru—the large and spirited Asses of Malta, the beautiful cattle of Tuscany, celebrated since the days of Virgil—the sheep of Spain and Barbary—the Llamas and splendid specimens of the feathered tribes from South America—the prolific swine, and fine poultry of the East Indies, and the delicious melons of the Mediterranean, with grain, grass, and vegetable seeds from all quarters of the world. And we have no doubt but the Constellation, just returned to her native shores, has brought her contributions from the fertile coasts of the Pacific. It is true these offerings to agriculture are silently and unostentatiously made; but they are surely not the less valuable, or worthy of record and remembrance on that account. It would give us pleasure, could we do more particular justice to the individual contributors, amongst whom may now be named, without intentional slight to such as may be omitted, Bainbridge, Stewart, Chauncey, Downes, Jones, Morris, Henley, Ballard, Booth, Hambleton and Lattimer. That all these have within a few years made valuable presents to their country, in the way we have described, has fallen in most instances, casually, and in others *more directly,* under the Editor's notice—at our very elbow, more than two dozen volumes by different authors, on Italian agriculture, serve to remind us of the well applied zeal of the present gallant commander of our Naval forces in the Mediterranean, and it was owing to his accomplished predecessor in command that several pens at our late Cattle Show, were filled to the admiration of the Society.

The course of reflection in which we have here indulged, was occasioned by turning over our file of communications to the following notices and extracts respecting the agriculture of Italy—for which we are indebted to S. Hambleton, Esq., Purser of the Columbus, on her late cruise in the Mediterranean, under command of Commodore Bainbridge.

# MORUS MULTICAULIS AND MERINO SHEEP

[Contribution by a subscriber to the *Farmers' Cabinet*, III (Oct. 15, 1838), 80–82.]

THE INTRODUCTION into our country of the valuable animals, fruits, and vegetables of other parts of the globe, is an object worthy the attention of patriots and philanthropists. But it is the nature of man to run into extremes in whatever he undertakes, and when an enterprise of any kind assumes the character of a wild speculation, it is the duty of those who wish to promote the general good, and also to place that enterprise on a permanent foundation, to check every attempt of mere speculators, who would force the business far beyond a sound and healthy growth. Many instances might be cited of very plausible speculations, in which numerous individuals have been ruined both in property and character; for the injurious effects of the spirit of speculation on the mind, are not amongst the least of its evil consequences. It produces a feverish, unsettled state of mind, which prevents the pursuit of a regular and useful business; for how can a man whose brain is filled with projects by which he is to become suddenly possessed of his tens of thousands, pursue a steady manual labor employment which will produce him little more than a respectable living?

At the present moment, when a very remarkable speculation is carried on in Mulberry trees under the plausible calculation that the raising of silk is to become suddenly *the* business of the whole population of the United States, it may be considered rather presumptuous to call in question the propriety of this speculation, and to exhibit some of its probable effects. I hope, however, that the right of *free discussion* will not be denied even by those who are most deeply affected with the mulberry mania. . . .

Many of the readers of the Cabinet will doubtless remember the ruinous speculation which followed the introduction of Merino sheep, a few years ago. It was the general opinion that wool was to become immmediately the grand staple of the country—every hill and valley was to echo with the bleating of sheep—every farmer was to raise the greatest possible number, and beyond all doubt, sudden riches were to be the portion of every inhabitant of these fortunate states. But it was only a certain kind of sheep, viz: the Merino—that was to accomplish this great object—and consequently the demand for them was to be almost unlimited until every farmer in the country became supplied. The most active and enterprising individuals who were caught with the fever, were of course anxious to secure the earliest supply, and make contracts to pay the most extravagant prices for all the merino lambs that could be furnished them within certain periods. The bubble, however, soon burst, and ruin followed to the infatuated speculators and the duped farmers.

Now, instead of wool, we are told that silk is to be the grand staple of our country; every female and all the children in the land are to be engaged in feeding the silk worms, and reeling the silk; we are all to become suddenly rich, and casting away our rough woollen garments, and our vile cotton and linen fabrics, our precious bodies are to be wrapped in soft, silken folds! And as a particular kind of sheep was to work the wonders of wool, so it has been discovered that no kind of mulberry but the *multicaulis* is worth planting for the purpose of raising silk, notwithstanding the experience of other countries to the contrary. From this great discovery has resulted one of the most remarkable speculations of the age. It is rapidly extending amongst all classes, and in my view, is assuming an alarming character. We hear daily reports of individuals who have *made* their thousands and tens of thousands of dollars, rising suddenly from poverty to wealth, and the storekeeper, the farmer, the mechanic, the clerk, the teacher, are dazzled with the golden vision, and rush from their useful employment into the grand speculation. What, let me seriously inquire, is to be the end of all this? Is it not time to pause and examine the foundation upon which this new Babel is attempted to be built? That the raising of silk is an object of great importance, and also attainable in this country, will, perhaps, be generally admitted; but that it is to be universally adopted this year or next, and that it will be so profitable as to justify the present extravagant prices of the multicaulis, is not quite so clear. We are told, indeed, that the fact of three legislatures having offered a premium on all cocoons raised within their several states within certain periods, is proof enough that the *silk business* will be permanently and profitably established. It must be recollected, however, that the members of those law-making bodies are men of frail judgment, like ourselves, and may possibly be influenced by other motives than those purely patriotic, in the exercise of their legislative duties. It is well known that a number of active mulberry speculators were busily engaged during the last winter, in *boring* the members of one of these legislatures, and it seems highly probable that they were pretty thoroughly inoculated with the mulberry fever, for it is said that after passing the premium law, many of the members hurried home to plant mulberries and make their fortunes—not by raising cocoons, but by selling trees!

. . . It seems evident then that where all are planting and propagating trees to sell, and none are purchasing to feed silk worms, the trade must soon come to a close; and happy, I believe, it would be for the community, and for hundreds now engaged in the speculation, if this were speedily to happen. There are individuals within my knowledge who wish to venture all they have and more in the speculation, hoping it will continue another year at least, and richly repay their risk; thus they and thousands more will be led on to ruin. I have met with very few speculators who believe that the present extravagant prices can be long maintained; but they say we will make our fortunes while the fever is up. . . .

In throwing my views on this subject before the readers of the Cabinet, I wish to call their serious attention to it, hoping the great and rapidly increasing evil may speedily be arrested. I am opposed to all speculations—in the common acceptation of the word—believing them to be highly injurious to those engaged in them, ruinous to the community, and in the end, destructive to the interests and success of whatever may be made the subject of them.

# REMARKS ON AGRICULTURAL HOBBIES AND HUMBUGS

[Editorial in the *Farmers' Register*, VI (April 1, 1838), 47–48.]

WE HAD DESIGNED, long ago, to submit, more at length than heretofore, remarks on a class of subjects which may be comprised under the general designation of "agricultural hobbies and humbugs;" to which might be added, "agricultural frauds," if it were not improper and offensive to place in juxta-position the designing deceivers, and the deceived; that is, some of the most knavish of the agricultural profession, with many of the most disinterested, honorable, and public-spirited. But so it is—by the united operations of the cheats and their dupes, (who, being deceived themselves, innocently aid in deceiving others,) there is a perpetual succession, in the agricultural and other papers, of agricultural humbugs and deceptions, by which a few make large profits, and many find disappointment and loss. We have heretofore, though concisely, and as incidental to particular subjects, expressed opinions designed to oppose the progress of this evil; and similar views were more generally expressed by our friend, James M. Garnett, esq. in his last 'Address,' published in this work. Still, there remains an ample field of humbug to treat of, and to expose.

There are various modes of practice, in this branch of business; but the most usual is the following: Some new variety of corn, wheat, or other crop, is announced, and recommended as being superior in production and value, to any thing known before. This statement may be partially true, or altogether mistaken—and in either case, the publication may have been made with honest and praiseworthy intentions. On the other hand, the first motive of the publication, as well as the final effect, may have been to make dupes, and to get hold of their money. According to these, or other circumstances, the early publication either contains, or is afterwards followed by, a notification that Mr. Such-a-one offers to sell seed or plants of this rare and valuable variety, at a price three or four times, or it may be ten times, as high as is usual, or as would be a sufficient remuneration for the raising. The enthusiastic and credulous, among the readers of these highly-wrought panegyrics, are put all agog to obtain seeds which, as represented, alone will add 30 per cent. to their crops. The regular seedsmen hasten to give their orders, that they may profit by the new demand, just as milliners would for bonnets of the latest fashion, and knowing well that the fashion will be as likely to be transient in the one case as the other. The editors and publishers of agricultural journals, who are not themselves seedsmen, find that articles on these wonderful new products are interesting to many of their readers, and very convenient to fill their columns; and therefore, they help on its way every successive humbug, and the more wonderful the account, the more sure it is of being selected for republication. But if the publisher is himself engaged, directly or indirectly, in the selling of seeds, &c. (which is a usual, because a very convenient and profitable combination of trades,) then he can best profit by his readers' appetite for novelties and wonders; for he can not only start and direct the puffs, but such recommendations are, beyond comparison, the most effective of advertisements, because not suspected to be such, nor to be otherwise than the honest and disinterested opinions of the writers. . . . If we were to discard our scruples, and, instead of neglecting or opposing, were fully to sustain the puffing system, our publication would have many more articles of temporary interest to readers, and in more ways than one, would be productive of much more profit to the publisher. . . .

Do not let us be understood as denying all value to these overrated novelties—nor as wishing to discourage trials of new plants, or new (supposed) improvements of any kind. On the contrary, we would encourage trials, if made carefully and accurately, as being calculated not only to amuse and gratify all cultivators of inquiring minds, but also as sometimes leading to results which, if correctly appreciated, will be of important value to agriculture. But we *do* mean to avow a very general distrust of these many newly discovered values in particular varieties of seeds, &c. and especially when it is manifest that he who makes or sustains the recommendation of the article, has a private and pecuniary interest in raising its reputation, or maintaining for it a high selling price. Let all the attendant and connected circumstances be borne in mind, and the trials be economically, as well as cautiously and accurately made, and we would urge the trial of every new thing that was even plausibly recommended. . . .

# CHEMICAL AGRICULTURE

[*Prairie Farmer*, XV (July, 1855), 201–2. This editorial is characteristic and indicates the reaction of the period to the overemphasis on chemical agriculture.]

It is a notable fact that there is in the Agricultural papers a considerable waning in regard to the quantity of Chemical literature applying to agriculture; as compared with the same journals a few years ago. There was a time, when in most of the leading papers of this class—and in some which did not lead—the body of their teachings, was but the iteration of Chemical lore. Liebig, Johnston and the like, were the men in authority then; and if one did not believe all they said, he had better be silent.

The thing seems somehow to have considerably changed though in a very silent way; and now with the exception of the Working Farmer—J. J. Mapes' paper—no one of them accords to chemical teaching any other than a proportionate space or consideration. *Professor* Mapes indeed blazes away at it as usual, seeming to think that the Moon *is* of green cheese whatever outsiders may say and that you have only to cut with your knife, and cut away to your liking. Let him cut; he will do no harm and possibly some good.

Eight years ago and there abouts, Chemical theories were the burden, not only of professional men, but of multitudes of writers, who took up the subject on a mere reading, and then followed these theories into their various application to Agricultural practice. According to some of these farming was the easiest thing imaginable. All one had to do, was to take a pound of soil from a hundred acre field, give it a run through the crucibles—a thing which anybody could do—see what it was made of, and then fill up the deficiencies of the field short hand. It was just as easy as tailoring, when the tailor takes your measure, and then proceeds to give you a fit. This was a most beautiful theory; and had the practical results been at all equal to it, we had now been getting one hundred bushels per acre of wheat, and five hundred of corn, just as sure as Christmas.

Our people cannot be fooled long. They always want some cider for considerable talk; and after this thing had been sufficiently preached, some crops were expected to correspond with the preaching. Alas the crops would not come. Like Owen Glendower, who "called the spirits from the vasty deep," Chemistry called for the crops from the tinkered earth; but as the spirits did their liking toward Glendower's invitations, so did the corn and potatoes, when summoned across lots. Field manufacture was a failure. Analysis did its best, and could tell precisely what the earth wanted; but when Synthesis took hold to supply the want, somehow it came out wrong. Nor are we aware to this day, that analysis has been able to do much in this work of supplying the deficiencies of barren lands. We are not aware that any deficiency has been remedied by it which mechanical examination has failed to do. We know that *Professor* Mapes tells large stories of this sort; and so does *Professor* Comstock of Terra Culture; and we have as much faith in one class of stories as in the other. Were such cases common, somebody else would know it also. It must be admitted that the hopes a few years ago entertained, that chemical science would soon be able to remedy barrenness by analysis, have not been, nor are they likely to be, very soon realized. The difficulties have been already set forth fully in our columns, and we have no desire of repeating them now. We only call attention to the fact, as one of the varying phases of Agricultural progress worthy of attention. Chemical Science has doubtless many applications to Agriculture; but a remark made by us years ago in this journal, holds still true, that its benefits are to be looked for in detached and fractional instances, and not in any grand or general way; as if a railroad were opened up through the Agricultural kingdom, when any body might get on and ride at his leisure. Good land will not be made out of poor, by any empirical process, at a cent an acre. Observation and experience are worth something yet, and are likely to be though Charlatanism flouts them.

# THE LATER ARCHITECTURE

[Editorial in the *Prairie Farmer*, XIII (Jan., 1853), 2–3.]

IT IS a fact that within ten years or thereabouts, the style of our buildings, more particularly of our dwellings, has undergone an entire change. One need not go out of this city to ascertain that fact, but if he does travel in almost any direction, he will find it confirmed by a large proportion of the newly erected dwellings he sees. Fifteen years ago, the only style common was the Grecian, with its unvarying features of close cornices, smooth window-frames, and columnar verandas. There were some deviations, it is true, from this, in the more ambitious residences of such as were able to indulge the fancy; sometimes with due regard to correct taste, and sometimes without it. A readiness to adopt a new and more expressive style, began to grow up with the revival of Agricultural Literature, which in the very necessity of the case, prepared the way for whatever changes should be required by utility, or a better disciplined taste. Agriculture and Horticulture are cognate subjects. They are twined into each other, and are naturally treated in the same pages, and by the same instrumentality. Architecture is equally inseparable in connection with both. There is no such thing as a farm without a house, and a garden without its dwelling, is all but an inconceivable idea. Hence as one of the first things claiming attention in our new routine of topics was domestic Architecture. As quick as people began to find out that they had any taste in matters of building, they discovered that it was either unsatisfied or absolutely outraged with the prevailing styles. Hence when a writer competent to the task made his appearance, the country was ready for him. Had Downing arrived thirty years ago, he would have preached to "thin congregations." His books would have gathered dust on the shelves, awaiting, in neglected slumbers, the "good time coming." He came at the right time for himself, and for his countrymen, and his books at once were hailed as the help for which they had begun to look about them.

But as in the beginning of the application of new ideas, in all the world's history, so here, many of the first innovations were made in utter defiance of the laws under which the innovators had enlisted. The old and the new were often jumbled together in a way not much better than that condemned by the old Roman satirist, when the head of a fine woman is tacked to the body of a scaly fish. Good work was made of improper materials, and bad coloring spoiled good work. It took some little time to get newly awakened and unsettled ideas arranged in order, and in consistency with themselves. It took, and is taking longer yet to get the common idea sufficiently simplified as that it should not be striving after unattainable things; and thus to fail of what can be easily reached, if we will cease to look above it. These and other mistakes, all natural enough, have been made, and will continue to be made for some time to come.

But we have gained all that is implied in the first steps of improvement. The stupid and unvarying Grecians with that same gable to the street, has yielded, in the first place to—variety. The American is a mixed character, and personal independence is no where else so insisted on. It is, therefore, becoming that this truth should be expressed in our Architecture. We can tolerate uniformity in nothing else; why should we be condemned to dwellings of one particular cut and pattern? But we have got beyond that, and buildings of a simple and intelligent beauty, are appearing in almost any district of the country—buildings expressive of the uses to which they are put, and harmonizing with the beautiful in the surrounding scenery. These things perhaps carry to the mind of the uninitiated, an idea of a greatly enhanced expense; but such is not by any necessity the fact in regard to them. The change is made by an alteration of a very few particulars. One is the overhanging eave with its attendant bracket, which now is thought essential in almost every house of any pretentions, where the innovator has penetrated. Another is the style of the veranda, with its supports. These, instead of the huge and unsightly columns so ambitiously coveted in former years, consist now of simple standards of some neat and lightly wrought patterns. Another is the style of the windows with their frames, another is the frequent and expressive use of gables breaking up the monotony and furnishing occasion for more or less of ornament, when it can best be used; another, is the change of color. From the uniform and glaring white, we have travelled through a sea of dingy browns and hideous brindles, till we have at last got among quiet neutral hues, which all are bound to love as soon as we come to know them. That the subject will now take care of itself, with the hold it has on the public mind, we have no doubt.

# The Times

## COGITATIONS OF AN OLD FARMER

[*American Agriculturist,* XVII (Jan., 1858), 3–4. This represents a characteristic analysis of the depression of 1857.]

HARD TIMES! So, everybody says; and so say we—for the "times" do bother us, as everybody else is bothered. Possibly we may not be so badly damaged as some others, but we know enough about the hard scratching which they inflict upon us to wish they were otherwise. We have had "good" times too, and quite a run of them for several years, until a few months ago. So the same "everybody" told us time and again. Yes, they were good times. We had free-trade, and free-credit abroad; and we used it freely too, with a vengeance. We have built a long array of free rail roads, free to the select coteries of speculators who got them up for their own especial benefit, mind you, on bonds which were gobbled up by the usurers with decided *freedom.* The roads gave free passes to the legislators, and judges of the country, as well as to various editors, for which, the little share that we had in the riding we shall never cease to thank them. We imported millions of free goods that we did not need, but which we have made out to wear, and eat, and drink, and dispose of in one way and another; and the beauty of it is, those which are not paid for, or used up—and they are many—the owners are free to send back to where they came from, as many of them probably will, or let the goods lie a long while in the bonded warehouses, awaiting better times for sale and consumption.

The truth is, for the last eight or ten years we have built extravagantly, dressed nonsensically, lived lavishly, speculated wildly, trusted everybody about us, as we got trusted abroad, and "laid loose" around generally. Our farmers got great prices for their produce to feed the fools and tyrants who were doing up their own fighting in Europe; and they got such prices so long that they supposed they were always to have them. Our towns were so prosperous, and people in them got rich so rapidly that a vast many others, old and young, who were doing well enough on their farms and thought they could do a great deal better in town, left them to know little peace or quietude afterwards. Our women and girls quit spinning stocking yarn at home, and took to spinning street-yarn, and wearing crinoline abroad. Instead of thumping the clothes in the pounding barrel in the kitchen, they took to thumping the piano, and the melodeon in the parlor; while the boys, and "Young America," took to "fast horses," "long nines," "cock-tails," and a general "cut-up," all round the board, and so went the world. . . .

. . . In sober truth, we must "settle up," and again go to work. We must cease importing goods we do not want; we must abandon superfluities we do not need; we must stick to our farms, our workshops, and our trades, whatever they may be—if we can get a living by them—and if we can not do that, take to those at which we can. Instead of earning one, five, or ten hundred dollars a year, and spending more, we must earn all we can, and spend less. That is the only true and honest way to fortune. A great master of human life has said:

> "Sweet are the uses of adversity,
> Which, like a toad, ugly and venomous
> Hath yet a precious jewel in its head."

He did not know much about toads, however, for they are decidedly good things in a garden.

# THE PRESENT AND FUTURE OF SLAVERY—NO. 2

[A defense of slavery by a Southerner found in the *American Cotton Planter*, XIII (March, 1859), 73–75.]

NEARLY the whole exports of the country are products of slave labor and manufactures of slave products. Of these, the chief is Cotton.—Now, it has been demonstrated by experiment that the semi-tropical productions, Cotton, Sugar and Tobacco, can be produced in large quantities by slave labor alone. Abolish slavery and their production, as commodities, would cease at once. How often do we hear the remark that "cotton is the only thing that brings money into the country." It freights our steamboats and railroads; furnishes employment to thousands of men in its transportation and shipment, feeds the miner, the ship-builder, the machinist, and many another artesan besides. What then would be the consequence of the cessation of its production? Evidently this almost countless host would be thrown out of employment in our midst, and must necessarily come in direct competition with the labor of all "who have no interest in the continuance of slavery, because they do not own a slave, nor ever expect to." But these will not be your only competitors. The liberated slaves too must find employment of some kind, and there is no cotton to bring money to pay wages. You are, then, most certainly deeply interested in the continued production of cotton. You all know that the free negro will not work steadily and constantly. Can the whites take his place in the cotton field? A single week's cotton-picking during August and September in the Mississippi, Alabama, Red River, Arkansas, or Yazoo Valleys, would kill any white man. All the best cotton lands in the Union would of necessity be abandoned. Every inhabitant of a cotton-producing State is directly interested in the continuance of its culture, and since it can be cultivated only by slave labor, he is equally interested in the continuance of slavery. The same may be said of the inhabitants of the sugar and tobacco States.

If property in slaves were struck out of existence, would you be the richer? Nay, so far from it, you would partake in the ruin and desolation it would bring upon your country. It is clearly the interest of every man South of Mason and Dixon's line to uphold slavery, as indissolubly connected with the prosperity of the South.

The Constitution of the United States guarantees to the citizen of one of the States all the rights and privileges of a citizen of all the States; yet this principle was violated under the sanction of law, and with our own consent by the "Missouri Compromise." (Though why that should be called a compromise, where all is yielded on one side and nothing on the other, passes my understanding.) True, by it we yielded only an abstract right; and that for the sake of peace, and the Union; but our enemies used it as the stepping stone, the vantage ground, from which to attack other rights.

Then followed "the Omnibus," which gave the sanction of a statute for the performance of that, which was secured to us by the Constitution—the return of fugitive slaves. For this BOON(?) we consented to the abolition of the Slave Trade in the District of Columbia—another "Compromise," in which we gave all, and received nothing, for love of the Union.

What is the next blow? "The Kansas-Nebraska Bill"—which was to be the healing of sectionalism. Under this belief how eagerly did we swallow this Douglas pill, that gave only the *promise* of future good for what we surrendered. Still we were willing to yield for the sake of peace. What did we gain? A ten-fold fiercer war upon us and our rights.

Thus, one by one, has the North demanded our rights under the Federal Constitution, and thus has the South yielded them, till emboldened by success the great leader of the crusade against us does not hesitate to declare that he will not cease his efforts till he has deprived us of the rights we hold under our State Constitutions, till slavery is abolished. This was no idle vaunt. This very declaration secured the election of the Black Republican candidates in the Empire State, by nearly 20,000 votes.

Are we still to go on yielding? If not, where is to be the stopping point? "Are we to lie supinely on our backs till our enemies have bound us hand and foot?" The past is irrevocable; the future is our own.

I love our glorious Union for the unnumbered benefits it has heaped upon us. It has fostered our growth and progress, till we have become a mighty people, both in arts and arms. I venerate the memory of those sages who framed our Constitution, and wonder at the wisdom displayed in devising the most perfect instrument of government the world ever saw, or ever will.—They designed it to secure the greatest amount of happiness to all the citizens of our country. How admirably was it adapted to that end? . . .

When the British Parliament essayed to rob our forefathers of their rights and property, what did the revolutionary heroes—how did they compel the repeal of those odious acts which taxed them without their consent? What measure thwarted the whole British power for ten years? THE RESOLUTION NOT TO USE ANY ARTICLES SUBJECT TO DUTY. The same remedy is open to us. The ancestors of the very men, who now seek to crush us have pointed out the only effectual means of *peaceful resistance* to the aggressions of their degenerate descendants.

Form societies in every county throughout the entire South, pledged not to use a single article produced or manufactured in a Northern State, "till our grievances are redressed," and our National Legislature is purged of its crowd of "negro-worshippers." It may be said that the ladies will not be willing to lay aside their finery for such a purpose. When did the ladies of the South hesitate to make any sacrifice in the cause of patriotism? Never! nor will they now. Indeed, I am much deceived if they are not found foremost in such a course when the crisis arrives. . . .

Who shall set this ball in motion? Will our *prominent* men, our *politicians!* No! the fear of unpopularity will ever be the phantom to deter those whose life is the popular breath, from attempting anything out of the common track—*as if anything in the common track could meet the exigencies of our case.*

We must do our own work. The planter, the mechanic and the farmer, must both set it in motion, and keep it in motion, till our purpose is accomplished. Our politicians must be our mouth-pieces; they must utter our sentiments, and not lead us to adopt theirs. Those who are now in public life are too wedded to old ideas—too much fettered by precedents, to deal with this question properly. They have looked with veneration upon Compromises, but the day of Compromises is past.

Bonaparte committed his fortunes in his last conflict, to those who had grown grey in service, to those who had attained to wealth, fame and rank—and he fell—and so will it ever be. Choose the young and ambitious for new and bold exploits; *if we would succeed we must commit the conduct of this question to the rising generation.* Can we trust our Southern delegation in Congress with the task of stopping the wheels of Government if a "Negro-worshipper" is chosen President in 1860? I for one will not vote for a member of Congress for 1860, who will not pledge himself not to take his seat if a Black Republican is chosen President at the next election. I will never be disgraced by my Representative acting on an Abolition Message. Ours is emphatically a war of defense, and let us employ every means in our power to defeat our enemies. But says one, this is disorganizing the government. Is the North forever to say to us, "Heads I win, tails you lose"? They have in practice freed themselves from the obligations imposed by the Constitution, and shall we continue bound?

Away, with such sophistry! We are already free by their own acts, for the violation of a contract by one party frees the other.

Some may say, "After our fathers had thwarted the purposes of the British ministry for ten years, they were forced into a bloody war to maintain their rights, and our resistance may have the same result." Let us do our duty to ourselves, our children and our country, and trust the consequences to the All-wise Ruler; if war should result, it would not be of our seeking—and as Patrick Henry once said: "Three millions of people, armed in the holy cause of liberty, and in such a country as we possess, are invincible by any force our enemy can send against us."

A—

*Enon, Miss.*, Dec. 22, 1858

# THE MORAL CULTURE OF SLAVES

[Editorial in the *South-Western Farmer*, I (Feb. 3, 1843), 169.]

THE FANATICS of the North and of foreign countries have no idea of the deep anxiety entertained by our planters upon the subject announced at the head of this article. The slaves on our large plantations are not regarded by us as so many beasts of burden or dumb brutes, as is often falsely stated; but they are regarded by the owner, provided he be an intelligent and upright man, as moral beings whose welfare, both mental and bodily, are entrusted to his charge by an all-wise Providence. To him they look up, not only for food and clothing—but for such instruction, by precept and example, as will give a stamp, either for good or evil, to their characters. Principles, either salutary or pernicious, are at the foundation of all conduct—and even slaves may be taught to do right on principle. Let no one sneer at this—for it is true. We appeal to the experience of every intelligent man, born and reared among the colored race—to say whether we utter any thing absurd—nay, whether we tell not that which is as clear as a sunbeam—that a negro can, by proper instruction, be taught to act aright, *from principle.*

This being premised,—which, we have no doubt, will be admitted by all our readers—a question arises which, of course, will be solved in many different ways—that is,—how shall the minds of slaves be operated upon so as to instil into them correct principles of action?

On this subject we have reflected much, and we have inquired much of those who have had the best opportunities of judging; and the result is, that we are confirmed more strongly than ever in the opinion we have long entertained, that the best way to operate upon their minds, is, *through religious truth.* The reasons of our opinion it is hardly necessary to give, we presume; for we think that the most of our readers are men who believe and acknowledge the divine origin of the scriptures.

The great duty, then, to be performed, is—not to convince the Mississippi public of the importance of instructing their slaves in the principles of religion—but, to point out the best methods. Here, happily, we are not left in the dark, to grope our way with untried theories—for, many planters, both in this state and Louisiana, have long been pursuing a systematic plan in this matter, and, as we are happy to learn, with the most pleasing results. Indeed, so far back as 1835, when the lamentable excitement took place in this country, in consequence of the mad schemes of the negro's worst enemy, Abolitionism, measures were in extensive progress upon many of our plantations, which bade fair to produce highly beneficial effects upon our black population. But the benevolent schemes of our planters were nipped in the bud; and the melioration of the moral condition of their servants was set back at least ten years.

To instruct our negroes in the truths of the Bible, it is not necessary to teach them to read. By mere oral instruction, they have, in many instances, become well versed in the doctrinal, preceptive, and especially in the historical parts of the sacred writings—also in the catechisms—while, in the committing of hymns to memory, in learning sacred music, and in other devotional exercises, their quickness is proverbial. We well recollect hearing a negro preacher, a slave, many years ago, who went through all the usual clerical exercises with considerable cleverness—giving out his hymns, line by line—announcing his text, and directing his audience to chapter and verse—all this, too, without using any book, and without being able to read if he had had one.

To us, it would be a most gratifying task to travel around for a while from plantation to plantation in those sections where a systematic course of religious instruction is pursued; for we believe that a publication of the facts which might thus be obtained would be one of the most acceptable, because most profitable services we could perform to the agricultural community. Such a tour, however, is impracticable—and we must therefore appeal to others, who are conversant with facts, to correspond with us, and furnish us, for publication, with such information and such suggestions, in regard to this subject, as may benefit the public.

In Adams county, especially, the good effects of religious instruction upon the slave population have been experienced in an eminent degree. In one particular neighborhood, known by the name of *Second Creek*, we have heard of a degree of attainment by mere oral teaching, which many would deem incredible. But, the enlightened planters in that neighborhood, not less distinguished for their intelligence than for their moral worth and their ardent piety, have patiently pursued their system for a series of years. They have employed ministers of the gospel, thoroughly furnished for the undertaking, not to excite the mere feelings of their slaves, but, by a course of persevering effort, to inform their minds—to make them understand the responsibilities of their state as moral beings—and to teach the young children the rudiments of Christianity by means of catechisms and hymns. The result is, that the slaves are more tractable than if they were not instructed—and, as a proof of this, it is stated that it is very unusual for the planters who have adopted this mode to employ an overseer. Nor is this to be wondered at; for, if the master and the servants are alike brought to that happy state of mind whereby they act continually with reference to religious accountability, the necessity of an overseer is well nigh superseded.

In indulging a lively wish for the religious instruction of our servants—in the desire to see a solemn *duty* performed—it is gratifying to know, at the same time, that we are only urging a measure of *interest*—the interest of the owners and the community at large. And we earnestly hope that self-interest, if no other consideration, will open the way for ministers to enter more generally, and more frequently, and with more devotedness, upon this interesting field of labor.

Will some friend in Adams county, who is well informed on these topics, furnish us with a communication upon the subject? In Madison, Yazoo, Warren and Claiborne, as also in Louisiana, we believe that considerable attention has been directed to this subject —and from those points also we should like to hear.

# RULES FOR OVERSEERS

[*Carolina Planter*, I (Feb. 26, 1840), 49–50.]

Mr. Editor:—

When I employ an Overseer, I read to him the rules of my plantation, and give him to understand at once that he is to be governed by them. It is part of my contract. Those which relate more particularly to the Overseer's duty, I send you for publication. I cannot help thinking that if such rules were generally adopted and enforced, this useful and necessary class of men would be improved.

Franklin.

## RULES OF THE PLANTATION

1st. A good crop means one that is good, taking into consideration every thing—negroes, land, mules, stock, fences, ditches, farming utensils, all of which must be kept up and improved in value—the effort therefore must not be merely to make a given number of bales of Cotton, but as much as can be made without interrupting the steady increase in value of the rest of the property.

2d. The Overseer will never be expected to work in the field, but he must always be with the hands when not otherwise engaged in the employer's business—and will be required to attend on occasion to any pecuniary transaction connected with this plantation.

3d. The Overseer must never be absent a single night or an entire day without permission previously obtained. Whenever absent at church or elsewhere, he must be on the plantation by sun down without fail.

4th. He must attend every night and morning at the stalls and see that the mules are watered, cleaned and fed and the doors locked. He must keep the stable key at night, and all the keys in a safe place, and never allow any one to unlock a crib but himself. He must endeavor also to be with the plough hands at noon.

5th. The Overseer must visit every negro house at day light in the morning and see that all are out. Once a week or more he must visit them after horn blow at night, to see that all are in. Once a week he must visit every negro quarter after night. The horn will be blown in winter at 8, in summer at 9 o'clock, after which no negro must be seen out of his house.

6th. The Overseer will be expected not to degrade himself by charging any negro with carrying news to the employer. There must be no news to carry. The employer will not encourage tale bearing, but will question every negro indiscriminately whenever he thinks proper about all matters connected with the plantation, and require him to tell the truth. When he learns any thing derogatory to the Overseer he will immediately communicate it to him.

7th. The Overseer must ride but one horse unless he obtains permission to do otherwise. And as the employer's business will require his whole attention he is expected to see but little company.

8th. He will be expected to obey strictly all instructions of the employer. His opinion is requested on all questions relative to plantation matters as they arise, and will be treated with respect, but when not adopted he must cheerfully and faithfully carry into effect the views of the employer, and with a sincere desire to produce a successful result. He must carry on all experiments with fidelity, and note the results carefully, and he must, when instructed by the employer, give a fair trial to all new methods of culture and new implements of agriculture.

9th. The whole stock will be under the immediate charge of the Overseer, and he will attend to them personally, with the assistance of any negro he may choose from time to time. He must see and feed every hog at least twice a week, and salt and count the cattle once a month. The Overseer must without being asked inform the employer of any thing going on that may concern or interest him.

10th. The negroes must be made to obey and to work, which may be done by an Overseer who attends regularly to his business, with very little whipping; much whipping indicates a bad tempered or an inattentive manager, and will not be allowed. The Overseer must never on any occasion, unless in self defence, kick a negro, or strike him with his hand, or a stick, or the but[t] end of his whip. No unusual punishment must be resorted to without the employer's consent.

11th. The sick must be treated with great tenderness, and visited at least three times a day, and at night if necessary. The greatest attention must be paid to all the children, and they must be kept clean, dry and warm by the nurses. Suckling and pregnant women must be indulged as much as circumstances will allow, and never worked as much as others. Sucklers must be allowed time to suckle children, and kept working as near the house as possible. No lifting, spinning or ploughing must be required of pregnant women.

12th. The use of spirits is absolutely forbidden on this plantation, unless when prescribed by a physician—not even a Christmas dram is allowed the negroes. Should the Overseer get drunk he must expect to be instantly discharged.

13th. The Overseer is particularly enjoined to keep the negroes as much as possible out of the rain and from all kinds of exposure, and to see that they make good fires in cold weather and after rains.

14th. The Overseer must not punish the driver except on some extraordinary emergency that will not allow of delay, until the employer is consulted. Of this rule, the driver is, however, to be kept in entire ignorance.

15th. It is distinctly understood in the agreement with every Overseer, that whenever dissatisfied he can quit the employer's service, on giving him one month's notice in writing—and that the employer may discharge him at any time, by paying him for his services up to that period, at the same rate as he agreed to pay for the year.

# STICK TO THE FARM

[Henry F. French, in the *Country Gentleman,* III (April, 27, 1854), 268–69.]

. . . You ARE TEMPTED to exchange the hard work of the farm, to become a clerk in a city shop, to put off your heavy boots and frock, and be a gentleman, behind the counter! You, by birth and education, intended for an upright, independent, manly citizen, to call no man master, and to be no man's servant, would become at first, the errand boy of the shop, to fetch and carry like a spaniel, then the salesman to fill the place which at best, a girl would fill much better—to bow and smile and cringe and flatter—to attend upon the wishes of every painted and padded form of humanity—to humbly suggest to rakes and harlots, as well as to starched and ruffled respectability, what color and fabric best becomes the form and complexion of each—and finally, to become a trader, a worshipper of mammon, as Carlyle says, "a kind of human beaver that has learned the art of ciphering," compelled to look anxiously at the prices current of cotton and railroad stocks, in order to learn each morning, whether you are bankrupt or not, and in the end, to *fail,* and compromise with your creditors and your conscience, and sigh for your native hills.

Or, perhaps, your party being in power, you would obtain a clerkship at Washington, and remove your little family from the north to a more genial climate, to live at your ease, and grow rich on twelve hundred dollars a year! You give up your little farm, your New England privileges of schools and churches, your independent and influential membership of parish, and district and town and church, the woods and playgrounds for your children, your friends and kindred and *home.* . . .

You have left your home. At the end of a single year in "the city of *magnificent distances,*" you have bitter realizations of the meaning of that phrase. It has proved indeed to be full of magnificent distances, for you, from happiness, from independence, from advantages of every kind. For the first time, you have felt how sore a thing it is, for a northern freeman to be dependent, to labor at stated hours, at the bidding of a superior officer, to feel that the office you fill, on which depend your very means of living, for yourself and family, is held at the arbitrary will of another, who may, if he please, make a servile conformity of your views with his own, on political or what you may deem moral questions, the condition, by which you retain your place. . . .

Look now, at the prices of necessary articles of food. On your farm, however small, your cellar was always filled with an unlimited supply of all such vegetables, as you desired, and barrels of beef and pork of your own slaughtering. . . .

We might follow this train of thought into further details, did time allow it, but enough it is hoped, has been said, to induce an independent Northern farmer to hesitate long, and consider well, before he exchanges his position for any place, where any *master* comes between him and his Maker. . . .

# THE AMERICAN FARMER
# A PORTRAIT

[*Franklin Farmer*, III (Oct. 12, 1839), 61.]

THE FOLLOWING beautiful description of our most worthy and happy class of citizens, is from the address of the Hon. Mr. Rowan of Kentucky, to the citizens of Louisville. If any one would enquire why farmers are more worthy than any other class, this description will answer the question. They follow that profession which is most essential to the public good, and which is the foundation and support of all other useful employments, and the situation of farmers leads him to a higher state of moral worth than any other class will attain.

—*Yankee Farmer.*

"Who is there among us that beholds the condition of our farmers and does not exult in the consciousness that he is an American citizen, and pant to superadd the character of a farmer? The house of the farmer is the abode of the virtuous.—It is a school in which lessons of practical wisdom are taught. It is a temple in which the precepts of our holy religion are inculcated. It is the castle of sovereignty, for it is owned by its occupant and he is a freeman. It is the residence of peace, order, harmony, and happiness. Patriotism and piety unite in consecrating the place, and in suffusing every countenance with their unction.—Indeed, what condition in life is so likely to produce that patriotism which will stand the country in stead upon emergencies, or that piety which will afford solace in extremity, as that of the farmer? He occupies a constant, intimate, and sensible relationship with Heaven. His mind is subdued with a love of order, by constantly beholding that which prevails around him. The regular successions of the seasons, of day and night, and of seed time and harvest, admonish him to the observance of regularity and order in all his conduct. He perceives that the sun and moon perform their circuits without loitering on the way; and learns from them that industry is required at his hands. . . . Matron chastity, and infantile innocence sweeten and religion hallows the atmosphere of his home, and render it irresistibly attractive. . . . His patriotism is an essential part of his conscious identity. Connected by his affections with the soil, and by his piety with heaven, it partakes of the stability of the former, and the purity of the latter."

# November

## DUTIES OF THE MONTH

[*Indiana Farmer*, VII (Nov., 1858), 225–26.]

THE PRINCIPAL WORK of the farm during this month is to gather the corn. The late and unfavorable spring put the corn back so that it will be unfit to crib until quite late; when it is sufficiently hardened all the available force should be set at it.

The best way to gather corn is to have two teams, one unloading while the other is loading; two men and a boy to gather and one to unload; take five rows at a time—drive over the middle row—the two men to gather the side rows while the boy gathers up the middle row behind the wagon; then throw it into the wagon, not on the ground to be picked up afterward, which makes double work and fills the corn with dirt, so that stock does not like to eat it.

Potatoes should be taken up and put in the dry before this time. They should always be dug and put away before the cold fall rains; they are easier dug, dryer and less dirt will stick to them, and they will keep better.

Now, too, is the time to see that all the stock is well furnished with shelter.—Cold nights are coming on, and stock feel the effect of and suffer as much from the cold rains of November as from the dry cold blasts of January. And if you are not careful, your stock will be losing flesh before you are aware of it, and in consequence will be harder to winter.—Most farmers provide shelter for their horses, cattle and sheep, while their hogs have to take care of themselves. Now, this is wrong. Hogs suffer as much and you lose as much by not providing a good warm place for them to sleep as by any other stock. To convince you of this fact, all we want is that you divide your hogs into two equal parts—provide for the one a good warm place for them to sleep, and let the other lot run out; feed both lots alike, weigh both in the fall and again in the spring. If this does not cure you of letting your hogs run out, then we must say you are past cure. The one lot will look thrifty in the spring, and have gained say from five to eight pounds of pork for every bushel of corn fed, and you perhaps have not lost a single hog from this lot, while one third of the lot that had to take care of themselves have perhaps died, or been killed by overlaying, while the remainder are in much poorer condition than they were in the fall, and weigh much less.

Now, too, is the time to plow for oats, barley, and in many cases for corn. The advantages of plowing in the fall for spring crops are well established. In the first place the alternate freezing and thawing of newly plowed land pulverizes it better than harrows and rollers can do it. In the second place it enables the farmer to get his spring crops into the ground as early as nature requires them to be to avoid drouth, rust and frost, and to obtain the best yield.

Whilst preparing for your stock and your farm, don't neglect your family.—They, too, must be warmly clad if you expect them to pass safely through the winter season. Especially should their feet be well protected from cold. Though your farm may be all underdrained, subsoiled and manured—though your barns are full of grain, and "the cattle on a thousand hills" are yours—what praise is due to you if your children, for whose welfare principally you claim to do all this, are consumptive from exposure of their feet, lymphatic in their temperaments from want of external warmth during the winters of their youth, and scrofulous from both these causes and from a want of healthy diet? The fact that you and your parents lived mostly on hog meat, bread and potatoes, is no reason why more than these—and especially beef and mutton, with cheese and good fruits—should not be allowed your children. Indeed, the fact that this generation is less in size and less enduring than our pilgrim fathers, is a strong argument against the *"hog and hominy"* regimen of Americans. If you will do justice to your family, as well as to your stock, now is the time to prepare supplies of vegetables and fruits in places convenient for access during winter.—The beef, mutton, fowl and cheese, can be obtained when needed,

# NOTES OF TRAVEL IN THE SOUTHWEST, NO. VII

By Solon Robinson

[*Cultivator*, n.s., II (Oct., 1845), 303–4.]

ONCE AGAIN, my friends, I come with my monthly greeting. Well, where parted we company last? Let us reflect. We had just visited Mr. Leigh, and given a slight sketch of his method of *farming*, which I have italicised to give the term a contradistinction from that of planting—the latter term meaning only the cultivation of cotton. But before leaving Mr. Leigh's neighborhood, I must notice that I was on President Polk's plantation, and earnestly hope that his cultivation of Uncle Sam's big plantation will be as well managed under the overseership of Mr. Polk, as his Mississippi cotton plantation is reported to be. The next point of interest that I visited was the plantation of Captain Wm. Eggleston, of Holmes county, who is one of the good farmers of Mississippi. He is a Virginian, from Amelia county, and having an introduction from his friend, Mr. Leigh, I met with a very hearty reception.

The 17th of February was an uncomfortably warm day. The peas in Captain E.'s garden several inches high, lettuce in full head, and other things in proportion. Captain Eggleston has about 1,400 acres of land under cultivation, and upon which live 20 whites, and 150 blacks, 70 of which are field hands; about one-third of his land is kept in corn and oats, the proportion of corn being as two to one. He keeps up a continued rotation of crops, and puts all the manure that he can upon the corn, which averages about 25 or 30 bushels to the acre; plants corn and sows oats in February. He is now working 43 mules and horses, and 28 oxen, and makes 560 bales a year, which he has to haul 10 or 12 miles.

He also raises all the grain and meat required upon the plantation, feeding his negroes at the rate of 3½ lbs. clear bacon per head per week, with about a peck and a half of corn meal, besides vegetables and fruit, melons, &c. Like Mr. Leigh, he gets his flour from Virginia, and asserts that no other will keep well through the summer.

I saw in his garden some very fine fig trees, which as far north as this produce remarkably well. Peaches are unfailing, but with grapes he has not been successful. Apples are not a southern fruit, yet many are attempting their cultivation. And now a word of Captain Eggleston's system of cultivation. His place is all hilly, thin, oak land, very light soil, that melts away in water not *quite* so easy as salt or sugar; and yet he has scarcely a gully upon the whole farm; but he has more than 20 miles of side hill ditches, which are so constructed that they take up all the surface water before it passes far enough over the ground to form gullies.

While riding over the plantation, I found one of the overseers engaged, with a large force of hands, laying off and making ditches upon some new ground, it being a rule never to put in a second crop until the land is ditched. . . .

. . . .

As before remarked, the rows have to conform to the ditches, however crooked, and the manner of plowing is to lay off the rows in the first instance, the middles often being left unbroken until after the corn is planted, and perhaps up. Captain Eggleston's plan is to plow deep directly under the corn, and plow shallow while tending the growing crop. His motto is to plow deep for all crops. He assures us that since he has adopted the level system of ditching and plowing, that in addition to the advantage to the land, that his crops are better and the soil improving instead of deteriorating.

All of his mechanical work is done by his negroes upon his plantation. He has two negro carpenters that he occasionally hires out to others at the rate of $40 apiece per month. He estimates that he has ten miles of plantation roads, and 20 miles of rail fence, more than half of which is to fence against other folks' cattle instead of his own; and this fence has all to be renewed once in seven years, as in this humid climate that period is the length of durability of rails. What an enormous tax! And with the enormous waste of timber going on, how long will it be before all the rail timber is exhausted? What is to be done then? What is to be substituted? It is time this matter was thought of even amid the forests of Mississippi. There is another matter that ought to be thought of too by every cotton planter. What are they going to do when the supply of basket timber is exhausted, as it already is in some parts of the state? Will they send to the north for these indispensable articles? Well, so be it. We are ready to furnish you, and we will soon learn that you cannot pick cotton without baskets. I advise you to commence immediately the cultivation of OZIER WILLOW. It will grow upon all your creek banks, and it will make a more handsome and valuable fringe than many that I have seen in the middle of your fields. There is another article that grows almost spontaneously upon some of the rich bottoms and waste corners of your plantations, that would bring money if sent to market; and that is red pepper, the grinding of which you can do in your own mills, and pack in your empty flower barrels—try it. You can get the willow from New-York; I don't know in particular from where, but I will venture to name my friend, Charles Downing, nurseryman, Newburgh, whose honesty I have great faith in.

And begin in time to husband your resources for fencing. Don't pursue a course that I witnessed a few days ago. Deadening good rail trees within the proposed enclosure to stand and rot down, and going outside among the standing timber and cutting down the trees for rails, and for the reason that by so doing it saved the trouble of clearing up the tops within the field—those outside could lay undisturbed to rot.

Leaving Captain Eggleston on the 18th, the first plantation I passed was one that once had been a very fine one, of comparative level and rich soil, now in utter ruins: Cause—debt, law and taxes. Fences, buildings and land all in ruins; the former rotted and fallen down, and the latter gullied away. In the midst of all this desolation, an ancient mound reared its lofty head, looking still more the lonely monument of an extinguished race than it would when met upon the wild waste where *civilization* had not yet set its more enduring mark. Even here upon this monument the hand of the white man had been, and exposed to view the interior, "full of dead men's bones."

After passing Lexington, the county seat of Holmes, which is rather a pleasant-looking town, we begin to leave the hilly country, and find one, though of the same kind of soil, much more level and showing more good farms, upon several of which I saw large forces busy planting corn. Cotton seed is much used for manuring corn, sometimes spread broadcast and sometimes put in the drill with the seed, which is generally planted in drills and covered with the plow.

From a Mr. Adams, whose hospitality I partook of this night, I learned that hot ashes are a very effectual remedy for what is generally called "the damps" in wells. They appear to absorb and neutralize the gas—so he says. It is easily tried. Mr. A. is a great economist of manure, and plows his land upon the level system, but without ditches, which Captain Eggleston says, upon side hills is worse than straight up and down. Mr. Adams' land is, however, comparatively level.

February 19th, I passed through the town of Benton, the county seat of Yazoo, and which is so superior to its namesake in Missouri, both in appearance and character of its inhabitants, that one or the other ought to change its name, and principally though for the reason that papers directed to one often get astray to the other. I regret that the anxiety that I began to feel to reach Log Hall, prevented me from making a stop at this town and forming more close acquaintance with some of the many friends and readers of the *Cultivator* that in a very short visit I found here. It was then my intention to return, which circumstances prevented. Although I would not make distinctions among friends, yet I may be permitted to signalise Mr. Jenkins, the P. M., and Wm. Battel, Esq., whom I found most active and anxious to encourage the reading of agricultural papers.

A few miles west of Benton I called upon John M. Cullen, who has invented, as he thinks, an improved cotton scraper—it being a small piece of steel attached to a plow in such a manner that he can "bar off" and "scrape" at the same time. I witnessed a trait in Mr. Cullen's character that I desire to mention, together with the wish that others would do likewise.

He owns a pond, which is the only watering place for teams upon the road for a long distance, and which he necessarily had to enclose; but instead of shutting the public out, he has gone to considerable expense to provide for their accommodation, and has put up a sign of "Bethesda," the meaning of which Bible readers will understand.

But let us go on with our wonders. To-day I first met with the "Spanish moss" regions, which, contrary to the opinion entertained by many, that it only grows upon trees in swamps, is found equally abundant upon the hills. I don't know that it shows any preference in the kind of tree it grows upon, for it is not a parasite; that is, so far as I could observe, it appears to have no connection with the tree, but hangs loosely upon the limbs, sometimes hanging down two or three feet. Its color is silvery grey, and when all the trees in the forest are thickly covered, it gives a curious appearance. Although at the north we esteem it valuable for mattresses, &c., it is here but little used.

This evening I crossed the "Big Black," a stream large enough for steamers in high water, but for want of improvement but little used. It runs through a wide, rich, overflowable bottom, entirely uncultivated. During two days ride I passed land that was not yet clear of timber, that had been worn out and thrown out of use. This bottom land would be more enduring.

In this region of the state there is great difficulty in getting wells, while streams and springs are few and subject to dry up; and though every body ought to have cisterns and artificial ponds, yet every body has not, and none that I have met with seem to be "fixed," but are ready to sell out and hie away to Texas, or some other place "further west."

February 20th I travelled on a very broken and poorly cultivated part of Madison and Hinds counties; passed several "gone to Texas" plantations, the appearance of which give the country a desolate look.

Enquiring for Dr. Phillips [Martin Wilson Philips, agricultural editor], I found that "a prophet is not known in his own country," and that if a man wishes to distinguish himself "among some folks," he must turn politician, instead of becoming a writer for agricultural papers. However, most that I inquired of seemed to know that the Doctor lived somewhere, though the exact where they could not tell, and for which latter piece of ignorance I did not much blame them after I knew myself, for a more out of the way place can't well be thought of. Knowing that his post-office address was "Edwards' Depot," I easily found that, but I cannot say that the seven miles from there was so easy to find in the night, or so pleasant to drive over; but perseverance accomplished the task, and I found the Doctor and his family so much more pleasant than the route to his place, that with the reader's permission who has traveled thus far with me, we will tarry awhile and partake of heartfelt hospitality while resting from the fatigue of our thousand miles journey. And now for another short month, dear reader, a kind adieu from your old friend and fellow traveller.

# A Day in the Country

## VISIT TO LINDENWALD

[*Cultivator*, n.s., I (Aug., 1844), 236.]

WE LATELY PASSED a beautiful summer's day in the vicinity of Kinderhook. Among other places of interest, we visited *Lindenwald*, the seat of ex-president Van Buren.

Lindenwald, formerly the residence of Judge Van Ness, is pleasantly retired, and commands a very agreeable landscape view, the most prominent features of which are the Catskill mountains, whose elevated summits are often veiled by the shadowy cloud.

We found Mr. Van Buren at home, and accompanied him in a walk over the farm. When he entered on the occupancy of this place on his retirement from the Presidency, three years since, it was much out of order: the land, having been rented for 20 years, and been under cultivation for a period of 160 years. Several of the buildings had become poor, the fences were old and were rotting down, and bushes and grass of wild growth had taken possession of much of the farm. During the short time it has been under Mr. Van Buren's management, the place has been greatly improved, and a course is now fairly begun by which a handsome income may be derived from it. The garden and pleasure-grounds have been enlarged and newly laid out—hot-houses have been erected—and a large number of fruit and ornamental trees, shrubbery, &tc. have been planted. The greenhouse contains a collection of exotic fruits and plants, among which were some fine grapes just ripening. In the garden we noticed fine samples of all the fruits of the season, and some of the finest melons we have ever seen, (so early in the year,) in this latitude.

Among the objects which give beauty and interest to the grounds, are two artificial ponds in the garden. They were easily made by constructing dams across a little brook originating from springs on the premises. Soon after they were made, (three years ago,) some fish were put into them, and they are now so well stocked, with trout, pickerel and perch, that Mr. Van Buren assures us they will afford an abundant supply for his table. This is a matter well worthy of consideration. There are many situations where such ponds may be made, and, with a trifling expense, the luxury of *catching* and *eating* a fine trout or pickerel may be had at any time.

Several of the fields have been enclosed with new fence, and several buildings erected; among which are a very tasty farm-house, and a barn calculated for storing 150 tons of hay after being pressed.

But perhaps the most important improvements which have taken place on the farm, have been made on a tract of *bog land*, thirteen acres of which have been thoroughly reclaimed, and are covered with luxuriant crops of grass or oats. Three years ago, this land was almost worthless. It was first drained by ditches. The stumps, bushes, &tc. were then cut out and burned, and the ashes spread on the land. It was afterwards sown to grass—using a mixture of timothy and red-top seed—three pecks to the acre. The whole cost of reclaiming was thirty-eight dollars per acre, and the land will now pay the interest of a hundred to a hundred and fifty dollars per acre. In this Mr. Van Buren has set a good example, which we hope will be followed by other farmers in the neighborhood who have lands similarly situated.

The *potatoe crop* is one of considerable consequence on this farm, as well as on others in the vicinity. Mr. Van Buren raises the variety called *Carters*, produced from the *ball* a few years ago by the Shakers. He considers these far the most profitable kind known. They yield well, and their quality is thought equal to any. Mr. Van Buren assured us that all which could be raised would readily command fifty cents per bushel, by the quantity, in New-York. All the crops appear to be well managed, and are promising. Leached ashes were tried here last season with excellent success. Great benefit has also been found from plowing in clover.

Mr. Van Buren keeps but little stock, a considerable object being the sale of hay, which a large portion of the farm is well calculated to produce—the horses for carriages and farm-work, with a yoke of oxen, and a sufficient number of cows to afford milk and butter for the family, comprising about all. We did not see the cows, but were informed that they were grade Durhams, and were excellent for the dairy. We were shown a good three years old Durham bull, whose head and limbs denote good blood, and whose mellow skin indicates that he is a thrifty animal.

All the improvements of which we have spoken, have been planned and executed under the immediate supervision of Mr. Van Buren, who finds in these useful enterprises a salutary exercise for the faculties of the mind and body, which seems to be highly enjoyed. In this pleasant retreat, removed from the cares of state, and the turmoil of political wars, he,

—"With a choice few retired,
Drinks the pure pleasures of a rural life."

# THE BENEFITS AND THE BEST MODE OF APPLYING SHELL AND OTHER MARLS *

[A short prize essay]

MARL has been used as a manure from the earliest ages. "When I marched an army," says VARRO, "to the Rhine, I passed through some countries where I saw the fields manured with fossil clay." All parts of England are dotted over with old marl pits from which marl has been drawn out on the land as manure. This marl consists principally of clay combined with carbonate of lime and more or less phosphoric acid. It will effervesce on the application of strong vinegar, which sets free the carbonic acid of the lime. The phosphoric acid, of course, is of great value; and the lime, as it slowly becomes available, is also beneficial. There are also many other valuable constituents of plants in the marl. But since the advent of artificial manures, the use of marl in England is far less common than formerly. The extensive cultivation of the turnip, which delights in light land, has also helped to render marl less needed. The turnips are consumed on the land, by sheep, and their pointed hoofs consolidate the soil; and this I regard as one of the principal effects of the old-fashioned practice of marling.

The usual amount of marl applied to the sandy lands of Norfolk, is from forty to seventy loads per acre—a load being about forty bushels. It is drawn out in autumn and winter, and spread upon the land, so that the frosts can pulverize it. Its effect is slow but very

* The above is representative of hundreds of short prize essays published in the *Genesee Farmer*. The committee of awards, while accepting the above, felt it was "not as full or as practical as the importance of the subject" required. *Genesee Farmer*, ser. 2, XX (Jan., 1859), 25–26.

permanent, and invariably beneficial. Lands worth only $1.25 per acre (rent) have been raised in value to $5 per acre, by the process of marling.

Shell marl is usually found under bogs and mosses, and is composed of the remains of small testaceous fish, which, dying in their shells, become converted into calcareous earth, their bodies, when decomposed, furnishing a rich mould of animal matter.

It consists principally of carbonate of lime, and is frequently burnt and converted into quick-lime. It is said that lime made from shell marl is more valuable for manure than that made from limestone; but of this I can not speak from experience. I can not see why it should be any better. Lime is lime, and it can be nothing else. The organic matter, which is undoubtedly very valuable as manure, would be all consumed by a heat sufficient to drive off the carbonic acid.

Where shell-marl is plentiful, I think it would be better to apply it in the crude state in considerable quantity, rather than to burn it. The organic matter would decompose in the soil, and furnish the ammonia so much needed by wheat and other cereals, while the carbonate of lime would become gradually available from the influence of the atmosphere and the acids of the soil.

It usually proves very beneficial on the mucky land near which it is found. It would seem as though nature had placed it there for use on such land.

M.R.

# DIALOGUE BETWEEN A FATHER AND SON, PART I

[*Farmers' Cabinet*, IV (Oct. 15, 1839), 78. This is a part of a series of the famous "Father and Son Dialogues" by James Pedder.]

*Supposed Conversation between a Provident and Improvident Farmer, and their respective crops and stocks, &c.*

*Frank.*—Father, which is the most profitable breed of sheep for the farmer? I should suppose the largest, as a sheep is a sheep you know, and a large one is of more value than a small one.

*Father.*—A prudent man will advise with his land on that subject.

*Frank.*—But can his land advise with *him?*

*Father.*—Yes, and the lessons which a farmer is taught by his land, are not soon forgotten, as, according to the old adage, "*bought wit is best.*" I sometimes fancy that my crops converse with me, when I visit them of an evening, and if I could do justice to those fancied dialogues which I seem to hear, and could commit them to paper, they would, I think, make a pleasant addition to your book.

*Frank.*—O, do try, "*nothing is impossible to a willing mind,*" you know.

*Father.*—Most opportunely quoted the *text*—now for the SERMON.

We will suppose then, that a slovenly *procrastinator* is visiting his fields on just such a glorious evening as the present, in just such a fruitful season as we are now blest with.—He goes up to the field, No. 1, which is wheat, and begins—

*Grabb.*—Good evening; fine weather this: but I don't think you look quite so well as you did the last time I visited you.

*Wheat.*—I wonder how I should—do you not see how I am

choked with weeds? How the thistles are goading me with their spikes, and the rag weeds are taking the food out of my mouth, while the bind weeds are dragging me down to earth; and how that I am smothered with evils innumerable?

*Grabb.*—But I allowed you a fallow and plenty of manure; you ought at least to have been able to cope with the weeds.

*Wheat.*—You forget that "the earth is own mother to the weeds, while she is only mother-in-law to the crops that are planted in her bosom": besides, you talk of a fallow—why this great thistle on my right, and which has one of his spikes fixed in my side, has just informed me that he is one of the progeny which was reared in this same fallow of yours,—his parent being the identical thistle under which the farmer sat on horseback and escaped a drenching, while his neighbors were wet to the skin! You seem to have forgotten that "*one year's seeding is seven years' weeding.*"

*Grabb.*—Ah well! I'll get these weeds pulled.

*Wheat.*—As you said a month ago, and will say again, and never do it!

*Frank.*—Excellent! But you never fallow or dung for wheat.

*Father.*—Nor have I ever such fine thistles. I always dung for green crops, and insure two things at the same time—more food for the cattle, and of course, larger dunghills. My object is, to retard the growth of the wheat, that it might be strong in the stalk, and I therefore do not encourage its lavish growth by manure and fallow. Now for No. 2.

2. Corn. *Grabb.*—Why you look very sickly; I thought you would do better, judging from the appearance you put on at first coming up—how's this?

*Corn.*—Ask yourself! You thought you were cheating me, when you sowed without manure—a favor you always promised me; I relied upon that promise and came up, with the expectation that I should find it when I needed it; but after sending my roots below in search for it, I find your promises are false—you complain of my sickly look! I can only say, if you had no more to feed upon than I have, you would not have shelled the three lower buttons on your waistcoat! Grabb tucked the shucks into the holes, and walked on.

*Frank.*—I now find that crops can advise, and admonish too: but could not the farmer still do something in the way of top dressing to remedy a part of the evil?

*Father.*—Yes; but he had no manure.

3. Barley. *Grabb.*—Ah! You'll come to nothing.

*Barley.*—I thank you, and return the compliment. But what did you expect when you sowed me after once ploughing, on a stiff and wet soil? "Nothing venture, nothing have." I only wish that *you* had to work so hard as I have for a living. You would then feel for *me.*

4. Oats. *Grabb.*—Well, I think you might do a little better than you do, if you would try; why, I shall not get the value of the seed back—that's too bad!

*Oats.*—Now, that's t h r i c e bad of you!—You know that you have had six grain crops in succession from the land on which I am sown, with not a spadeful of manure of any kind for the last six years? . . .

# THE EXECUTION OF GEORGE H. LAMB

[A letter from Mrs. Gage, *Ohio Farmer*, VIII (July 2, 1859), 214.]

I TAKE up my pen this morning, to mark passing events for the readers of the *Farmer*. Today, Friday, June 17, 1859, Geo. H. Lamb will end his career on the gallows, in this city [St. Louis], for the murder of his wife, Sarah Lamb, by taking her out upon the Mississippi in a skiff, and drowning her. Terrible are the details of this murder, and terrible is public indignation against the offender.

I do not purpose, now, to comment upon the crime, but upon the mode which a civilized and enlightened people—as ours boast of being—have chosen, all over the States, to punish murder in the first degree. A man of vile passions chooses to destroy a wife, of whom he is weary. He takes her out upon the waters, in the darkness of the night, and despite her tears, and prayers, and pleadings, puts an end to her life. "Human life is sacred," is the cry of the people, "and this death must be avenged"; and with calm deliberation, judges and jurors, witnesses and prosecutors, enter into the work of investigation; and if the culprit is found guilty, he is condemned "to hang by the neck till he is dead." The whole public mind is shocked with the announcement, if it be not already hardened and callous. The hanging of this fellow creature is talked over by all the loose, the low, and the depraved. Little children catch the thought, and the coming hanging is made a matter of jest or fear. Two human beings are cut off from the earth, the second one because he caused the violent severing of the life-cords of the first, and the death of the last is made as diabolical as possible. . . .

All punishment should be for the amendment of the offender, to prevent the repetition of the crime, or for the reformation of society.

Hanging a man certainly puts an end to all amendment, in this life at least. It is as certain that the second reason for the death of criminals is valid. The same man will not repeat the offense. But is society reformed? Does the reckless desperado, who is ready to give up his life any day, in a reckless brawl upon the street, care for this exposition of the end of the life of a murderer? Alas, no! but exactly the reverse; and the continual and minute accounts of all these tragedies, done up in the jocular and coarse tone of the American press, must have a tendency to lower the standard of public morals.

"What shall be done with the murderer?" is the question asked. Wiser ones than I must answer; but this I will say: do not commit murder, to avenge murder, thus paining and perhaps violating the highest and best feeling of humanity, and stirring into indignation, brutality and active life, the very worst. . . .

The sun has not shone out upon our city this day. It is fitting; for a vail of cloud and gloom should hang over a people, where such violence and wrong are being done. My arguments may be old and stale, and I may be accused of a false, sentimental sympathy with a wretched, depraved man. I do sympathize with every sufferer; but, in this case, not with the man hanged, as with the thousands that will be brought lower for the hanging—with the one hundred and fifty thousand people, who are thinking about it, and talking about it—with the aged father, who must stand by his erring child, and receive his mutilated body from the hands of the officers—with the relatives and friends—with that little girl, seven months old, who must bear through life the stinging pain, not only that her father was a murderer, but that the world murdered him.

The papers of this morning announce that he has been converted, and *has fully made his peace with God*, and expresses his calm resignation and willingness to die. Strange inconsistency! horrible Christianity!—God has forgiven him; yet man "hangs him by the neck till he is dead."

—F. D. Gage.

# DRESSES FOR TRAVELING, ETC.

[Anna Hope, in the *American Agriculturalist*, XVIII (June, 1859), 183.]

WE ARE a nation of travelers—farmers as well as others. Families are so scattered by emigration, that if we were all keepers at home we should, many of us, be compelled to bid a last farewell to some that we love, long before they bid adieu to earth.

An appropriate dress for the road is of no small importance, although it need not be of any expensive material. It should be of some plain color, drab or brown, or any other that will not attract attention. Bright colors are entirely out of place. Many of the India silks are suitable for traveling dresses—so are merinos and delaines. There are at the present time a great variety of cheap goods made of worsted and linen, or of worsted and cotton, that answer well for this purpose. A dark gingham is not amiss.

A traveling dress should be simply made—the waist buttoned up to the throat, and the skirt without flounces. A cloak of the same material as the dress, is, in most cases, in good taste. A gray flannel cloak is never unsuitable. The bonnet should be as simple as the dress. A colored straw, with but little trimming, is in good taste; so are shirred bonnets of plain colors. White straws are objectionable only because they are so soon soiled by the dust. The coarse "Rough and Ready" is much the fashion. Dress bonnets should not be worn except on dress occasions. For gloves, I prefer the doe-skin gauntlet, or the undressed kid; lisle thread are the best of low-price gloves; avoid soiled light colored gloves. Wear a linen or Marseilles collar, or an embroidered cambric—not lace or muslin.

I have just taken a journey of several hundred miles, and have seen examples of various styles of dress, which were not all of them in the best taste. One young miss, not far advanced in her teens, traveled in a low-neck dress, as it was very easy to see when she removed her cloak for her greater comfort. Another wore a many-colored chenille shawl, with a straw bonnet profusely trimmed with a ribbon in which red was one of the colors. The face trimming was a bright rose-color and black, and the strings another shade of rose-color without the black—the rose-color itself was beautiful, but its proper effect was ruined by the red, and the different shade of the same color. Another was still more marked in her style. She displayed prodigious hoops, wore no collar, but did wear an immense bloomer hat streaming with blue ribbons. She was excessively deficient in beauty, and should not have attracted attention by a peculiar dress. In the seat back of me sat a very neat little woman, in a drab dress and cloak, wearing a straw bonnet, with the cleanest of quilled lace for a face trimming. Her dress displayed both good sense, and good taste; good sense is always an element in good taste.

It is well for ladies to provide themselves with a lunch, as it is otherwise impossible for them to be comfortable in the hurry and scramble of railroad traveling. I would also recommend them to take a small tumbler with them, as it is not particularly agreeable to drink after others, especially after the victims of tobacco.

May I not offer a hint to gentlemen, to which I wish they would lend a listening ear. It is that they should leave their tobacco at home, and not bring it into the cars to annoy others. I pity the wives of these spitters, but as they were taken for worse, as well as better, perhaps there is no other way than to bear with them. I do not know a more disgusting practice than that of defiling cars, and public rooms, and private parlors even, in this way. If these men must chew, let them resort to the smoking car and enjoy their tobacco.

# OUR TRIP

[*Valley Farmer*, V (Nov., 1853), 406 ff. Mrs. Mary Abbott, editor of the "Family Circle" of the *Valley Farmer*, here reports her experiences on a trip across the State of Missouri.]

OWING to indisposition we cannot give so connected an account of our journey as we could desire; but we will give what we recollect. Our kind reception among our friends we shall *never forget*. . . .

We left St. Louis on the Kate Swinney, which we think to be one of the best managed boats on the Missouri river. Everything is done in quietness and order, and the officers understand their business, and are not too officious in their attention to the ladies, pushing themselves in the ladies cabin at unseasonable hours, as *some* clerks are apt to do. We arrived at Boonville in season to spend a quiet Sabbath and attend church. We heard a good sermon from Rev. Mr. Bell, which well paid us for going. On Monday, after our day of rest, we again embarked for Brunswick. On board were a number of gentlemen who were going to that place to attend the dedication of an Odd Fellows' Hall. . . . We made but a short stop at Brunswick, and then set off for Independence. There we were very sick, and cannot say much in favor of the place, as we did not see much of it. We took a short walk from the hotel to a dry goods store, and were nearly suffocated by ungentlemanly men smoking in our face, as we attempted to pass them. Dry goods are more than a third higher here than in St. Louis, and because we were strangers they made us pay more than double for the little we purchased. We were glad to leave there.

Our next stopping place was Weston. We spent about a week at the residence of Dr. Beaumont, five miles from town, where we were hospitably entertained and kindly cared for, by the Dr. and his wife, who did all in their power to make us comfortable. . . .

On Monday we set off for Columbia, to attend the *Boone* County Fair. We were surprised and pleased to see the good order and taste displayed in preparing the grounds for the exhibition. It was gratifying to see so great a display of domestic articles—the industry of the ladies. Here were blankets of the finest texture, jeans, carpetings, hosiery, flannel, besides numerous quilts, with such an enormous amount of work in them that we cannot say a word in their favor. We consider the making of them a useless and even sinful waste of time—time the makers had better employ in cultivating their own minds or those of their children, or in darning and patching. For our part we had rather have seen one pair of well darned stockings, than all the fine quilts that were displayed. But the ladies did well in their *useful* articles, and they deserve praise. There was some excellent butter, preserves and pickles. This Fair was as well, if not better conducted than any of the fairs we have attended this fall. The President and Marshal did their speaking in so loud and clear tones that we not only saw but understood everything that was going on. Major Rollins very politely and generously recommended the *Valley Farmer*, not forgetting our own humble efforts, for which kindness we feel very grateful. . . .

We left Major Rollins' on Saturday morning for Rocheport, in hopes of getting a boat for *Boonville* so as to rest there over Sunday, but there was no boat that we could get till Sunday, and on *that* day we would not, nor will not travel. We rested at Rocheport till Monday morning, the first day of the Boonville Fair. We started off early in a coach, and arrived there in as good season as those who went up on Sunday, much refreshed by our invigorating and quiet rest, and were comfortably entertained at the house of Mr. H. M. Myers, which really seemed another home to us while we attended the State Fair. We were at the fair grounds in good season, and saw a fine display of ladies' industry, including such articles as we saw at the Columbia Fair, besides bread and boiled hams, and socks knit by little girls, which pleased us much. Mr. Myers' little girl of *seven* got the certificate, we thought she deserved the premium, considering her age, so we gave her the premium, and she gave us the socks, which we keep to show what *little* girls can do. We could not tell the difference between the certificate and premium butter, except there was rather more fancy work in moulding it. Mrs. Porter of Boonville, got the premium for butter, and Mrs. Perry for bread. We hope there will be more competition in these articles next year.

While at Boonville and Columbia we met with many warm friends, some that we had wanted to see, but had never before had the opportunity. It pleased us much to see how our poor endeavors to be useful are appreciated; and to hear our department spoken of with so much approval was very encouraging to our humble self. We had rather know that we are kindly spoken of by both sexes, than to be known as the head and leader of the unscriptural Woman's Rights party, and have our name received abroad as the great one among them, and appear as an enemy to those whom God in his wisdom and kindness has appointed to be the defenders and guardians of the gentler sex. . . .

After having been absent from home—'sweet home'—for more than four weeks, we set off on the Clendenen for St. Louis.

The Clendenen is a boat we cannot recommend to our friends. The ungentlemanly conduct of the first clerk—the noise and confusion all over the boat, especially in the ladies' cabin, occasioned by the intrusion of gentlemen who had no business there, led there by the example of the clerk—these *men,* we do not call them *gentlemen,* sought out the immodest, rude *girls,* we will not call them *ladies*—kept up such a noise, with unbecoming behavior, as utterly to astonish and confound the rest of the respectable and orderly passengers, preventing their rest and comfort. The men staying in the ladies' cabin till the doors were closed—keeping up a loud and boisterous noise with these girls. Their mothers were with them, encouraging them in their misdemeanors, and even joining in their noise.

While we were on our trip home, there was an awful and wicked deed committed, resulting in the death of the steward, a free and intelligent mulatto, who was cruelly beaten for taking the part of his sister, or near relation, whom the clerk had insulted and beaten without any just cause. It was *said* that he jumped overboard to escape being beaten to death. We look upon the deed as an actual murder. . . .

In out next number we will give some account of the other fairs we have attended this fall.

# ADDRESS BY ABRAHAM LINCOLN

[*Wisconsin Farmer*, XI (Nov., 1859), 389 ff. Given before the Wisconsin State Agricultural Society at its annual fair in Milwaukee on September 30, 1859.]

*Members of the Agricultural Society and Citizens of Wisconsin:*
AGRICULTURAL FAIRS are becoming an institution of the country; they are useful in more ways than one; they bring us together, and thereby make us better acquainted, and better friends than we otherwise would be. From the first appearance of man upon the earth, down to very recent times, the words "*stranger*" and "*enemy*" were *quite* or *almost* synonymous. Long after civilized nations had defined robbery and murder as high crimes, and had affixed severe punishments to them, when practiced among and upon their own people respectively, it was deemed no offence, but even meritorious, to rob, and murder, and enslave *strangers*, whether as nations or as individuals. Even yet, this has not totally disappeared. The man of the highest moral cultivation, in spite of all which abstract principle can do, likes him whom he *does* know, much better than him whom he does *not* know. To correct the evils, great and small, which spring from want of sympathy, and from positive enmity, among *strangers*, as nations, or as individuals, is one of the highest functions of civilization. To this end our Agricultural Fairs contribute in no small degree. They render more pleasant, and more strong, and more durable, the bond of social and political union among us. Again, if, as Pope declares, "happiness is our being's end and aim," our Fairs contribute much to that end and aim, as occasions of recreation—as holidays. Constituted as man is, he has positive need of occasional recreation; and whatever can give him this, associated with virtue and advantage, and free from vice and disadvantage, is a positive good. Such recreation our Fairs afford.

They are a present pleasure, to be followed by no pain, as a consequence; they are a present pleasure, making the future more pleasant.

But the chief use of Agricultural Fairs is to aid in improving the great calling of *Agriculture*, in all its departments, and minute divisions; to make mutual exchange of agricultural discovery, information, and knowledge; so that, at the end, *all* may know everything, which may have been known to but *one*, or to but *few*, at the beginning; to bring together especially, all which is supposed to not be generally known, because of recent discovery or invention.

And not only to bring together, and to impart all which has been *accidentally* discovered or invented upon ordinary motive; but, by exciting emulation, for premiums, and for the pride and honor of success—of triumph, in some sort—to stimulate that discovery and invention into extraordinary activity. In this, these Fairs are kindred to the patent clause in the Constitution of the United States; and to the department, and practical system, based upon that clause.

One feature, I believe, of every Fair, is a regular *address*. The Agricultural Society of the young, prosperous, and soon to be, great State of Wisconsin, has done me the high honor of selecting me to make that address upon this occasion—an honor for which I make my profound and grateful acknowledgement.

I presume I am not expected to employ the time assigned me, in the mere flattery of the farmers, as a class. My opinion of them is that, in proportion to numbers, they are neither better nor worse than other people. In the nature of things they are more numerous than any other class; and I believe there really are more attempts at flattering them than any other; the reason of which I cannot perceive, unless it be that they can cast more votes than any other. On reflection, I am not quite sure that there is not cause of suspicion against you, in selecting me, in some sort a politician, and in no sort a farmer, to address you.

But farmers, being the most numerous class, it follows that their interest is the largest interest. It also follows that that interest is most worthy of all to be cherished and cultivated—that if there be inevitable conflict between that interest and any other, that other should yield.

Again, I suppose it is not expected of me to impart to you much specific information on Agriculture. You have no reason to believe, and do not believe, that I possess it—if that were what you seek in this address, any one of your own number, or class, would be more able to furnish it.

You, perhaps, do expect me to give some general interest to the occasion; and to make some general suggestions, on practical matters. I shall attempt nothing more. And in such suggestions by me, quite likely very little will be new to you, and a large part of the rest possibly already known to be erroneous.

My first suggestion is an inquiry as to the effect of greater *thoroughness* in all the departments of Agriculture than now prevails in the North-West—perhaps I might say in America. . . .

The successful application of *steam power*, to farm work, is a *desideratum*—especially a steam plow. It is not enough that a machine operated by steam, will really plow.—To be successful, it must, all things considered, plow *better* than can be done with animal power. It must do all the work as well, and *cheaper;* or more *rapidly*, so as to get through more perfectly *in season;* or in some way afford an advantage over plowing with animals, else it is no success. I have never seen a machine intended for a steam plow. Much praise and admiration are bestowed upon some of them; and they may be, for aught I know, already successful; but I have not perceived the demonstration of it. I have thought a good deal, in an abstract way, about a steam plow. . . .

The world is agreed that *labor* is the source from which human wants are mainly supplied. There is no dispute upon this point. From this point, however, men immediately diverge. Much disputation is maintained as to the best way of applying and controlling the labor element. By some it is assumed that labor is available only in connection with capital—that nobody labors, unless somebody else owning capital, somehow, by the use of it, induces him to do it. Having assumed this, they proceed to consider whether it is best that capital shall *hire* laborers, and thus induce them to work by their own consent, or *buy* them, and drive them to it, without their consent. Having proceeded so far, they naturally conclude that all laborers are naturally either *hired* laborers or *slaves*. They further assume that whoever is once a *hired* laborer, is fatally fixed in that condition for life; and thence again, that his condition is as bad as, or worse, than that of a slave. This is the "*mud-sill*" theory. But another class of reasoners hold the opinion that there is no *such* relation between capital and labor, as assumed; and that there is no such thing as a freeman being fatally fixed for life, in the condition of a hired laborer, that both these assumptions are false, and all inferences from them groundless. They hold that labor is prior to, and independent of, capital; that, in fact, capital is the fruit of labor, and could never have existed if labor had not *first* existed—that labor can exist without capital, but that capital could never have existed without labor. Hence they hold that labor is the superior—greatly the superior—of capital.

They do not deny that there is, and probably always will be, *a* relation between labor and capital. The error, as they hold, is in assuming that the *whole* labor of the world exists within that relation. A few men own capital; and that few avoid labor themselves, and with their capital, hire, or buy another few to labor for them. A large majority belongs to neither class—neither work for others, nor have others working for them. Even in all our slave States, except South Carolina, a majority of the whole people of all colors, are neither slaves nor masters. In these free States, a large majority are neither *hirers* nor *hired*. Men, with their families—wives, sons and daughters—work for themselves, on their farms, in their houses and in their shops, taking the whole product to themselves, and asking no favors of capital on the one hand, nor of hirelings or slaves on the other. It is not forgotten that a considerable number of persons mingle their own labor with capital; that is, labor with their own hands, and also buy slaves or hire freemen to labor for them; but this is only a *mixed*, and not a *distinct* class. No principle stated is disturbed by the existence of this mixed class. Again, as has already been said, the opponents of the "*mud-sill*" theory insist that there is not, of necessity, any such thing as the free hired laborer being fixed to that condition for life. There is demonstration for saying this. Many independent men, in this assembly, doubtless a few years ago were hired laborers. And their case is almost if not quite the general rule. . . .

The thought recurs that education—cultivated thought—can best be combined with agricultural labor, or any labor, on the principle of *thorough* work—that careless, half-performed, slovenly work, makes no place for such combination. And thorough work, again, renders sufficient, the smallest quantity of ground to each man. And this again, conforms to what must occur in a world less inclined to wars, and more devoted to the arts of peace, than heretofore. Population must increase rapidly—more rapidly than in former times—and ere long the most valuable of all arts, will be the art of deriving a comfortable subsistence from the smallest area of soil. No community whose every member possesses this art, can ever be the victim of oppression in any of its forms. Such community will be alike independent of crowned-kings, money-kings, and land-kings.

But, according to your programme, the awarding of premiums awaits the closing of this address. Considering the deep interest necessarily pertaining to that performance, it would be no wonder if I am already heard with some impatience. I will detain you but a moment longer. Some of you will be successful, and such will need but little philosophy to take them home in cheerful spirits; others will be disappointed, and will be in a less happy mood. To such, let it be said, "Lay it not too much to heart." Let them adopt the maxim, "Better luck next time;" and then, by renewed exertion, make that better luck for themselves.

And by the successful, and the unsuccessful, let it be remembered, that while occasions like the present, bring their sober and durable benefits, the exultations, and mortifications of them, are but temporary; that the victor shall soon be the vanquished, if he relax in his exertion; and that the vanquished this year, may be the victor the next, in spite of all competition.

It is said an Eastern monarch once charged his wise men to invent him a sentence, to be ever in view, and which should be true and appropriate in all times and situations. They presented him the words: "*And this, too, shall pass away.*" How much it expresses! How chastening in the hour of pride!—How consoling in the depths of affliction! "And this, too, shall pass away." And yet let us hope it is not *quite* true. Let us hope, rather that by the best cultivation of the physical world, beneath and around us; and the intellectual and moral world within us, we shall secure an individual, social and political prosperity and happiness, whose course shall be onward and upward, and which, while the earth endures, shall not pass away.

# AN EVENING DISCUSSION

[*Prairie Farmer*, XX (Sept. 22, 1859), 177 ff. Report of a farmers' discussion group at the Illinois State Fair, Freeport, September 7, 1859.]

. . . THE MEETING this evening was organized by the election of K. K. Jones as chairman. Plowing and drainage was the subject for discussion. A large gathering of farmers attested the interest felt in the matter.

Mr. Mills, of Marion county, related his experience with deep culture. Much better crops and affected less with drouth, were the result of deep tillage. He thought Fields' rotary plow, which we described last week, and which was on exhibition, was likely to be what is needed. At least it was a step towards it.

D. J. Powers, of Wisconsin, had experience with subsoil plowing; used one of Mapes' subsoil plows; subsoiled with it 16 or 18 inches; had been very dry. On the subsoiled land, plants and vegetables had grown continuously. The best time to commence deep culture subsoiling is now. Farmers of Wisconsin and Illinois are now scratching the soil like an old hen. We must have a different system of culture, and now is the time to commence.

Jno. Periam, of Cook Co., had used the Michigan double plow. Good effects resulted. He had used the subsoil plow with other good results, but he was not satisfied. Last spring he bought a plow with a subsoil attachment, with which he plowed twenty-two inches deep. His soil is dry, sandy land, and sensitive to the influences of drouth, and yet on that soil the crops have not suffered, although the drouth was so terrible that from ground plowed in the ordinary manner no crops have been taken.

Joseph Carpenter, of Union Co., thinks in Egypt [Illinois] subsoiling should be done in the fall. Grapes had done finely on land prepared with Mapes' Subsoil Plow; said that when he first went to Egypt, in conversation with some of his neighbors, he asked, "how deep do you plow in Egypt?" The answer was, "mighty deep." "How deep is that?" asked Carpenter. "Four inches!" was the reply.

B. G. Roots, of Perry Co., said that in Egypt, soil trench plowed —that is the subsoil turned on the surface—gave a poor crop the first year.

Dr. Ostrander, of Livingston Co., said that if we are going to deep plowing we must look out for draining in a wet season.

Atkinson, of Fulton Co., entered into an elaborate disquisition of principles which affected the introduction of a new invention of his. One of the principal accomplishments of his implement as we gather from his description was, that it "cuts and covers" beautifully. This remarkable process is without doubt well known to too many farmers in the West, and we have no desire to encourage it.

Hind, of Whiteside Co., thinks we do not sow wheat enough per acre; should sow two and a half bushels per acre; is farming on an old farm; it is pretty well reduced; believes he can make it perfectly as good as new prairie, by using the plow properly. He made some bold assertions about Frye's plow, now being manufactured in Detroit; says he believes it will be the plow in general use within two years. Subsoil must be got to the surface. It has become necessary to drain the surface and save the crops. This is the alternative presented us. It will pay us to adopt the preventive, rather than lose our labor and seed.

Mills, of Marion, says much of the land in Egypt is plowed with a single horse, so that when you pick the nubbins of corn the stalks are often pulled up, so short are the roots and so little their hold upon the soil. Teams are needed there. Mr. Mills is not in favor of any implement that "cuts and covers."

Col. Harris, of Ohio, says that cutting and covering is practiced by some of the Sciota farmers. It is easy to put grain in in this manner, and it has another advantage, *it is easily harvested!* Deep plowing alone will not prevent the soil from becoming worn, or the crops from deteriorating in quantity and quality. The soil must be fed.

Mr. Allen, of ——, is in favor of deep plowing for planting trees.

His experience is positive on this subject. On ground plowed 15 inches deep the growth of trees and plants has been constant.

Preston, of De Kalb Co., hopes there will be as much practice as talk about deep plowing. Has had experience with this kind of culture; gets double the corn per acre; crops on land plowed deep do better in wet seasons as well as dry. He knows instances where the recent frosts have made sad havoc in corn on shallow plowing, while that beside it on the deep plowing has not been injured. This is an important fact, and if our readers know any similar instances, we should like to be informed of them, and of the peculiarity of the soil, situation and difference in the depth plowed.

Hoyt, of Wis., says Wisconsin farmers—some of them—skim the surface as though the devil was below it. Thinks in poor soil the trench plow should be used, but in much of the soil of Illinois and Wisconsin, the subsoil plow is to be preferred. Mr. Hoyt spoke at some length of the importance and philosophy of deep plowing, of the action of the atmosphere upon the soil, &c.

Dunlap, of Champaign, would plow stubble land shallow in the fall, if for corn, and then deep in spring. But for wheat and potatoes would plow it much deeper in the fall, and shallow in the spring. In this manner the weeds are not turned to the surface at all, and exposed to the action of the atmosphere. Stubble plowed for corn in the fall and plowed two or three inches deep is again turned over in the spring eight or ten inches, thus burying the seeds of weeds beneath the surface at the second plowing. The deep plowed stubble for wheat and potatoes is not disturbed in the spring except at the surface. Gave his mode of putting in potatoes, which has already been described in the *Prairie Farmer*.

Roots of Tamaroa has noticed a marked change in the appearance of grass land where under drains have been made.

A long discussion followed, with reference to the use of the mole ditchers, and some important facts were elicited. . . .

# PLOUGHING MATCH

[*New England Farmer*, IV (Oct. 14, 1825), 92. This is an official report of a match held at Concord, Mass., October 5, 1825.]

THE COMMITTEE on the *Ploughing Match* ask leave to make the following Report, viz.

The ground allotted for ploughing was a smooth, even piece of light grass land, free from all stones or other substances to impede the plough. It was divided into lots of one eighth of an acre each. These lots were assigned to the several competitors, being fourteen in number, by lot, and they commenced at the same time. By a vote of the trustees of the society, your committee could not take into consideration the time in which any plot was ploughed, if the work was done in thirty minutes. As by this arrangement each competitor except one was enabled to complete his piece within the time prescribed by the trustees, your committee was mainly confined in forming an opinion who were entitled to the prizes, to the quality of the work. And here it hardly need be remarked that there was some difficulty in coming to a result. Much attention was paid by your committee to the style and manner of ploughing, the docility of the oxen and the skill of the ploughman and driver.

The several lots having numbers attached to them, and those who entered the lists being unknown to most of the Committee, after a careful and minute examination, the Committee came to the conclusion, that the premium should be awarded as follows:

| | | | |
|---|---|---|---|
| To No. 1, | the first premium of | | $21 |
| No. 13, | the second | do | 17 |
| No. 5, | the third | do | 10 |
| No. 9, | the fourth | do | 7 |

On a recurrence to the names of the individuals who had drawn these numbers, they found them to be Samuel Hoar, 2d, Darius Hubbard, Silas Conant, jr. and Abiel H. Wheeler.

Your Committee would further remark that so well did the others perform, that we can only regret that there were not other premiums which could be awarded them. They however may be assured that they did well, if there were found those who could do a little better. The Committee were highly gratified in witnessing so much regularity in all concerned, during the Ploughing Match, and in conclusion would express a wish that those who have not come off conquerors and won the prizes this day will not put their hand to the plough and look back, but that they will look forward to another year, and to other prizes.

For the Committee,
John Keyes.

# FARMING LIFE IN NEW-ENGLAND

[*Country Gentleman*, XII (Aug. 19, 1858), 114. This delightful sketch, which was pronounced, in the main, accurate by the editor of the *Country Gentleman*, was copied from the *Atlantic Monthly*. It is quoted in part.]

. . . We propose to present the portrait of such a home [a New England farm in the fifties], and, while we offer it as a just outline of the farmer's home generally, in districts removed from large centers, we gladly acknowledge the existence of a great multitude of happy exceptions. But the sketch:—A square, brown house; a chimney coming out of the middle of a roof; not a tree nearer than the orchard, and not a flower at the door. At one end projects a kitchen; from the kitchen projects a wood-shed and a waggon cover, occupied at night by hens; beyond the wood-shed a hog-pen, fragrant and musical. Proceeding no farther in this direction, we look directly across the road to where the barn stands, like the hull of a great black ship-of-the-line, with its port-holes opened threateningly upon the fort opposite, out of one of which a horse has thrust its head for the possible purpose of examining the strength of the works. An old ox-sled is turned up against the wall close by, where it will have the privilege of rotting. This whole establishment was contrived with a single eye to utility. The barn was built in such a manner that its deposits might be convenient to the road which divides the farm, while the style was made an attachment of the house for convenience in feeding its occupants.

We enter the house at the back door, and find the family at dinner in the kitchen. A kettle of soap-grease is stewing upon the stove, and the fumes of this, mingled with those that were generated by boiling the cabbage which we see upon the table, and by perspiring men in shirt-sleeves, and by boots that have forgotten, or do not care where they have been, make the air anything but agreeable to those who are not accustomed to it. This is the place where the family live. They cook everything here for themselves and their hogs. They eat every meal here. They sit here every evening, and here they receive their friends. The women in this kitchen toil incessantly from the time they rise in the morning until they go to bed at night. Here man and woman, sons and daughters, live, in the belief that work is the great thing,—that efficiency in work is the crowning excellence of manhood and womanhood, and willingly go so far into essential self-abasement, sometimes, as to condemn beauty and those who love it and glory above all things, in brute strength and brute endurance. . . .

# Part III

## SKETCHES OF CERTAIN IMPORTANT JOURNALS

## 1819–1860

IN DESCRIBING the size of the journals discussed in the following pages, the writer has adopted the scale of the American Library Association. According to this method a periodical measuring more than twenty centimeters and not over twenty-five centimeters in height is termed an octavo, more than twenty-five centimeters and not over thirty centimeters a quarto, and more than thirty centimeters a folio.

# *The New England Farmer* (1822–1846)

Boston, Massachusetts

THE *New England Farmer* made its appearance August 3, 1822, in Boston and was the second important agricultural journal to be established in the United States.[1] Thomas W. Shepard, the proprietor, appointed Thomas G. Fessenden as the first editor.[2] In the prospectus, attention was directed to the services already rendered to agriculture by the *American Farmer* at Baltimore and the *Plough Boy* at Albany. It was pointed out that the aforesaid agricultural papers could not be circulated with as great convenience in New England as a similar publication might be, if printed in Boston. Furthermore, the material contained in these periodicals was not always adapted to the soil and climate of the New England region.[3]

The *New England Farmer* was a weekly of eight quarto pages selling for $2.50 per annum in advance. In general approach and content it was not dissimilar to the *American Farmer*. The Baltimore editor promptly suggested an exchange of papers thereby sending a complete file of his current numbers to Fessenden, who was delighted with this friendly gesture and responded in kind.[4]

Thomas Green Fessenden, poet, journalist, and inventor, was born April 22, 1771, in Walpole, New Hampshire. He spent his youth on his father's farm and amid these rural scenes acquired

[1] For a graphic description of home-life on a New England farm in the late fifties, see pp. 316–17.

[2] No attempt has been made to enumerate the changes in proprietorship of this journal or any of the journals that follow. A complete list of its proprietors may be found in *New England Farmer*, XXIV (June 24, 1846), 414.

[3] *Ibid.*, I (Aug. 3, 1822), 1.

[4] *Ibid.*, I (Aug. 17, 1822), 22.

the general practical details of husbandry. By means of teaching and conducting singing schools, he worked his way through Dartmouth College, where he was graduated valedictorian in 1796. The study and practice of law followed, to be shortly abandoned for far more congenial literary and scientific pursuits. In 1822 Fessenden moved to Boston to assume the editorship of the *New England Farmer*, a position he held until his death fifteen years later.[5]

Although a tireless worker, Fessenden realized that he was not an agricultural authority, with years of practical and scientific training. Naturally he relied greatly on men so qualified and consistently urged the advantages gained through "reciprocal communication." While he used the shears freely, he did not blindly approve every fad and frill advanced by the agricultural press. "We had better [he wrote] by half follow the beaten track of our ancestors, if it be a little rugged and circuitous, than strike out at once into a wilderness of whim-whams, and theories not sanctioned by actual and repeated experiments." [6] Since in many parts of New England, farmers were compelled of necessity to be their own cattle doctors, he made provision for articles by "approved authors" on the diseases of animals. Likewise, he hoped to meet the demand for discus-

[5] *Dictionary of American Biography*, VI, 347; *Cultivator*, V (July, 1848), 207; *New England Farmer*, XVI (Jan. 3, 1838), 201; Perrin, *Thomas Green Fessenden*, pp. 160 ff. In the Foreword of Perrin's study, H. M. Ellis says that Fessenden was probably the most noteworthy American satirist in verse between the time of Trumbull and Lowell. In his later years, Fessenden was editor of two other agricultural papers, *Horticultural Register and Gardeners' Magazine* and the *Practical Farmer and Silk Manual*. He also edited several agricultural handbooks and was a member of the state legislature.

Nathaniel Hawthorne, an intimate friend of Fessenden describes for us an interesting picture of him in his library during these last busy years. "The table, and great part of the floor, were covered with books and pamphlets on agricultural subjects, newspapers from all quarters, manuscript articles for the 'New England Farmer.' . . . On my part, I loved the old man because his heart was as transparent as a fountain; and I could see nothing in it but integrity and purity, and simple faith in his fellow men, and good-will towards all the world." Hawthorne, *Works*, XII, 246 ff.

[6] *New England Farmer*, I (Aug. 10, 1822), 15.

sions relating to topics such as the "best mode of laying down ploughed lands to grass," the best manner of "reclaiming salt marshes," and the latest improvements in the manner of "cultivating hops, curing and preparing them for market." [7]

The value of the *New England Farmer* was greatly increased in 1832 when it became a vehicle for publishing the results of discoveries and improvements made by members of the Massachusetts Horticultural Society.[8] The curator of the Botanic Garden at Cambridge likewise made known the details and results of his experiments in various horticultural subjects through the *New England Farmer*.[9] In 1839 the *Horticultural Register*, a Boston monthly, was merged with the *New England Farmer*.[10] It was the aim of the editor to emphasize the subjects of especial interest to New England.

The journal was strongly humanitarian and reformatory in its tone. Its crusading zeal is exemplified in the numerous attacks on the use of tobacco, criticisms on lotteries, articles against cruelty to animals, and propaganda for temperance reform. Pride in the high moral tone of the *New England Farmer* is evident in the editor's boast that its pages had "never been stained with a statement, sentiment, or expression, which would raise a blush of shame on the cheek of modesty, or infuse a poison in the uncorrupted mind." [11]

Many social-economic problems of a sectional nature were aired in the columns of the *Farmer*. For instance in 1831, as a result of the great number of New England farms under mortgage, a writer suggested that farmers' daughters might well do their part in easing the financial burden by serving as "hired help." [12] This immediately

[7] *Ibid.*, I (Nov. 16, 1822), 127. [8] *Ibid.*, XI (Jan. 2, 1833), 199.

[9] *Ibid.*, X (Oct. 12, 1831), 104. [10] *Ibid.*, XVII (Jan. 30, 1839), 238.

[11] *Ibid.*, XVII (Jan. 2, 1839), 207.

[12] *Ibid.*, X (July 20, 1831), 1. This writer suggests that the financial embarrassment on the part of the farmers of New England was due in a large measure to the decay of the household industries. He says that "the cheapness of the fabrics that are sent out from our manufactories, has superseded the labors of the loom and the distaff. Spinning and knitting are obsolete toils. And the labors of only one or two can contribute anything to the support of the family." This whole story is part of the industrial and

brought forth a farmer's reply: "I love my children; I am happy, yes, I am proud to see them earn their bread by honest labor; yet if I know my own heart, I would sooner, infinitely sooner, follow my daughters to the grave than see them '*go out to service.*' "[13] "Oliver" who felt that pride was at the root of such an attitude, begged the editor to give his "cogitations" on the subject of domestics, which he did.[14] "If farmers' daughters are willing to become farmers' wives, let them not hesitate when occasion requires, to go into the service of other respectable farmers or into any other good, moral, respectable families; and instead of suffering their spirits to be wounded, or feeling chagrin or mortification, let them make themselves as cheerful, useful and agreeable as possible."[15] And so the lines were drawn to be expanded and clarified by other correspondents.

After Fessenden's death in 1837, Joseph Breck, the proprietor, became editor, although Henry Colman was engaged "to afford such aid, advice, and correspondence as he could render" without sacrificing his official duties to the state.[16] (Colman was at this time making an agricultural survey of Massachusetts at the request of Governor Everett.) This arrangement continued until January, 1841, when Allen Putnam of Danvers, a practical farmer who had been a Unitarian clergyman, became editor.[17] Finally in 1843,

---

agricultural revolutions in New England and their effects on household economy. In the period after 1800, especially after 1820, the transfer of the textile industry from farmhouses to factories is part of the transition from self-sufficient to commercial agriculture. See Bidwell, "The Agricultural Revolution in New England," pp. 693 ff.

[13] *New England Farmer*, X (Aug. 3, 1831), 19.

[14] *Ibid.*, X (Aug. 10, 1831), 27.

[15] *Ibid.*, X (Aug. 24, 1831), 46.

[16] Henry Colman (1785–1849) was graduated from Dartmouth College in 1805, and later became a minister of the Congregational Church. He published several volumes of sermons which were widely circulated in the United States and reprinted in England. Later, partly due to impaired health, Colman turned to farming, an occupation for which he had always felt much enthusiasm. *Dictionary of American Biography*, IV, 312; *New England Farmer*, XVI (Jan. 17, 1838), 217.

[17] *New England Farmer*, XIX (Dec. 23, 1840), 198; XIX (Dec. 30, 1840), 206; XIX (Jan. 6, 1841), 214.

Breck again took over the editorial duties and retained this office until the *Farmer* was discontinued.[18]

After twenty-four years of service, the *New England Farmer* came to an end, rounding out a career of unusual length. It had pursued the "even tenor of its way," with less change and greater steadiness of purpose than had attended most other periodicals in the country.[19] It possessed a reputation for sound, practical, common sense and was considered a wise counselor of the farmers throughout the United States and the Canadas. However, as the subscription list gradually diminished, the journal became a financial burden to the owner.[20] Unfortunately, in the thirties the paper had placed too much emphasis upon the "silk business," later known as the "silk mania," and this reacted unfavorably. Competition from journals springing up in all parts of the United States dealt a keen blow. Many subscribers were dissatisfied with the editor because he failed to include a greater variety of miscellaneous nonagricultural reading. "We felt [wrote the editor] that an important trust had been committed to our hands—that it would be akin to sacrilege to mutilate a work whose foundations were laid by such men as Fessenden, Lowell, Prince, Pickering, Buel and others of a kindred spirit." [21] A reduction in subscription rates resulted in no material benefit, so in 1846 at the close of the twenty-fourth volume,[22] Breck terminated the *New England Farmer*, while continuing his duties as

[18] *Ibid.*, XXII (Sept. 27, 1843), 102. Breck, author of several books on horticulture, after the discontinuance of the *New England Farmer*, acted as agent for the *Horticulturist* (Albany) and was its Boston correspondent. *New England Farmer*, XXIV (June 24, 1846), 409.

[19] *Cultivator*, n.s., III (July, 1846), 227.

[20] The actual circulation figures of the *New England Farmer* have been well concealed by its editors.

[21] *New England Farmer*, XXIV (June 24, 1846), 414.

[22] In 1848, two years later, another *New England Farmer* was launched in Boston with J. Nourse as publisher and S. W. Cole as editor. This journal continued until 1913 running through a number of series with minor changes at various times. After a period, it carried the notation "established in 1822" and numbered its volumes consecutively from that date. This has led to some confusion with regard to the ancestry of the (new) *New England Farmer*. Fortunately, the record is clear. In 1852 Breck wrote an article which was copied in the *New England Farmer*. He said in part, "I think it is out-

proprietor of the New England Agricultural Warehouse and Seed Store.

---

rageously gross and ungenerous to take the credit, honor and labor of others to build up *themselves*. The public as well as the publisher of the real *New England Farmer* have a right to complain of the confusion which is introduced by one person's assuming the name and using the reputation of another." The editor of the (new) *New England Farmer* said, ". . . we cannot, by any strict apostolic succession therefrom, trace our right to edit, print or publish our paper. We do distinctly deny that we are in any way connected by purchase, descent or otherwise, with Joseph Breck & Co., as publishers of the *New England Farmer*." *New England Farmer*, IV (Feb., 1852), 66.

# *The Maine Farmer* (1833–1924)

Winthrop, Hollowell, and Augusta, Maine

THE *Maine Farmer*, another early, important, and long-lived New England periodical, began in 1833 at Winthrop, Maine, and continued publication until 1924.[1] This weekly closely resembled the *New England Farmer* in appearance and general content selling at a subscription rate of only $2.00.[2] It carried the subtitle, "Journal of the Useful Arts," and was "devoted to Agriculture, Mechanics, and General News." In the first number, the editor explained that the journal was projected because the needs of the people in the "Siberian portion (as it has been churlishly called) of the Union" were not being met by the periodicals already in circulation. He stated that the main objective was the "mutual improvement of the Farmer and the Mechanic" giving the usual promise to keep "aloof from the melancholy jarrings of party and sectarian zeal." [3]

Ezekiel Holmes, the first editor, held this office until his death in 1865, a unique record of continuous service in the agricultural press, which also explains the unvarying policies of the *Maine Farmer* during the period. Holmes was born in Kingston, Massachusetts, in 1801. At the age of twenty, he completed his course at

[1] For the first two months this journal was called the *Kennebec Farmer*. Its place of publication was changed to Hollowell in 1837, where it remained until the following year, when it was again moved to Winthrop. Finally in 1844, the *Farmer* found its permanent home in Augusta. It was first issued by William Noyes. A list of proprietors before the Civil War may be found in N. T. True, "Biographical Sketch of Ezekiel Holmes," p. 212.

[2] Later, modifications were made in the size of sheet, number of pages, and price. See Griffin, *History of the Press of Maine*, pp. 97–98.

[3] *Kennebec Farmer*, I (Jan. 21, 1833), 4.

Brown University and three years later was graduated Doctor of Medicine at Bowdoin College. He practiced medicine in a small way throughout his life; his main interests lay in other fields. After serving a number of years as instructor of agriculture and later as principal of the Gardiner Lyceum, Holmes in 1833 became associated with this periodical.[4] In the same year he was appointed lecturer on chemistry and natural history in Waterville (now Colby) College, continuing in this capacity until 1837. During his editorship, he also served a number of years in the State Legislature and Senate, and in 1852, was nominated for governor by the Liberty Party.[5]

While the *Maine Farmer* contained many items on general agriculture, those subjects of particular interest to Maine and New England were stressed. A great deal of attention was directed to articles on potatoes, oats, barley, corn, hay, wheat, livestock, and dairying. In addition, numerous agricultural extracts were copied from the contemporary press throughout the United States.

The zeal for reform so marked in the *New England Farmer* was even more conspicuous in the *Maine Farmer*. This is not surprising since Maine gave to the world Elijah P. Lovejoy, Dorothea Dix, and Neal Dow. Articles against the use of tobacco, snuff, and alcohol were common. The magazine did its part in bringing about, in 1851, the passage of the first prohibition law in our history. Attacks

[4] The Gardiner Lyceum, the first American institution devoted exclusively to the teaching of the science of agriculture, was established in 1821 at Gardiner, Maine. Its founder was Robert Hallowell Gardiner (1782–1864). The school began with 20 students and received financial aid from the state. In 1832 the school was closed, owing to the financial embarrassment of its founder, together with other causes. A. C. True, *A History of Agricultural Education in the United States*, pp. 35 ff.; Bidwell and Falconer, *History of Agriculture in the Northern United States*, p. 194.

The *New-England Farmers' and Mechanics' Journal* (1828) of Gardiner was conducted by Doctor Holmes.

[5] N. T. True, *op. cit.*, pp. 207 ff.; *Dictionary of American Biography*, IX, 163–64. In 1838 he was employed by the state to make a survey of the natural resources of Aroostook Territory. His influence aided the establishment of a state Board of Agriculture in 1852 and he served as its secretary for the next three years. He was also active in the founding of the Maine Agricultural Society in 1855 and served as its secretary until his death. Frequently he spoke before agricultural societies throughout the state.

on slavery were frequent; land speculation and "stock gambling, as carried on in the great cities," received much criticism. In 1834 the reformers greeted with joy the editor's enthusiastic announcement of the organization of an Anti-Slandering Society by the young ladies of Waterville "for the prevention of evil speaking."[6] Correspondence on this subject brought about the demand for Anti-Tobacco Societies. When a bill for the abolition of capital punishment came before the State Legislature in 1835, the journal office was flooded with correspondence. As usual, the editor was finally compelled to call a halt on the debate.[7] In 1838 Holmes opened the pages of the *Farmer* to those interested in the subject of peace. There was for a time a "Peace Department," mainly set aside for those who wrote upon the principles of the advocates of non-resistance. The editor stood aloof on this subject but allowed each correspondent to have his turn.

As we might expect, the periodical reflected the experiences and interests of the editor. As Holmes had taught school for many years, educational subjects were stressed. His years in the Maine House and Senate further expanded his interests and no doubt suggested a column on "Proceedings of the Legislature." In order to familiarize the people with state jurisprudence a "Legal Department" was introduced and continued for a number of years. Included in this unique department was a series of articles which resembled a course in state law. Problems of government were discussed and answers to specific legal questions were printed. A gentleman, "eminently qualified" was in charge of this section. Holmes's medical training naturally turned his interest to current medical practice and theory; so articles on hydropathy, mesmerism, and phrenology received a good share of his attention.

The *Maine Farmer* was popular from its early days and was considered one of the most useful and entertaining journals of the period; the editor uniting good sense with much good humor. He

[6] *Maine Farmer*, II (Dec. 5, 1834), 362; II (Dec. 12, 1834), 371; II (Dec. 19, 1834), 380; III (Feb. 6, 1835), 5.

[7] *Ibid.*, V (March 4, 1836), 38. This act was finally approved in 1837 and brought about the practical abolition of capital punishment in Maine. *Ibid.*, V (Aug. 15, 1837), 214.

was thoroughly competent and highly respected by his fellow editors.[8] Commencing with about 200 subscribers in 1833, its patronage increased to 2,200 by 1845. From this time its expansion was rapid, rising to 4,000 in 1848, and 6,000 in 1852.[9] Although this journal circulated throughout the United States, it was far less national in scope than the *American Farmer*, since it contained many items of purely local interest, such as marriages, births, and deaths. Holmes's contribution to Maine's agricultural history was well summarized by his successor who wrote, "To him must be rightfully accorded the honor of being the founder of systematic and intelligent farming in Maine." [10]

[8] Holmes has been given credit for the introduction and popularization of the Jersey breed of cattle in Maine about 1855. Stackpole, *History of Winthrop, Maine*, p. 181.

[9] *Maine Farmer*, VIII (Jan. 11, 1840), 1; XIII (Dec. 25, 1845), 205; XVI (March 2, 1848), 36; XX (Jan. 1, 1852), 4.

[10] N. T. True, *op. cit.*, p. 222.

June 7, 1924, was the date of the last independent issue of the *Maine Farmer*. From June 15, 1924, to July 26, 1925, it was issued as a supplement to the Sunday edition of the *Portland Press Herald*.

# *The Farmer's Monthly Visitor* (1839–1849)

Concord, New Hampshire

ON January 15, 1839, six years after the founding of the *Maine Farmer,* Isaac Hill issued at Concord, New Hampshire, the first number of the *Farmer's Monthly Visitor;* he continued as editor during the eleven years of its existence. When Hill established this journal, he was fifty years of age, and had achieved success as an editor and politician. After serving in New Hampshire as state representative and senator, he was, in 1830, elected to the United States Senate where he became a member of Jackson's famous "Kitchen Cabinet." [1] He played an important part in the establishment of the spoils system and was "the Marat of the Kitchen Cabinet, the fanatic, calling for heads—more heads—and unseemly in his mirth as they fell." [2] In 1836 he was elected governor of New Hampshire, and served till 1839. The thought of entering the editorial field of agriculture occurred to Hill when he made a thorough study of the *Genesee Farmer,* and was profoundly impressed by its beauty and utility.[3]

[1] Isaac Hill (1789–1851) was born in Cambridge, Mass. When he was about nine, his mother purchased a farm in Ashburnham, where he spent the next four years. Lameness and a slight physique handicapped him as a farmer, so in 1802, he was apprenticed to Joseph Cushing, a printer in Amherst, N.H. In 1809 he published the first number of *Hill's New Hampshire Patriot,* a paper destined to have great influence in New England. It was an organ of the Republican (Jeffersonian) party and it took no small amount of courage in those days to oppose the Federalists in New Hampshire. *Dictionary of American Biography,* IX, 34–35; Appletons' *Cyclopaedia of American Biography,* III, 204–5; Bradley, *Biography of Isaac Hill of New Hampshire,* pp. 23 ff.

[2] Bowers, *Party Battles of the Jackson Period,* p. 158.

[3] *Farmer's Monthly Visitor,* XI (Dec. 31, 1849), 182.

The *Farmer's Monthly Visitor* was a small folio of sixteen pages, slightly larger than the *American Farmer*. It sold originally for seventy-five cents per annum, but this was reduced to fifty cents, effective with Volume V. As the editor estimated that each number contained about fifty-three thousand words, the *Visitor* was advertised as the "cheapest agricultural paper in the United States." [4]

The first number announced that the journal aimed to "elevate the character and calling of the practical farmer—to contribute to his intellectual improvement—to induce him to cultivate the kindly and social affections and to be content with his occupation and calling"—and to impart to him such information as would "enable him to cultivate his ground to the best advantage and the greatest profit." [5] The editor further explained that the periodical would include

> suggestions of improvements in cultivation, in agricultural implements . . . , in the construction of farm-buildings, in the application of chemistry to the general purposes of agriculture, in the destruction of insects injurious to vegetable life, in the eradication of weeds; the discovery of new varieties of grain, and other vegetables useful to man, or for the food of domestic animals; useful information in regard to the management of woods for fuel and timber; information relative to the best breeds of live stock, and the various methods of preventing and curing the diseases of cattle, sheep, pigs and poultry.[6]

As a true Jeffersonian Democrat, Governor Hill adopted a motto for the *Visitor* from the writings of his political prophet: "Those who labor in the earth are the chosen of God, whose breasts he has made his peculiar deposite for substantial and genuine virtue." [7] This motif runs throughout the volumes of the magazine.[8] The

[4] In July, 1840, the *Farmer's Monthly Visitor* absorbed the *Cheshire Farmer* of Keene, N.H. *Farmer's Monthly Visitor*, II (July 31, 1840), 108.

[5] *Farmer's Monthly Visitor*, I (Jan. 15, 1839), 2.

[6] *Ibid.*, I (Jan. 15, 1839), 1.

[7] Thomas Jefferson, quoted in the *Farmer's Monthly Visitor*, I (Jan. 15, 1839), 1.

[8] This same theme is conspicuous in most of the journals, but received special emphasis in this paper. Editor Hill included on his title page as a part of the purpose of the *Visitor* these words: "to defend the dignity of the agricultural profession."

following are typical statements: "The farmer is the most noble and independent man in society," "They are little kings," "It is admitted that agriculturists are the most honest, upright, industrious and peaceful class of mankind."

Aside from items on agriculture, two main subjects were emphasized: history—American and European—and education. James Buchanan once remarked in the Senate, that he had never known a man with a wider range of information on American affairs than Senator Hill.[9] This knowledge and interest in American affairs was reflected in many articles typical of which are the following: "Soldiers of the Revolution," "John Randolph," "The Capture of Andre," "Washington's Farewell to his Army," and the "Capture of Burgoyne's Army." Items on European history abounded. Educational subjects also received a liberal allotment and in 1849 the "Common School Department" was inaugurated as a special feature conducted by Professor Rust, commissioner of common schools in New Hampshire.[10]

During the last fifteen years of his life, Hill was extensively engaged in agriculture on his own account.[11] However, according to his personal testimony, he considered himself a "novice in agriculture," and, as an agricultural editor, a "mere gatherer" when compared in practical knowledge and experience with such editorial giants as Buel of the *Cultivator*, Skinner of the *American Farmer*, Ruffin of the *Farmers' Register*, and Holmes of the *Maine Farmer*.[12] While never considered an outstanding agricultural expert by his colleagues, he was nevertheless a veteran editor, and had the faculty of engendering enthusiasm among his contributors, and the good sense to select valuable articles from contemporary agricultural journals. Hill asserted that the *Visitor* would recommend no new plan or system of agriculture the practicability about which its

[9] *Dictionary of American Biography*, IX, 34. Hill, as a youth, was an omnivorous reader. It was said that he read completely through the Bible before he was eight years of age. Bowers, *op. cit.*, p. 156.

[10] *Farmer's Monthly Visitor*, XI (May 31, 1849), 69.

[11] Stearns, *Genealogical and Family History of the State of New Hampshire*, IV, 1982.

[12] *Farmer's Monthly Visitor*, IX (Dec. 31, 1847), 190; III (Dec. 31, 1841), 190.

editor felt any uncertainty. He claimed the "merit of holding back in the Morus Multicaulis speculation," that is, the silk craze that caused financial losses to so many farmers throughout the United States.[13] Furthermore he felt that his journal had no controversy with any of the religious or political dogmas of sects or parties and had "presented no matter calculated to provoke the hostility of the mendacious, to awaken obstinate prejudice, or to raise a blush upon the cheek of modesty." It was "fitted for the readers of both the parlor and the kitchen." [14]

While this periodical in 1847 boasted of the hearty approval of men such as ex-President Van Buren, who placed the *Visitor* at the top "of the numerous agricultural journals" he received, nevertheless it was not a financial success.[15] In the first two years, the circulation was about 5,000; in its fifth year 4,000; in its seventh year 2,500; and the following year 2,400.[16] At the end of Volume IX, Hill announced that by force of circumstance "the labor of the editor all the time" had been gratuitous. He attributed the declining circulation to the fact that it had "at all times been published as secondary and subsidiary to another newspaper," *Hill's New Hampshire Patriot* and as a result there had been no regularity of publication and delivery. Pressure on the editor from other business duties and his increasingly frail health contributed to the neglect of the farm journal.[17]

Therefore, when in 1849 the *Farmer's Monthly Visitor* [18]

[13] A detailed discussion of this speculation may be found on pp. 60–61.

[14] *Farmer's Monthly Visitor,* X (Dec. 31, 1848), 190.

[15] *Ibid.,* IX (March 31, 1847), 43.

In 1848 there were, however, more than 650 agents for the *Visitor* scattered over the United States. *Farmer's Monthly Visitor,* X (Nov. 30, 1848), 174–75.

[16] *Farmer's Monthly Visitor,* V (Dec. 30, 1843), 189; VII (Nov. 30, 1845), 176; VIII (Dec., 1846), 184.

The *Cheshire Farmer* (Keene, N.H.) was united with the *Farmer's Monthly Visitor* in 1840. *Cheshire Farmer,* n.s., II (May, 1840), 182.

[17] Hill served as head of the Boston subtreasury, 1840–41. *Farmer's Monthly Visitor,* II (July 31, 1840), 111; *Ibid.,* IX (Dec. 31, 1847), 190.

[18] Two years later, in January, 1852, another *Farmer's Monthly Visitor*

came to an end with the completion of Volume XI, little surprise was manifested at its demise.

---

was inaugurated in Manchester, N.H., under the editorship of C. E. Potter. An octavo of thirty-two pages, it was priced at $1.00 a year. This Manchester journal was actually entirely distinct and unrelated to the Concord *Visitor*, although the editor retained the title and volume numbering of the older journal and wished the new work to be considered a continuation of the old. In 1854, after merging the Manchester *Visitor* with the *Granite Farmer*, the title was changed to *Granite Farmer and Visitor*, and a year or two later to *New Hampshire Journal of Agriculture*. Other changes followed. The complicated history of this Manchester journal may be found in Hurd, *History of Hillsborough County, New Hampshire*, p. 61. This account is very detailed and, in the main, accurate.

# *The Genesee Farmer* (1831–1839)

Rochester, New York

IN THE PERIOD under review, New York furnished more outstanding agricultural journals than any other state in the Union.[1] One of the best was the *Genesee Farmer*, which appeared January 1, 1831, in Rochester and was published and edited by Luther Tucker.[2] He was of the opinion that while other branches of science had been progressing, aided by the exertions of men of learning and invention, agriculture and horticulture had been neglected. Wherever he traveled, he witnessed a lack of intercommunication among the farmers, the tendency of an all-prevailing

[1] The first agricultural journal in New York State was the *Plough Boy* which was established on June 5, 1819, in Albany. This quarto weekly was under the editorship of Solomon Southwick, who for the first few months used the pen-name, "Henry Homespun, Jr." His early training was in the printing business; in 1812 he became regent of the state university and later ran for governor. At the time he inaugurated the *Plough Boy* he was postmaster at Albany. In 1820 this journal became the official organ of the State Board of Agriculture. Southwick seems to have had no practical experience in farming and lacked a vital interest in it. (The books he wrote dealt with other subjects.) These facts, coupled with ill health and extreme carelessness in his financial records, account for the failure of this pioneer journal after four years of publication. *Appletons' Cyclopaedia of American Biography*, V, 614; *Dictionary of American Biography*, XVII, 413–14; *Plough Boy*, III (April 20, 1822), 374; IV (Oct. 15, 1822), 158.

[2] The first volume of the *Genesee Farmer* carried the names of Tucker and Stevens as proprietors. However, Stevens's connection with the enterprise was nominal and his name was soon dropped. While the first two volumes indicate that N. Goodsell was the editor, Tucker seems, nevertheless, to have been the leading spirit. In later volumes, David Thomas, Willis Gaylord, and John J. Thomas assisted in the editorial department.

practice toward deterioration of the soil, and the almost universal absence of agricultural reading.[3] The journal was inaugurated contrary to the advice of many of his friends, who felt it was "hazardous almost to madness to attempt such a work." [4]

Luther Tucker was one of the great figures in early agricultural journalism in America. He was born in Brandon, Vermont, on May 7, 1802. Although his formal schooling was meager, he acquired a good education through his own efforts and became adept as a writer. He was apprenticed to a printer in Middlebury, Vermont, at the age of fourteen, then served as a traveling journeyman, and before his twenty-fifth birthday, established the *Rochester Daily Advertiser* in 1826, the first daily paper west of Albany.[5]

The *Genesee Farmer*, which received most of his attention, was unique as one of the first journals to be written directly from the standpoint of practical experience. Indeed, early in Volume II, the editor boasted that he was enabled to present to the public a number composed entirely of original articles, many from the pens of the most outstanding agriculturists in the country.[6] In the first two volumes over a hundred men contributed original articles on almost every phase of agriculture and horticulture. In nearly all cases these were practical treatises by practical men.[7] The contributions on horticultural subjects were famous in this field.

One of the subjects receiving especial attention was agricultural education. Tucker advocated that agriculture be taught to farmers' sons as a science as well as an art. He felt that in order for this training to be of any permanent good, it should be taught concretely, not merely from books. Practical application in farm management as a required part of an agricultural school course was his plan.

[3] *Genesee Farmer*, I (Jan. 1, 1831), 1; *Cultivator and Country Gentleman*, XXXVIII (Jan. 30, 1873), 72.

[4] *Genesee Farmer*, II (Jan. 7, 1832), 1. The journal was a weekly quarto of eight pages the price $2.00 in advance. During the years 1836–39, Tucker published a *Monthly Genesee Farmer and Horticulturist* consisting of selections from the weekly.

[5] Ogilvie, *Pioneer Agricultural Journalists*, pp. 19 ff.; *Dictionary of American Biography*, XIX, 35–36.

[6] *Genesee Farmer*, II (Jan. 7, 1832), 1.

[7] *Ibid.*, III (April 20, 1833), 121.

These hopes were finally realized on a grand scale through the Land Grant Act of 1862.

In October, 1839, Tucker's editorial career was influenced and his activity expanded through the death of Judge Buel, who had conducted the *Cultivator* at Albany. Accepting an offer to purchase the *Cultivator*, Tucker consolidated it with the *Genesee Farmer* and, under the title of the *Cultivator*, issued it at Albany. The *Genesee Farmer*, in its last year, had reached a circulation of 18,000.[8] Before considering this new and larger *Cultivator*, a brief review of a journal that later claimed to be the successor of the original *Genesee Farmer* should be made.

In January, 1840, a *New Genesee Farmer* was published in Rochester with John J. Thomas and M. B. Bateham as editors.[9] This *New Genesee Farmer* at first claimed no relationship to Tucker's discontinued *Genesee Farmer*.[10] However, with Volume VI the word "New" was dropped from the title and to the volume number was added "second series," suggesting a continuation of the first *Genesee Farmer*. In 1858 Thomas and Bateham asserted that the journal had been inaugurated in 1831 and not in 1840, a claim unsupported by Luther Tucker, editor of the original *Genesee Farmer*.[11] Despite this confusion in titles, the *New Genesee Farmer* was destined to be one of the most important and influential jour-

[8] *Ibid.*, IX (Nov. 9, 1839), 353. In the same year the *Monthly Genesee Farmer* had a circulation of about 14,000. *Ibid.*, IX (May 4, 1839), 137.

[9] M. B. Bateham and E. F. Marshall were the proprietors of this quarto monthly of sixteen pages, which sold for the small sum of fifty cents a year. Bateham, who furnished the real initiative in the undertaking, had been proprietor of a seed store in Rochester for five years and his business depended to a large extent upon the *Genesee Farmer*. *New Genesee Farmer*, I (April, 1840), 52–53.

[10] *New Genesee Farmer*, I (Jan., 1840), 1.

[11] *Ibid.*, XIX (Jan., 1858), 36; XIX (Feb., 1858), 67. To simplify the footnote references to this journal founded in 1840, the title and series numbering have been followed exactly as found in the journals. Possibilities of further confusion were added on July 4, 1833, when N. Goodsell published in Rochester a new journal, *Goodsell's Genesee Farmer*, a journal which seems to have lasted only a year. In 1843 a *True Genesee Farmer* was launched in Rochester, but fortunately for the cataloguers after a few months it was merged with the *New Genesee Farmer*.

nals in the country. Among the agricultural and horticultural authorities who served in an editorial capacity prior to 1860 should be mentioned Henry Colman, orator, and famous for his *European Agriculture* and other agricultural writings; Doctor Daniel Lee, nationally known agricultural lecturer and one of the best agricultural chemists and educators of the period; James Vick, florist and seedsman of world-wide reputation; Joseph Harris, a chemist by profession, author of several farm books as well as a research worker in agricultural fields; and perhaps the best known, Patrick Barry.

Patrick Barry, noted nurseryman and horticultural authority, conducted the famous "Horticultural Department" of the *New Genesee Farmer* from its commencement in 1845 until 1853. This journal was one of the first to inaugurate such a department under the supervision of an expert. Barry's wide experience was further enriched by horticultural tours over the United States and Europe. His department retained its leadership until Barry resigned to become editor of the *Horticulturist,* following the death of the editor, Andrew Jackson Downing.

Possibly some indication of the extensive influence of this journal may be gained from its circulation figures. In 1841 it listed more than 17,000 subscribers; in 1851 the number increased to approximately 40,000; and in 1859 still remained at about that figure.[12] Fifty cents a year was considered extremely cheap for such an excellent paper. It is not surprising that patrons were drawn from every state and territory in the Union, and also from Great Britain, Ireland, and Canada. "Canada West" alone furnished 3,000 subscribers in 1859.[13] Finally, in 1865, this popular magazine went the way of many another agricultural periodical, when it was merged with the *American Agriculturist* of New York city.[14]

[12] *New Genesee Farmer,* II (June, 1841), 88; XII (May, 1851), 124; ser. 2, XX (March, 1859), 94.

[13] *Ibid.,* XII (April, 1851), 100; ser. 2, XX (May, 1859), 164.

[14] *Ibid.,* ser. 2, XXVI (Dec., 1865), 361.

# *The Cultivator* (1834–1865)

Albany, New York

As WE have already noted, the *Genesee Farmer* was closely associated with a contemporary New York periodical, the *Cultivator*, founded at Albany in 1834 by Judge Jesse Buel. In a very real sense, Buel was the *Cultivator* during its first six years, and to understand him is to appreciate fully this remarkable journal. Like Luther Tucker, he was an experienced journalist and printer, also a son of New England, born at Coventry, Connecticut, on January 4, 1778. Buel was a self-made man, for he had received a scant six months of formal schooling, when he was apprenticed, at the age of fourteen, to a printer in Rutland, Vermont. His rise was rapid. By 1821 he had initiated a total of five newspapers and served as judge of the Ulster County Court. Having achieved a competence, he disposed of his business interests and turned to agriculture, as so many successful men in all walks of life were doing at this period. He was now forty-three years of age, physically sound, financially independent, and possessed of a mind which had attained the full maturity of its powers.[1]

He purchased an eighty-five acre farm near Albany, lying in the "Sandy Barrens," which was "unimproved, covered with bushes, and apparently doomed to everlasting sterility." Here he farmed, experimented, and wrote until his death eighteen years later.[2] It

[1] *Farmer's Monthly Visitor*, II (March 31, 1840), 33; Ogilvie, *Pioneer Agricultural Journalists*, p. 13; *Dictionary of American Biography*, III, 238.

[2] Judge Buel continued actively in many fields other than agriculture. During this period he served several times in the New York State Legislature and was the Whig candidate for governor in 1836. (It is noteworthy that no mention of this fact appeared in the *Cultivator*.) At the time of his death

is safe to say that no man did more for the advancement of farming than Jesse Buel in the two decades before 1840. He was not only one of the most influential writers, but serves today to typify the movement for agricultural improvement in the East during the period.[3]

It is as editor of the *Cultivator* that Buel will be longest and most favorably known, for this journal had few equals among farm papers in any part of the world. Inaugurated as the organ of the New York Agricultural Society, it was, for the first year, under the immediate direction of a committee of publication with Jesse Buel as the leading spirit. Since the editors served without pay, this quarto monthly of sixteen pages was sold for the ridiculously low price of twenty-five cents a year. After the first year of experiment, the *Cultivator* was turned over to Buel as sole editor and conductor, and under his direction it was gradually enlarged and the price increased to $1.00 per annum.[4]

Buel was a man of prominence in the agricultural world long before he became editor of the *Cultivator*. He officiated for several years as secretary of the New York State Board of Agriculture, served as president of both the State Agricultural and the Albany Horticultural Societies, and issued a number of agricultural tracts. He was nationally known through his contributions to the *American Farmer*, *New England Farmer*, *New-York Farmer*, and his editorials in the *Genesee Farmer*. As evidence of respect and admiration for his life's work, he was made an honorary member of agricultural and horticultural societies in all parts of the United States, as well as in Lower Canada, England, and France.[5]

The periodical under Buel was quite successful in breaking down

---

he was regent of the University of the State of New York. *Dictionary of American Biography*, III, 238.

[3] *Farmer's Monthly Visitor*, II (March 31, 1840), 33; *Genesee Farmer*, IX (Oct. 26, 1839), 337; Bidwell and Falconer, *History of Agriculture in the Northern United States*, p. 319.

[4] Judge Buel personally paid the debt of $500 incurred during the first year of publication. Buel, *The Farmer's Companion*, p. xvi. The circulation during this first year was 12,000. *Cultivator*, I (Feb., 1835), 177.

[5] Buel, *op. cit.*, p. xvii; *Cultivator*, V (Jan., 1848), 25; Hedrick, *A History of Agriculture in the State of New York*, p. 321.

the prevalent opposition to book-farming. The Judge was a book-farmer in the strictest sense of the term and his scientific methods brought such remarkable results as to convince the "dullest unbelievers." The many interested visitors to his farm bore witness to the conversion of his land into one of the most fertile and valuable spots in the state.[6] The fact that its value had increased sevenfold in less than twenty years spoke in language that farmers everywhere could understand. The editor had been shocked with the evidence of soil exhaustion throughout the country's settled regions, a result of the contemporary attitude that the greatest immediate return from the smallest amount of labor was the important consideration. In the pages of the *Cultivator*, this farmer-editor continued to teach his advanced system of agriculture—sustaining and strengthening the soil by manuring, draining, good tillage, crop alternation, root cultivation, and the substitution of fallow crops for naked fallows.[7]

Over a period of twenty years, Buel emphasized in the New York Assembly, preached from the public platform, and stressed in the *Cultivator*, the need for professional schools of agriculture as well as such instruction in the district schools.[8] His journal became the medium of propaganda for his ideas on agricultural education. Indeed, he outlined an educational program which has been in a large measure adopted in the last fifty years.[9] He further urged legislative encouragement and appropriations and did all in his power to promote the organization of both state and county agricultural societies throughout the country.[10]

To express the design of the *Cultivator*, Buel selected the motto: "To improve the soil and the mind." He enlisted more than two

[6] *Genesee Farmer*, IX (Oct. 26, 1839), 337.

[7] *Farmer's Monthly Visitor*, II (March 31, 1840), 33.

[8] A. C. True, *A History of Agricultural Education in the United States, 1785–1925*, p. 47.

At the request of the Massachusetts Board of Education, Buel wrote *The Farmer's Companion* for use in school and rural libraries. This little volume designed as a textbook went through at least six editions and contained much material from the pages of the *Cultivator*. Buel, *op. cit.*, p. xxii.

[9] *Dictionary of American Biography*, III, 238.

[10] *Cultivator*, V (Feb., 1839), 197.

hundred correspondents of rich and varied experience from all parts of the country to aid him in this process. However, the real success of the magazine was due, in large part, to Buel's editorial tact, his great practical experience, his extensive and varied information, as well as his personal popularity. Twenty thousand families throughout the Union and in foreign lands received his messages each month.[11] The announcement of his death in October, 1839, came as a great shock and loss to the agricultural world. Isaac Hill, editor of the *Farmer's Monthly Visitor*, expressed the general tenor of the farm press when he wrote, "To no man is the country so much indebted for improvement in Agriculture as to this gentleman; as a useful, practical, safe farmer, he probably excelled all others." [12]

It was at this time that Luther Tucker purchased the *Cultivator*, consolidated it with the *Genesee Farmer* and issued the new journal under the title of *Cultivator* without substantially changing its form, content, or emphasis. For the next five years this periodical was edited by Willis Gaylord and Luther Tucker. Gaylord had served as contributor and later as editor of the now extinct *Genesee Farmer*. It was said that he had no equal in the state of New York "in his ability to discuss clearly and correctly every department of agricultural science." [13] After Gaylord's death in 1844, Tucker was assisted in the editorial department by such agricultural leaders as Sanford Howard and John J. Thomas. Frequent contributions from agricultural and horticultural giants such as Solon Robinson, Andrew Jackson Downing, Professor John P. Norton, Henry Colman, and Frederick Holbrook kept the *Cultivator* in the leadership of the farm press.

In many respects this periodical was ahead of its time. An early appearance of the now familiar 10–1 corn-hog ratio is found in the August, 1847, issue.[14] Four years later, Tucker advocated the

[11] *Ibid.*, VI (May, 1839), 49.

[12] *Farmer's Monthly Visitor*, I (Dec. 20, 1839), 182.

[13] *Dictionary of American Biography*, VII, 199–200. For twenty years, Gaylord had contributed to American and European scientific magazines. He wrote articles for the newspapers in this country on literary, religious, and scientific subjects and occasionally he wrote poetry. *Cultivator*, n.s., I (May, 1844), 138.

[14] *Cultivator*, n.s., IV (Aug., 1847), 244.

establishment of agricultural experiment stations.[15] To meet the needs of the changing times, he introduced new departments such as "Horticulture," "Veterinary," and "Poultry Yard." Rural architecture received much attention and the paper was noted for the variety and excellence of its illustrations. Indeed Tucker was ever attentive to the demands of the public and in 1853, when he realized that rural life was broadening beyond the scope of the farm publications then existing, he inaugurated a weekly called the *Country Gentleman.*[16] This new journal took the center of the stage, and the *Cultivator,* with its price reduced to fifty cents, henceforth was composed of articles selected from this new, popular weekly. Both contemporary and present-day opinions agree quite generally that the *Cultivator* was the best agricultural periodical published in the East, if not the whole United States, during the period.[17] It continued to be published until 1866 when it was merged with the *Country Gentleman.*

[15] *Ibid.*, n.s., VIII (June, 1851), 194.

[16] In 1843 Tucker had inaugurated a monthly at 50 cents per year called the *Farmer's Museum* which was made up from material in the *Cultivator.* This journal lasted slightly more than a year. In the same year the *Cultivator Almanac* was issued. It had a circulation of 30,000 within the first six months. *Cultivator,* n.s., I (March, 1844), 73.

In 1846 he commenced a new journal called the *Horticulturist.* Under its first editor, Andrew Jackson Downing, it was noted for articles on architecture and later, under Patrick Barry, for hand-colored plates. In 1855 Tucker issued the *Illustrated Annual Register of Rural Affairs,* a condensed encyclopedia of rural matters. This appeared each year thereafter.

[17] Bidwell and Falconer, *op. cit.*, p. 470. *Dictionary of American Biography,* III, 238; Hedrick, *op. cit.*, p. 321; *Genesee Farmer,* IX (Oct. 26, 1839), 337; *Farmer's Monthly Visitor,* IV (Jan. 31, 1842), 16; *New England Farmer,* XXII (Sept. 27, 1843), 102; *Farmer's Monthly Visitor,* XI (Dec. 31, 1849), 182; *Prairie Farmer,* V (Feb., 1845), 33.

Frequently as many as three hundred different contributors appeared each year in the *Cultivator.* Its circulation was often 20,000 and more.

# *The Farmers' Cabinet* (1836–1848)

Philadelphia, Pennsylvania

THE *Farmers' Cabinet,* established in Philadelphia on July 1, 1836, differed in minor aspects only from the journals previously considered.[1] In the first place it was published semimonthly and contained sixteen octavo pages.[2] A sheet of this small size was selected, as the numbers could be then more readily and more inexpensively preserved and bound. The little journal cost $1.00 per year. Another difference might be noted: the strong and engaging personality of the editors, so keenly felt in the periodicals already mentioned, was almost totally lacking. Indeed, frequently, it is difficult to determine the identity of the editor.

The first important editor of the *Farmers' Cabinet* was James Pedder, who took over his duties in April, 1840.[3] Born in England on July 29, 1775, Pedder spent most of his life in his native land. While little is known of his early years and education, it is certain that for a long time he was an assistant to the celebrated chemist and writer, Samuel Parks. In 1832 he emigrated to Phila-

[1] Commencing July 16, 1838, the *Farmers' Cabinet* was published simultaneously in Philadelphia and Wilmington to accommodate the patrons principally in Delaware and the Eastern Shore of Maryland. *Farmers' Cabinet,* II (June 15, 1838), 352. Earlier it had been published in both Philadelphia and Pittsburgh for a short time.

[2] A year and a half later the journal was converted into a monthly of thirty-two pages in order to allow more time for the collection and preparation of materials for each number. *Farmers' Cabinet,* II (Feb. 15, 1838), 193.

[3] *Ibid.,* IV (April 15, 1840), 265.

John Libby was the first "official" editor, although Francis S. Wiggins "sustained the relation of editor" to the journal from its commencement until shortly before his death in February, 1840. *Ibid.,* IV (Feb. 15, 1840), 226.

delphia where he spent a number of years in the sugar manufacturing business.[4] Pedder, as a subscriber and contributor, had been closely identified with the *Cabinet* since its inauguration.[5] In fact, a great number of the best articles which appeared in its pages emanated from his pen. When Pedder assumed the editorial duties, his readers had been assured that their new editor was bred a farmer, was a practical man, and had made the best of his extensive opportunities for study and observation.[6] He was known to his contemporaries "as a good, plain, practical writer."

The *Farmers' Cabinet* was closely associated with the Philadelphia Agricultural Society and provided the medium of communication between that body and the rural community.[7] This society passed the following resolution in 1840:

> *Resolved*, That the Farmers' Cabinet, published in this city, and edited by James Pedder, a member of this society, merits our decided approbation, inasmuch as it is a means of diffusing much valuable information amongst the agriculturists of our country; and also *Resolved*, That this society being desirous of extending its patronage to said periodical, pledge its exertions to procure and present communications for publication in its columns.[8]

In 1842 Editor Pedder was elected librarian of the society.[9]

Commencing with Volume V, the title of the *Farmers' Cabinet* was changed to the *Farmers' Cabinet and American Herd-Book*. The readers were assured that the paper would "continue to be devoted to whatever was of importance in the pursuits of agriculture, horticulture, and rural affairs." However, a marked change in emphasis occurred at this time. The *Cabinet* now became the medium of communication between the breeders of fine stock in Pennsylvania and the adjoining states. In addition, the editor offered a "portion of its pages, in which to portray, in an improved style

[4] *Dictionary of American Biography*, XIV, 387–88.

[5] *Farmers' Cabinet*, IV (April 15, 1840), 266.

[6] *Ibid.*, IV (April 15, 1840), 265. In 1842 Editor Pedder published a book entitled *The Farmers' Land Measurer, or Pocket Companion*, which supplied easy methods for measuring grain in bulk. *Ibid.*, VI (July 15, 1842), 384.

[7] *Ibid.*, IV (July 15, 1840), 384.

[8] *Ibid.*, IV (May 15, 1840), 328.

[9] *Ibid.*, VI (Feb. 15, 1842), 232.

of engraving, animals of high blood—the Horse, Cattle, Sheep, Hogs, etc." [10] Among the "portraits" of famous animals were the Durham Bull "Prince of Wales," the Ayrshire Cow "Ellen McGregor," and the Bakewell Sheep "Major"; with each was given the pedigree of the animal and the name of its breeder or owner.

Although the magazine served as a valuable agricultural paper and exerted considerable influence throughout the United States, this did not spare it from criticism.[11] For instance a friend at Gettysburg begged leave to suggest the propriety of giving "more practical facts, rather than filling the paper with speeches at agricultural fairs, and controversies about ploughs," which were only known to particular sections of the country. "What we want to know, is how to raise the greatest amount from a certain quantity of ground, with the least labour and expense. Give us the facts, my dear sir, with the *modus operandi.*" [12] On the other hand, there were grateful acknowledgments for benefits received. A writer from Delaware, after explaining that he was a constant reader of the *Cabinet* and followed its instructions, admitted that he received the ridicule of some of the "old fashioned farmers" who laughed at book-farming. However, he added, "I have the satisfaction to know, that two years labor has about doubled the product of my farm, without any additional outlay." [13]

In 1848, with the close of Volume XII, the editor announced that the *Cabinet* would be discontinued.[14] It was sold to and merged with the *American Agriculturist* of New York City.

[10] *Ibid.*, IV (July 15, 1840), 384.

[11] The first volume of the *Cabinet* had a circulation of nearly 10,000, and later went through at least two more editions. The editor claimed his paper circulated in almost every state in the Union and some foreign lands. *Farmers' Cabinet*, I (Announcement Sheet, p. 3); *Ibid.*, III (April 15, 1839), 295. The average circulation for the first four years exceeded 7,000 annually. *Ibid.*, IV (April, 1840), 265.

[12] *Ibid.*, X (April 15, 1846), 293.

[13] *Ibid.*, X (March 16, 1846), 261.

[14] Josiah Tatum became proprietor and editor of the *Cabinet* in August, 1843, which offices he held to the conclusion of the paper. During this time he was also actively engaged in the coal, poudrette, and nursery business. After the sale of the *Cabinet*, he served as agent for the *American Agriculturist. American Agriculturist*, VII (July, 1848), 201.

# *The American Agriculturist* (1842– )

New York, New York

THE *American Agriculturist,* inaugurated in April, 1842, by the brothers A. B. and R. L. Allen, in many respects had a more notable history than any of its rivals.[1] Its national scope was evident from the beginning, for the editors promised in the prospectus to suit the work to every latitude and longitude in America; its subtitle assured the readers that the contents would appeal to the planter, farmer, stockbreeder, and horticulturist. New York City was chosen as the place of publication with some hesitation; the editors feared that the selection of this commercial metropolis might be interpreted by farmers as indicating less than complete devotion to agricultural matters.[2] As first published, this periodical was an octavo of thirty-two pages, issued monthly, at $1.00 per year.

Impressed with the superiority of English livestock, A. B. Allen had earlier imported some of the best blood to start herds of purebred stock in this country.[3] At the time of the inauguration of this journal, the brothers were engaged in the livestock business and as a result the *American Agriculturist* conducted a campaign for education in better breeding methods. The advantages of improved farm im-

[1] A third brother, Lewis F. Allen, whose chief lifework was as a breeder and improver of cattle, contributed freely to the *American Agriculturist* and similar journals. *Dictionary of American Biography,* I, 201. R. L. Allen retired from the editorship at the end of the first year, although he continued to write for the paper. He again became editor in 1849. The first publishers were Saxton & Miles, after which frequent changes occurred until 1856, when Orange Judd became publisher.

[2] *American Agriculturist,* I (April, 1842), 3.

[3] *Dictionary of American Biography,* I, 185.

plements were also stressed. In 1847 the brothers created the firm of A. B. Allen and Company and opened an agricultural implement warehouse in New York and, soon thereafter, an extensive agricultural implement works in Brooklyn.[4]

Of the many notable contributors to the *American Agriculturist*, Solon Robinson, famous for his travel articles, was perhaps the best known.[5] In 1851 he became assistant editor, and while serving in this capacity, he also traveled a great deal. In the same year the Allens decided to suspend publication because of the demands of their warehouse and factory.[6] Therefore, early in 1852, it was succeeded by the *Plow*, under the editorship of the veteran Robinson and with offices at A. B. Allen and Company. This journal which was the same size as the *American Agriculturist* and sold for only fifty cents a year, received contributions from a corps of talented writers including the Allens. The new paper was, alas, financially unsuccessful; its suspension at the end of the first year was due in part, at least, to the general demand for a publication more frequent than a monthly.[7]

In September, 1853, the *American Agriculturist* was revived as a sixteen-page weekly, a small folio, selling at $2.00 per annum.[8] Most significant for the future of this paper was the announcement that Orange Judd would assist A. B. Allen in the editorial department. This newcomer probably proved to be the most original agricultural journalist of his day. Judd (1822–92) was graduated from Wesleyan University in 1847, then taught school for several years, and in 1850 entered Yale as a graduate student to pursue

[4] *Ibid.*, I, 205.

[5] Between 1849 and 1850, Robinson was principally occupied in reporting his travels to the *American Agriculturist*, but he also sold seeds, nursery stock, farm implements, and machinery for A. B. Allen and Company. Kellar, *Solon Robinson*, I, 29.

[6] *American Agriculturist*, X (Oct., 1851), 324.

[7] *Plow*, I (Nov., 1852), 357.

[8] In January, 1853, Allen and Company inaugurated the *New York Agricultor* (weekly) and the *Farm and Garden* (monthly). The monthly was made up from the *Agricultor*. *Plow*, I (Dec., 1852), 361; *Country Gentleman*, I (Jan. 13, 1853), 20. Both publications were suspended after a few months. *American Agriculturist*, XI (Sept. 14, 1853), 16.

work in agricultural chemistry under Professor John Norton. After three years' study Judd became associated with the *American Agriculturist* and in 1856 sole owner and editor.[9] As he had first-hand acquaintance with both the farm and laboratory, this scholar-farmer soon found himself editing one of the most popular and influential journals of the day. He occupied a position of considerable power and influence in agricultural journalism until his death thirty-six years later.[10]

From its earliest days in the forties, this magazine was unique. Particular emphasis was placed upon European agriculture as may be especially noted in the January, 1845, issue, where the following articles appear: "Packing Provisions for the English Market," "Agriculture in Scotland," "Horticulture in France," "Sheep Husbandry in Spain," together with two pages of "Foreign Agricultural News." [11] Aside from these agricultural items, general European news was also included. The column for boys and girls, by far the most popular of its type, was widely copied in other periodicals.[12] In 1848 the *American Agriculturist* absorbed the *Farmers' Cabinet* of Philadelphia and thereafter continued to devour competitors at a prodigious rate.[13] This paper helped to initiate a new industry in the United States in 1856 when the editor distributed gratis great quantities of imported sorghum seed.[14] Two years later an edition of the *American Agriculturist* was issued in German to meet a

[9] *Dictionary of American Biography*, X, 231–32; *American Agriculturist*, XX (Jan., 1861), 1.

The *American Agriculturist* had been changed to a monthly in September, 1855.

[10] Ogilvie, *Pioneer Agricultural Journalists*, p. 33.

Judd served as agricultural editor of the New York *Times* from 1855 to 1863. *Dictionary of American Biography*, X, 231. Most of the important newspapers carried agricultural departments by the middle of the century.

[11] *American Agriculturist*, IV (Jan., 1845).

[12] See pp. 128–29 for a detailed consideration of this department.

[13] *American Agriculturist*, VII (July, 1848), 201. In the first thirty years of its history this journal absorbed a total of twenty-six periodicals. Mott, *A History of American Magazines, 1741–1850*, p. 731.

[14] *American Agriculturist*, XV (Oct., 1856), 305. In 1859 Judd said he had distributed to his subscribers 200,000 parcels of various kinds of seeds. *American Agriculturist*, XVIII (Sept., 1859), 282.

growing need.[15] In the same year Judd set another precedent when he was forced to copyright all articles as a protection against indiscriminate copying without proper credit. The following year (1859) an "extra early edition" was issued to accommodate subscribers in the Far West and foreign countries.[16] At this time a Western subscriber commented upon the "non-political, non-partisan, non-sectional—in short, the non-controversial character" of this farm journal.[17]

Editorial and advertising policies, particularly under Judd, maintained a high standard. Unusual vigilance was exercised over the material accepted for publication. The *American Agriculturist* was one of the first journals to examine, on the basis of sincerity and honesty, all advertising copy accepted. A column entitled, "What the Humbugs Are Doing," received Judd's personal attention and he ruthlessly exposed swindlers who preyed freely upon rural communities.[18] His periodical was also famous for its excellent illustrations; these were widely copied. Its popularity is attested by the circulation, which reached 10,000 in the third year of publication.[19] After the middle fifties, under Judd's supervision, these figures soared and in 1859 reached 45,125.[20]

[15] *Ibid.*, XVII (June, 1858), 192. This edition sold for $1.00 a year.

[16] *Ibid.*, XVIII (June, 1859), 188.

[17] *Ibid.*, XVIII (April, 1859), 102.

[18] Ogilvie, *op. cit.*, p. 33.

See pp. 150–51 for the attitude of this journal on advertising and humbugs.

[19] *American Agriculturist*, III (March, 1844), 65.

[20] *Ibid.*, XVIII (Nov., 1859), 349. Eighty thousand copies were printed in November of that year, but this was indeed exceptional.

# *The Country Gentleman*
# (1853– )

Albany, New York and Philadelphia, Pennsylvania

THE *Country Gentleman* was the last of the great triad of agricultural publications sponsored by Luther Tucker.[1] The title of this journal was suggested by Tucker's close friend, Andrew Jackson Downing, the famous landscape gardener, architect, and horticulturist.[2] As a true representative of the American country gentleman, a portrait of Washington was placed in the vignette.

The purpose of this periodical was outlined in the prospectus as follows:

> We were led to its publication, from the conviction that a large class of farmers and others interested in rural life wished something more in an agricultural journal than it was possible to furnish in the columns of a monthly. Horticulture and matters of rural taste, demanded more attention than the limits of the Cultivator would allow. The home education of farmers and their families, the social aspects of country life, the pleasures that slumber too often at the fire-sides of country residents, all present themselves as most desirable topics for discussion in an agricultural paper. A growing literary taste demands articles of a higher order than usually find their way into such journals. In a weekly journal all these subjects may be combined with strictly agricultural matter, in such way as to fit it for general perusal, and the improvement of the family circle, while the paper will be none the less valuable to the farmer. This combination, together with a concise summary

[1] The *Cultivator*, made up largely of articles drawn from the *Country Gentleman*, continued as a monthly and in 1866 was merged with the latter journal.

[2] *Country Gentleman*, I (Jan. 13, 1853), 20.

of news, will supply a want that is felt to a considerable extent by country gentlemen and their families.[3]

This weekly was a small folio of sixteen pages, beautifully illustrated and well printed on good paper selling at $2.00 per annum. The first volume was divided into departments devoted principally to the farm, garden and orchard, the fireside, current events, and produce market.

The original editors of the publication were Luther Tucker, John J. Thomas, and Joseph Warren. Thomas, who continued his services with the paper for forty years, was a pomologist, author, and an experienced editor, having previously served as such for the *Genesee Farmer* and *New Genesee Farmer*.[4] Warren assumed responsibility for the "Fireside" and "News Departments" for several years.[5] In 1856 Tucker took his son, Luther H. Tucker, into partnership.

Among the outstanding contributors to the *Country Gentleman* were John Johnston, the father of tile drainage in the United States; Samuel W. Johnson, professor of agricultural chemistry at Yale; Frederick Holbrook, later governor of Vermont; and Joseph Harris, favorably known for his scientific approach to farm problems, who also became associate editor in 1855.

The *Country Gentleman* in 1854 proposed what is said to have been the first test-plot experiments with corn in the United States.[6] The following year, the press printed a statement made by Robert L. T. White of Virginia declaring that most of the practical farmers were of the opinion that yellow corn was "more nutritious and better generally for stock than white corn,—" which is arresting in view of recent investigation. In 1922 the experiments of Steenback of Wisconsin University confirmed these earlier views, when

[3] *Cultivator*, ser. 3, I (Jan., 1853), 9.

[4] Thomas published *The Fruit Culturist* in 1846. This was expanded into *The American Fruit Culturist* in 1849, the content trebled, the information condensed, systematized, and illustrated. This book marks the beginning of systematic pomology in America. *Dictionary of American Biography*, XVIII, 439.

[5] *Country Gentleman*, IV (Nov. 9, 1854), 296.

[6] *Ibid.*, CI (March, 1931), 5.

large quantites of the fat-soluble vitamin were discovered in yellow corn.[7] In 1854, this periodical also carried a description of a new seedling grape developed by E. W. Bull of Concord, Massachusetts, today known as the famous Concord grape.[8]

The national and international character of the *Country Gentleman* may be observed in the four numbers issued in January, 1858. In this month alone there were 150 communications and extracts from correspondents in 20 counties of New York, "19 other states, and Canada, New Brunswick and India." [9] Correspondence from five or six hundred individuals annually from all corners of the earth proved to be the rule. Luther Tucker, the senior editor, made an extended agricultural tour of the West in 1857 and reported observations for his paper. It is interesting to note that he had sufficient vision to prophesy that Chicago, then a city of less than 30,000, would at no distant date rank as the second city of our land.[10]

Circulation figures during this period remain elusive. The editor stated in 1857 that they had not yet reached a number commensurate with the actual cost of production.[11] The "phenomenal" circulation of 250,000 so often quoted for the *Country Gentleman* as of 1858 appears to be the result of erroneously reading a statement made by the editor in January, 1859. (This estimate of a quarter of a million was made with reference to the aggregate circulation of the entire agricultural press.[12]) While others had a greater circulation, this journal was probably the most "modern" of its contemporaries.

Today the *Country Gentleman* carries the statement that it is the "oldest agricultural journal in the world," declaring that it is a "consolidation of the Genesee Farmer, 1831–1839, and the Cultivator, 1834–1865." The method of assuming priority by claim-

[7] *Ibid.*, VI (Nov. 15, 1855), 316; CI (March, 1931), 5.
[8] *Ibid.*, III (Feb. 23, 1854), 122.
[9] *Ibid.*, XI (Jan. 28, 1858), 64.
[10] *Ibid.*, X (July 16, 1857), 41; CI (March, 1931), 140.
[11] *Ibid.*, IX (June 25, 1857), 409.
[12] *Ibid.*, XIII (Jan. 6, 1859), 9.

ing the birthday of a periodical which has been absorbed is open to question.[13]

[13] A number of other journals make similar assertions; some of these, however, have not been published continuously. The *Prairie Farmer* (1840–    ) appears to have a good case. Claims have also been made for the *Southern Planter* (1841–    ) and the *American Agriculturist* (1842–    ). The difficulty in determining the "oldest journal" has arisen as a result of the various methods used for computing age. See Mott, *A History of American Magazines, 1850–1865*, pp. 435–36.

# *The Southern Agriculturist* (1828–1846)

Charleston, South Carolina

THE *Southern Agriculturist* was the first important farm journal of the South. Established in 1828 at Charleston, with J. D. Legare as editor and with the emphasis on regional subjects, this octavo of forty-eight pages sold at $5.00 per annum. For convenience it was divided into three parts. Part One consisted of original communications embracing agriculture, horticulture, botany, rural affairs, and domestic economy with particular emphasis on the chief objects of Southern culture—cotton, rice, Indian corn, and sweet potatoes. Part Two contained reviews of works dealing with Southern agriculture and selections from important books, both foreign and domestic. It was carefully pointed out that the journal took the place of an agricultural library and that the planter of small estate, not the wealthy landowner, was the object of the editor's solicitation. Part Three contained such miscellaneous agricultural items as proceedings of agricultural societies and descriptions of agriculture in other portions of the globe.[1]

Apart from its small size, the *Southern Agriculturist* differed in many respects from the journals already considered. Miscellaneous matter—items of interest to women and children, poetry, jokes, and news events—was less conspicuous than in the majority of other periodicals. The editor remained almost completely in the background; communications were not directed to him personally as

[1] *Southern Agriculturist*, I (1828), Introduction, pp. i ff.

In January, 1853, the *Southern Agriculturist* of Laurensville, S.C., with A. G. and W. Summers as editors, was inaugurated. This journal seems to have continued until September, 1854. (The writer has not examined this periodical.)

was true of the very popular papers. No editorial columns kept the readers regularly appraised of the editor's attitude on contemporary problems and the Buel type of leadership seems to have been utterly lacking. John D. Legare was the editor [2] during the greater part of its existence. He was not considered an authority by his contemporaries and wrote comparatively little for his paper. For a period he was proprietor of a repository of field and garden seeds and of agricultural implements located at Charleston. He was also greatly interested in the Grey Sulphur Springs in Virginia, which he hoped to convert into a health resort.[3]

After twelve volumes of the *Southern Agriculturist* had been issued, the name was changed in 1840 to the *Southern Cabinet.* The emphasis on agriculture, horticulture, rural and domestic economy was continued, but articles on the arts and sciences, literature and sports (under the heading "Sporting Intelligence") were added. Stories of love and adventure, items of history, and literary notices were found side by side with "Diseases Peculiar to Cattle—Remedies" and "Notes on European Agriculture." The following year, the paper was again called the *Southern Agriculturist,* a new series inaugurated, and the original emphasis upon content reëstablished.

The editor urged his readers to communicate results of personal experiments in all fields of agriculture and to report and describe failures as well as successes.[4] Subscribers were normally unwilling to make such contributions. One writer lamented that certain questions he had propounded to the paper for a possible solution from the field received no reply after an interval of twelve months. Such apathy on the part of subscribers forced the editor to rely upon other journals for much of his material. "It is to our reproach, [wrote one farmer] that a Journal, so exclusively devoted to the advancement of agriculture in the Southern States, should depend for so much of its reading matter, upon *selections* made from the pages

[2] From 1828 to 1834 and from 1840 to 1842, both inclusive. A. E. Miller and Bartholomew R. Carroll also served as editors. Miller was the first publisher.

[3] Legare, *Account of the Medical Properties of the Grey Sulphur Springs, Virginia.*

[4] *Southern Agriculturist,* V (June, 1832), 305.

of similar periodicals, in the more distant sections of the Union." [5] The paper never had a large circulation and frequently the necessity for increased subscriptions and prompt payment of old accounts was voiced.[6] Although the price was reduced to $3.00 in 1843, the number of subscribers was not materially supplemented. Finally, the competition from "so many cheap northern papers" made the continuance of the journal a problem of questionable financial wisdom.[7] As a result, in 1846, the journal was terminated.

[5] *Ibid.*, n.s., III (Feb., 1843), 66.

[6] *Ibid.*, n.s., III (Feb., 1843), 65; n.s., IV (Oct., 1844) cover; n.s., IV (Nov., 1844), cover.

[7] *Ibid.*, n.s., IV (Oct., 1844), cover.

# *The Farmers' Register* (1833–1842)

Shellbanks and Petersburg, Virginia

OF FAR GREATER influence than the *Southern Agriculturist* was the *Farmers' Register*, a journal which John Skinner called the "best publication on agriculture which this country or Europe has ever produced."[1] It was inaugurated in June, 1833, with Edmund Ruffin as editor and proprietor. Its appearance brought to an end Virginia's pioneer agricultural periodical, the *Virginia Farmer* (established at Scottsville in 1832), which was said to droop "like a harebell before the sun."[2] The new publication was an octavo of sixty-four pages, published monthly at "Shellbanks" (later Petersburg) at a subscription rate of $5.00 per year. In order to increase space for reading matter, the editor promised to exclude advertisements.[3] Although conscious of the "general and long existing apathy manifested with regard to the improvement of agriculture in Virginia," Ruffin launched upon the new venture without awaiting the assurance of a subscription list sufficient to defray the expenses of publication.[4]

Edmund Ruffin (1794–1865), was born in Prince George County, Virginia, and educated at home until his sixteenth year when he entered William and Mary College for a short and unprofitable period. After brief and inactive service in the War of 1812, he returned to his home to assume charge of the Coggin's

[1] Craven, *Edmund Ruffin*, p. 62.

[2] *Farmers' Register*, I (July, 1833), 128. No copy of the *Virginia Farmer* is known to be extant.

[3] *Farmers' Register*, I (June, 1833), 64. The *Farmers' Register* did, however, carry advertisements on its cover. *Ibid.*, I (Dec., 1833), 446–47.

[4] *Ibid.*, I (June, 1833), 62; I (July, 1833), 128.

Point farm left to him upon his father's death.[5] Here he began the exhaustive researches in agriculture which were to play a large part in the agricultural revolution in Virginia.[6] At the time of the inauguration of the *Register,* he already enjoyed a wide reputation for his valuable and scientific work on calcareous manures.

Ruffin early announced that the publication aimed to provide a medium of exchange for the farmers of Virginia and upon this feature, he laid great stress. He editorialized as follows:

> It would serve to farmers as an *Exchange* does for merchants—an institution which would enable every individual to obtain readily any and every kind of information that any other individual is able to furnish. Whenever one of our readers wishes to be informed on any point connected with agriculture, let him freely make the inquiry through the *FARMERS' REGISTER* and let those who can, as freely answer: and if no other than this humble purpose was served by the publication, its support would directly promote the pecuniary interest of every farmer who can afford to spend five dollars a year on anything beyond the absolute necessaries of life.[7]

While he believed that most farmers were able to express themselves intelligently on paper and he urgently requested them to do so, nevertheless he was forced, because of the paucity of communications, to write nearly half of the material which appeared in his periodical.[8] No editor so completely dominated his journal as did Edmund Ruffin, and in order to appear less conspicuous in its pages, he frequently resorted to the use of the *nom de plume.*[9] Particularly powerful were his editorials, which covered all phases of agriculture, with comparatively little emphasis, however, upon agricultural machinery.[10] It was with reason called Ruffin's *Farmers' Register.*

While discussion of the marl question—a favorite hobby of the editor—recurred with great frequency, almost every topic in which

[5] *Dictionary of American Biography,* XVI, 215.

[6] See pp. 65–66 for a more detailed discussion of Ruffin's investigations.

[7] *Farmers' Register,* I (June, 1833), 62–63.

[8] Craven, *Edmund Ruffin,* p. 62; *Farmers' Register,* I (June, 1833), 62; I (Sept., 1833), 193.

[9] *Farmers' Register,* V (Aug. 1, 1837), 249; IX (Oct. 31, 1841), 576. See p. 112 for an account of Ruffin's wide use of pen names.

[10] Hutchinson, *Cyrus Hall McCormick,* I, 415.

the farmers of Virginia might conceivably be interested received comment. Special attention was devoted to agricultural chemistry; slavery, particularly in its nonpolitical aspects; internal improvements; questions of political economy; and agricultural education. The practical operations of the farmer, however, received the greatest care.[11] Ruffin was always eager to expose agricultural humbugs of every description.

The *Farmers' Register* avoided jokes, poetry, and items of interest to women and children to a still greater extent than did the *Southern Agriculturist.* To many, this journal which largely turned its back on the "lighter side," had a scholarly, forbidding, and "scientific" appearance. Ruffin reprinted the seventh edition of John Taylor's *Arator,* the *Westover Manuscripts,* and the third edition of his own *Essay on Calcareous Manures* as numbers and supplements in the *Register.*[12] A particularly large number of articles dealing with Virginia history were published; extracts from Berkley's *History of Virginia,* "Scraps from Old Authors Respecting Virginia," "Clayton's Letters from Virginia, in 1688," and "Sketches of the Habits and Manners of Old Times in Virginia." The editor explained that this was done, in part, because of the "dearth of communications," and because he felt in this way he could supply the deficiency "with other original and good and interesting matter." [13]

On the whole, the *Register* articles were carefully written and edited, or selected from the best American and European authorities. An unusually large number of original contributions were included. Its high standard was maintained throughout its existence and during that time it was a leading factor for agricultural re-

[11] Cutter, "A Pioneer in Agricultural Science," p. 500.

The hot-headed editor, who was later proud to fire the first shot at Fort Sumter, included little material on the controversial side of slavery. In the ten volumes of the *Register* his references to this subject were, in general, restrained and moderate. See Craven, *Edmund Ruffin,* p. 108.

[12] *Farmers' Register,* VIII (Dec. 31, 1840), 703; IX (Supplement, 1841); X (Supplement, 1842).

This was the first publication of the *Westover Manuscripts.* See Boyd, *William Byrd's Histories of the Dividing Line betwixt Virginia and North Carolina,* p. xvi.

[13] *Farmers' Register,* IX (Oct. 31, 1841), 577.

form in this region.[14] Many present day authorities look upon it as the "greatest agricultural paper of the time." [15] Earl G. Swem in *An Analysis of Ruffin's Farmers' Register* calls attention to this neglected historical source and states that these ten volumes are "one of the most authoritative sources we have for the study of agricultural and other economic questions of the ante-bellum period in Virginia." [16]

Ruffin realized no financial gains from the publication of the *Register*.[17] The subscription list totalled more than 1,300 at the close of the first year; few figures thereafter are available.[18] During its last two years, the editor entered the forbidden field of politics in his fight for banking reform in Virginia. The hostility aroused by this move played a part in his final withdrawal from editorial work and is expressed by a correspondent as follows: "You will therefore oblige me by placing my account in the October number of the *Register*, and withdrawing my name from your subscription list, unless you shall see proper to keep every *political article* out of the paper." [19] Other factors added to the distressed condition. In August, 1841, Ruffin commented upon subscription arrears, which amounted to more than $4,000. He also stated that for a long time there had been insufficient income to pay for a capable clerk.[20] De-

[14] Craven, *Soil Exhaustion*, p. 139.

[15] Craven, "The Agricultural Reformers of the Ante-Bellum South," p. 310; Phillips, *Life and Labor in the Old South*, p. 132. Contemporaries of the *Farmers' Register* agree with this evaluation. See *Tennessee Farmer*, I (Dec., 1834), 7.

[16] Swem, *op. cit.*, Introductory Note, p. 42. See also Bruce, "Materials for Virginia Agricultural History," pp. 10 ff.

[17] Swem, *loc. cit.*

[18] As a supplement to Volume I, Ruffin attached a list of subscribers. While sixteen states and the District of Columbia were represented, making a total of more than 1,300 subscribers, over 80 percent were drawn from 85 counties in Virginia.

[19] *Farmers' Register*, IX (Oct. 31, 1841), 617–18; X (April 30, 1842), 179 ff.

Ruffin published the *Bank Reformer* (Sept. 4, 1841–Feb. 5, 1842), which also caused hostility. *Dictionary of American Biography*, XVI, 215.

[20] *Farmers' Register*, IX (Aug. 31, 1841), 507.

Early in 1841, the *Carolina Planter* of Columbia, South Carolina, was united with the *Farmers' Register*. *Carolina Planter*, I (Jan. 13, 1841), 409.

ploring the general apathy of the agricultural community of Virginia, the disillusioned editor added in disgust, "But with the close of this volume [X, 1842] will end the editor's labors for ten years of the best years of his life; and he will no longer obtrude, on the agricultural public, services which seem to be so little appreciated, and which have been so little aided by the sympathy of the great body of the members of the interest designed to be served." [21] The *Register* was sold to T. S. Pleasants who also became editor but the journal under his direction was abandoned in a short time.[22]

[21] *Farmers' Register*, X (April 30, 1842), 155.

[22] Ruffin, however, continued his work for improved agriculture in the South. In 1841 he had been appointed a member of the first State Board of Agriculture and was its first corresponding secretary. In the following year he was appointed agricultural surveyor of South Carolina. Always a strong advocate of agricultural societies, he played an important role in the organization of the Virginia State Agricultural Society (1845), serving as president in 1852. He was a prolific writer on agricultural subjects for newspapers and farm papers and often spoke before agricultural societies.

Ruffin had always been interested in politics. He served in the Virginia Senate (1823–26) and was from the first an ardent defender of slavery. He was one of the earliest secessionists in Virginia, and with the collapse of the Confederacy, in utter despair, he took his life. *Dictionary of American Biography*, XVI, 214–15; Craven, *Edmund Ruffin*, p. 1.

# *The Agriculturist* (1840–1845)

Nashville, Tennessee

TWO YEARS before the *Farmers' Register* was abandoned the *Agriculturist* was established by the Tennessee Agricultural Society to serve as the organ for the state and county societies. The following officers of the state society were appointed editors: Tolbert Fanning, Doctor John Shelby, and Doctor Gerard Troost.[1] These gentlemen served without compensation.[2] This monthly quarto of

[1] *Agriculturist*, I (Jan., 1840), 14.

Tolbert Fanning (1810–74) was a minister of the Disciples of Christ. In 1837 with Mrs. Fanning, he opened a boarding and day school for girls at Franklin, Tenn., which was moved in 1840 to "Elm Crag," a farm about five miles north of Nashville. Fanning was interested in agricultural advancement, took an active part in the Tennessee Agricultural Society, and imported some of the best breeds of stock. *Dictionary of American Biography*, VI, 268–69.

John Shelby (1786–1859) was an eminent physician in Nashville. He had served as an army surgeon under General Jackson, suffering the loss of an eye in one of the battles with the Indians. He was postmaster at Nashville for a number of years. The Shelby Medical College was named after him. He was vitally interested in all phases of agriculture. *The National Cyclopaedia of American Biography*, II, 151.

Gerard Troost (1776–1850), a native of Holland, received his medical degree from Leyden. Early in the nineteenth century he came to the United States. After a number of business ventures he spent a year at New Harmony and in 1827 went to Nashville. From 1828 until his death he was professor of geology and mineralogy and, for a time, chemistry at the University of Nashville. He was active in the founding of the Tennessee Agricultural Society and contributed freely to the *Agriculturist*. *Dictionary of American Biography*, XVIII, 647–48.

Cameron and Fall were the publishers.

[2] *Agriculturist*, I (March, 1840), 49; V (Feb., 1844), 26.

twenty-four pages sold for $2.00 per annum; its objective lay in the diffusion of information which might promote the cause of agriculture, and elevate and improve the condition of farmers and mechanics generally.[3] The wide range of subjects to be considered was suggested in the first number and included everything that would "improve mankind, intellectually, morally, in a domestic or pecuniary point of view." [4]

As one of the main interests of the Tennessee Agricultural Society was the establishment of an agricultural school, it is understandable that the *Agriculturist* should abound with material on this subject.[5] In the second year of publication, the editors announced that education and particularly agricultural education would be discussed more fully in this journal than had yet been done elsewhere in America.[6] In January, 1843, Fanning, the most active of the editors, opened a school on his farm at Elm Crag with six students.[7] While regular academic studies were pursued, practical agriculture was also taught; the editor claimed that it was the first institution of the kind in America.[8] A full discussion of this school including the course of study, disciplinary problems, costs to students, in fact all phases of school life, may be followed in the pages of the *Agriculturist*. In January, 1844, after much labor on the part of the Tennessee Agricultural Society and this periodical, the Legislature of the state chartered an agricultural school to be called "Franklin College." The board of trustees including John Shelby and Tolbert Fanning, held their first meeting at the *Agriculturist* office, where the agricultural society also held its meetings.[9] It was decided to lease Elm Crag as the site of the proposed institution. In January, 1845, Franklin College opened its doors with more than sixty students. Academic subjects were taught, although agriculture and manual training were important features of the

[3] *Ibid.*, I (Jan., 1840), 13. The price was changed to $1.00 a year with Volume V, and the number of pages per issue was reduced to sixteen.

[4] *Ibid.*, I (Jan., 1840), 2.

[5] *Ibid.*, I (Jan., 1840), 23; I (May, 1840), 97.

[6] *Ibid.*, II (Dec., 1841), 287.

[7] *Ibid.*, IV (Jan., 1843), 11.

[8] *Ibid.*, IV (Jan., 1843), 11; IV (Feb., 1843), 30; IV (May, 1843), 78; IV (Sept., 1843), 132–33.

[9] *Ibid.*, III (Aug., 1842), 189; V (Feb., 1844), 22.

curriculum.[10] In June, 1844, Fanning was compelled to relinquish his editorial activity, temporarily, because of arduous duties in connection with the new college.[11] The following year the college faculty became a part of the *Agriculturist* editorial staff.

This periodical was noted for its aggressive leadership in other directions. In an editorial entitled, "True Agricultural Policy of Tennessee," the editors outlined, in no uncertain terms, what they considered a wise policy to be followed in Tennessee. They deplored the waste of virgin forests and the resulting excess land brought to the mercy of the "burning sun and beating rains." They advised the termination of cotton cultivation as unsuited to the soil of Tennessee, while tobacco, because of its deleterious effects on the consumer, was also to be abandoned. Indian corn, the chief product of Tennessee (it raised more than any other state in the Union) was to be reduced by one half. On the other hand the cultivation of potatoes, beets, carrots, parsnips, and turnips was encouraged. The editors desired that more attention should be given to grasses (Red clover, Blue grass, Timothy, Orchard grass, and so on), and the dairy business was strongly recommended. Four years later the editors were gratified with the diminished interest in cotton and the greater diversification of crops. At this time they also began to stress the advisability of rearing sheep, hogs, cattle, horses, and mules.[12] As the lack of a seed store caused great inconvenience to the farmers and gardeners in this locality, the *Agriculturist* announced that all kinds of grass and garden seeds would be sold at the office of the journal.[13]

The magazine also provided a channel for airing the varied and ofttime conflicting ideas with regard to penitentiary reform in Tennessee. Editor Fanning objected to the competition of convict labor as opposed to the "honest" shoemakers, stonecutters, and cabinetmakers throughout the state and suggested that prisoners be employed in noncompetitive work such as internal improvements and silk manufacturing.[14] Solon Robinson, on the other hand, inaugu-

[10] *Ibid.*, V (March, 1844), 47; VI (Feb., 1845), 22.

[11] *Ibid.*, V (June, 1844), 86.

[12] *Ibid.*, II (June, 1841), 130 ff.; VI (Feb., 1845), 25.

[13] *Ibid.*, I (Dec., 1840), 287.

[14] *Ibid.*, IV (June, 1843), 82–83.

rated a heated discussion when he advanced the idea that the only way to prevent crime was to sentence convicts to the most terrible punishment, "*a life of idleness,* solitary and alone." [15] A writer promptly objected to taxing honest people "to support law breakers in idleness," while others felt that convicts should "pay their own way" in prison.[16] Capital punishment was also debated widely in this journal.

In January, 1841, the *Tennessee Farmer,* the pioneer agricultural journal of Tennessee, and the *Southern Cultivator* (Columbia, Tennessee) were absorbed by the *Agriculturist,* which now became the sole farm paper in the state. A few years earlier there had been five periodicals in the field, and with the expiration of the *Agriculturist* four years later, there were none.[17] The circulation of the *Agriculturist* was 2,500 in September, 1841.[18]

This publication was terminated in December, 1845. Fanning explained that the paper had been a source of "deep mortification" to him, because it lacked "spirit and accuracy" due to the fact that not one of the three editors could give sufficient attention to the work. A new publication called the *Naturalist and Journal of Natural History, Agriculture, Education and Literature,* which Fanning assured his readers was the "Agriculturist in a new form, under a new name, and different auspices," made its appearance in January of the following year. It was published at Franklin College under his editorship with the assistance of the faculty, but it continued less than a year.[19]

[15] *Ibid.,* III (March, 1842), 67.

[16] *Ibid.,* III (April, 1842), 84–85; III (May, 1842), 109; III (June, 1842), 136–37; III (July, 1842), 157–58.

[17] *Ibid.,* II (Jan., 1841), 23; II (March, 1841), 61.

[18] *Ibid.,* II (Sept., 1841), 193.

[19] *Ibid.,* VI (Nov., 1845), 171.

# *Southern Planter*
# (1841– )

Richmond, Virginia

ALTHOUGH Charles T. Botts (1809–84) regarded the *Farmers' Register* very highly, he felt that this scholarly, high-priced periodical with its profusion of "philosophical and theoretical essays" did not meet the need of a large class of farmers in Virginia. Thus in 1840 he decided to establish the *Southern Planter* at Richmond, to sell at $1.00 a year, a "price within the reach of all." The first issue appeared in January, 1841. He promised to reject "long, and even perhaps able essays" for he desired to make his paper the "medium for the promulgation, in a condensed form, of the observations and deductions of practical men." [1] It was first published as a sixteen-page octavo, by 1860 it was gradually increased to sixty-four pages, and the price raised to $2.00 per annum.

While the *Southern Planter* advocated popular policies such as crop rotation, deep plowing, and diversification, it also stressed the culture of tobacco and wheat, as well as agricultural education. Botts was particularly interested in farm machinery, consequently this subject received special emphasis. His practical bent along this line was demonstrated by the invention of "Botts' Straw-Cutter" which received favorable notice in the farm press.[2] While attempting to popularize agricultural machinery in the South, he was frequently rebuffed with the remark that a certain machine might "do well for a northern farmer" but it would "never do" for Southern Negroes. While he labored to extend the beneficial effects of machin-

[1] *Southern Planter,* I (Jan., 1841), 1.

[2] *Cultivator,* X (Dec., 1843), 193; n.s., I (Jan., 1844), 35; *American Agriculturist,* III (Oct., 1844), 309–10; *Southern Planter,* III (April, 1843), 92–93.

ery to Southern agriculture, he found it "an uphill business." [3] The importance he placed upon this subject is expressed in these words: "There is no greater bar to agricultural improvement at the South than the entire ignorance of our farmers and planters of the simplest mechanical details." [4] In 1842, in conjunction with L. M. Burfoot, who assisted in the editorial department during part of the year, Botts opened an agricultural warehouse and in addition, manufactured Botts' Straw-Cutter.[5] A land agency was undertaken in the following year.[6]

The editor announced that under the heading of "Miscellany," a page would be devoted to "polite literature," which would include selected anecdotes, poetry, and the like. This material appealed particularly to the women and younger members of the household.[7] In 1850 an horticultural department was inaugurated with A. D. Abernethy and later E. G. Eggeling in charge. Both were well-known horticulturists and florists in Richmond.[8]

Although Botts sold his journal to P. D. Bernard in 1846, he continued as editor for most of the following year.[9] John M. Daniel then assumed these duties for a short time and was succeeded by Richard B. Gooch, a recent graduate of the University of Virginia, who had also traveled in Europe.[10] His untimely death in 1851 brought another famous editor to the *Planter* in the person of Frank G. Ruffin, a "graceful, fluent, and forceful writer." It was well, for the journal had declined greatly, even as the subscription list.[11]

The new editor promptly disclaimed any near kinship to the more famous Edmund Ruffin but again proved the fact that a journal is as good as its editor; for immediately the *Planter* was revived. The publisher assured the readers that Ruffin was a farmer, engrossed in agricultural pursuits and wholly dependent on his land for a living,

[3] *Southern Planter,* III (Sept., 1843), 205, I (June, 1841), 91–92; II (April, 1842), 95; XCI (Jan. 1, 1930), 3 ff.

[4] *Ibid.*, IV (Nov., 1844), 253.

[5] *Ibid.*, II (Sept., 1842), 212.

[6] *Ibid.*, III (Jan., 1843), 24.

[7] *Ibid.*, I (Jan., 1841), 15.

[8] *Ibid.*, X (Dec., 1850), 354; XVII (Feb., 1857), 186.

[9] *Ibid.*, VI (Dec., 1846), 282; VII (Nov., 1847), 359.

[10] *Ibid.*, IX (July, 1849), 193; XI (June, 1851), 161.

[11] *Ibid.*, XI (July, 1851), 195; XVIII (Aug., 1858), 497.

thereby giving a guarantee of the practical character of his work.[12]

Ruffin, with his usual frankness, stated that he did not feel that the *Planter* was the sort of paper which would best promote progress in agriculture. His ideal was the more serious *Farmers' Register,* which appealed to a different class of readers, to men better able, for a variety of reasons, to introduce reforms, shake off prejudices, and correct bad habits. However, the fate of the *Register,* he said, warned him against a similar venture.[13]

His new duties soon presented a problem common to those early part-time editors. The full direction of his large plantation suddenly fell upon Ruffin's shoulders through the death of his overseer. For a time, the editor's office was under an old tree near his barn, where, with the aid of a little wheat straw, he arranged a "tolerable" lounge among the rocks. Surrounded by his papers and books, he directed his farm and edited his paper.[14] In 1855 he became both proprietor and editor. He paid a "high price" for the journal, recognizing the improved financial prospects which he attributed to the large advertising section. While admitting that he had made little or no money from it, he confessed that he enjoyed the work, since it afforded a "species of occupation and excitement." Then too it enabled him to "do some good." The *Planter* maintained a high standard during these years. Although advised by his friends to discontinue the constant attacks on humbugs, he felt he owed this protection to his readers.[15] In 1858 Ruffin disposed of his interests in the paper for "private" reasons and was succeeded by Dr. James E. Williams of Henrico County.[16]

The popular appeal of the *Planter,* rising from its emphasis on the practical, assured its success from the beginning. The first number in 1841 registered 1,200 copies and at the year's close the circulation soared to 3,000. Still more remarkable was the editor's statement that he would make a profit of two hundred dollars for that year.[17] By 1854 the circulation totalled 4,200 and was increased by 400 in the following year.[18] This periodical is still in pub-

[12] *Ibid.,* XII (Jan., 1852), 30; XII (Aug., 1852), 237.

[13] *Ibid.,* XI (July, 1851), 195.

[14] *Ibid.,* XI (Aug., 1851), 225.

[15] *Ibid.,* XV (Jan., 1855), 17–18.

[16] *Ibid.,* XVIII (July, 1858), 387.

[17] *Ibid.,* I (Nov., 1841), 223.

[18] *Ibid.,* XIV (Jan., 1854), 25; XV (Jan., 1855), 17.

lication and today carries the notation that it is the oldest agricultural journal in America.[19]

[19] Because of the war, this journal was suspended from 1862 to 1867. The hundredth anniversary of the *Southern Planter* was celebrated in 1940. Current issues carry many interesting articles dealing with the history of this old farm journal.

# *The Southern Cultivator* (1843–1935)

Augusta, Athens, and Atlanta, Georgia

THE LAST of the Southern journals to be considered is the *Southern Cultivator,* inaugurated at Augusta, Georgia, on the first of March, 1843. For the first two years it was a bimonthly quarto of eight pages, thereafter converted into a monthly of sixteen pages.[1] This pioneer periodical of Georgia was devoted exclusively to Southern agriculture and in order to place it within the reach of the "most humble tiller of the soil," it was priced at $1.00 per year. The editor announced that his primary objective was to restore the exhausted lands of the South, to introduce an enlightened system of agriculture, and to "afford an acceptable medium for the interchange of views between planters." He declared that the "fatal system" of agriculture prevalent in the South had been a source of deep anxiety to him.[2]

The first important editor of the *Southern Cultivator* was James Camak of Athens, Georgia, who assumed his duties with the beginning of Volume III, in January, 1845.[3] A former editor of the *Georgia Journal,* he had always felt a deep interest in agricuture.[4]

[1] As the first number of the *Cultivator* was two months late, this paper was published weekly for a short period. *Southern Cultivator,* I (May 10, 1843), 79.

[2] *Ibid.,* I (March 1, 1843), 6.

[3] J. W. Jones was editor of the *Southern Cultivator* for the first two years. It was published by J. W. and W. S. Jones.

[4] He was the first president of the Central Bank and was one of the builders and a director of the Georgia Railroad. He died June 16, 1847. Kellar, *Solon Robinson,* II, 477; *Cultivator,* n.s., IV (Aug., 1847), 257; *American Agriculturist,* VI (Aug., 1847), 260; *Southern Cultivator,* IX (Dec., 1851), 185.

Alive to the importance of instituting radical changes in Southern cultivation, he maintained that agriculture was a science requiring deep study and research. Among other things, he introduced a variety of choice fruits, which he planted around his home to demonstrate to his friends how such luxuries were perfectly adapted to the soil and climate of the region.[5] Pomology continued to be emphasized in the *Cultivator* long after Camak ceased to be editor.

At the death of James Camak in 1847, a still more distinguished editor, Daniel Lee, took over the direction of the publication.[6] The paper was under his editorship for well over a decade and became one of the most influential journals of the period. Lee's reputation was far-reaching even at this early date, mainly as the result of his activity in the editorial chair of a Northern journal, the *New Genesee Farmer*. When the new editor took over his duties in the South, he deplored his ignorance of Southern agriculture, climate and soil, as well as the habits and customs of the people; but this deficiency, if it existed, was soon overcome.[7] Southern agricultural leaders such as M. W. Philips and Dr. N. B. Cloud (the latter became editor of the *American Cotton Planter*) immediately coöperated and corresponded freely with Lee. Philips wrote: "I will try to send you before January twenty subscribers. Dr. Daniel Lee, not only merits such treatment at our hands in coming South, but we aid our country in putting his writings into the hands of our people." He added, "Dr. Lee shall have the aid of my little experience, when and where he may demand it, provided I can write enough between supper and bed time, as this has been penned." [8] Besides editing the *Cultivator*, Lee analyzed soils, gave lectures on agriculture and, for a period, was professor of agriculture at Georgia University.[9]

Before very long a number of innovations were introduced. Among the new features were a well-conducted "Horticultural Department," a column on "Domestic Economy," "Answers to Correspondents," and "Our Book Table." Lee, like many other

[5] *American Agriculturist*, X (Nov., 1851), 336.

[6] See p. 67 for additional material on Daniel Lee.

[7] *Southern Cultivator*, V (Aug., 1847), 120.

[8] *Ibid.*, V (Oct., 1847), 153.

[9] *Ibid.*, XVI (May, 1858), 145; A. C. True, *A History of Agricultural Education in the United States, 1785–1925*, pp. 71–72.

editors, had a hobby. His bent was agricultural chemistry. The "Agricultural Chemistry" department in his paper was one of the few of its kind. "Dr. Lee is an able editor; [wrote the editor of the *Prairie Farmer*] and though fond of riding with a vengeance, when he strides a hobby, he does know how to make a good paper." [10] In the same year Lee obtained the services of an accomplished engraver, whose labors were exclusively devoted to illustrating the pages of the *Southern Cultivator*.[11] No effort or expense was spared to render this journal of greatest utility. In order to increase the circulation, seven hundred and fifty dollars in premiums were offered for obtaining subscribers.[12]

Lee was born in New York State and lived there until the age of 45 when he assumed the editorial duties of the *Southern Cultivator*. The position he took on slavery is both interesting and illuminating. In candid and brilliant editorials this leader discussed all phases of this subject and permitted a free and open discussion in his pages. He wrote,

> It took the writer some time to make up his judgment to the effect that negro labor, as it exists at the South, is, upon the whole a good thing; and he has been much slower in suggesting to his readers that an increase of this good thing will tend to equalize the market value of land and labor in the planting States; and thus check the blighting sacrifice of land because it is cheap, and the misemployment of labor because it is dear.[13]

The editor claimed he did not advocate the revival of African slave trade as it existed previous to 1808, but voiced his position on this subject as follows: "If it is right to hold persons as slaves in Africa and America at all, then it cannot be worse to transport slaves from the Niger to Savannah." [14]

As the ethical side of slavery was not stressed, little acrimonious discussion appeared in the paper. "I need not tell you [wrote a correspondent] how much I am pleased with the *Cultivator* and especially with the characteristic fairness and impartiality which

[10] *Prairie Farmer*, IX (Jan., 1849), 37.

[11] *Southern Cultivator*, V (March, 1847), 40.

[12] *Ibid.*, VI (Oct., 1848), 160.

[13] *Ibid.*, XVI (May, 1858), 138.

[14] *Ibid.*, XVII (March, 1859), 84.

seems to govern its editorial conduct. You do right in making it the medium of free discussion for all questions of interest to the Southern farmer and planter, and in allowing each correspondent to 'have his say' in his own way." [15]

With the close of Volume I, the editor announced that he was delighted with the patronage extended to the *Southern Cultivator*, but that thus far it had "been barely sufficient to defray the actual cost of publication." [16] When, three years later he stated that expenses had not yet been met, the *Maine Farmer*, in an editorial, told the people of Georgia they should be ashamed of themselves for such neglect.[17] In 1848 while comparing the thriving journals of the North with the ill-supported Southern periodicals, Lee announced the meager circulation of 5,000 for the *Southern Cultivator*.[18] However, by 1850, under his direction, the popularity of the paper mounted and the editor announced "with confidence" that the *Cultivator* was "permanently established." [19] Two years later, the circulation rose to 10,000 and when Lee left the paper in 1859 he felt that the periodical possessed "vitality and strength." [20] In 1860 the subscription list presented a "very flourishing condition." [21]

At the time of Lee's retirement in 1859, D. Redmond, who had assisted him for a number of years, continued with the editorial duties. In this same year, the *South Countryman* of Marietta, Georgia, was absorbed by the *Southern Cultivator*, and its late editor, Reverend C. W. Howard, became Redmond's associate in the editorial work.[22] The journal was now on a firm foundation; it continued in publication to 1935.

[15] *Ibid.*, XVII (May, 1859), 133; XVIII (July, 1860), 217; XVI (May, 1858), 137.

[16] *Ibid.*, I (Nov., 1843), 190.

[17] *Ibid.*, IV (May, 1846), 72.

[18] *Ibid.*, VI (Dec., 1848), 185; III (May 1, 1845), 72.

[19] *Ibid.*, VIII (Nov., 1850), 168.

[20] *Ibid.*, XVII (June, 1859), 176.

[21] *Ibid.*, XVIII (March, 1860), inside cover.

[22] *Ibid.*, XVII (Aug., 1859), 240.

# *The Prairie Farmer*
# (1840– )

Chicago, Illinois

In October, 1840, the *Union Agriculturist* (later *Prairie Farmer*), one of the first important "western" journals, was inaugurated in Chicago by the Union Agricultural Society. This monthly folio of eight pages was edited by the corresponding secretary of the society, John S. Wright. The editor immediately announced that the paper would be published "at cost," that is, $1.00 per year.[1] "We hope in time [wrote the editor] to make the Union Agriculturist to the West, what the Cultivator [Albany] is to the East." [2] In January, 1841, the subtitle, "Western Prairie Farmer," was added to make it more indicative of the character of the magazine. (A journal bearing the title *Western Prairie Farmer* had been recently discontinued in Springfield.) It soon became the organ of the Illinois State Agricultural Society. In January, 1843, the proprietorship was transferred to John S. Wright, who changed the title to *Prairie Farmer*. He was assisted by J. Ambrose Wight in the editorial work. No change was made in the general plan of the journal, although the size of the sheet was reduced and the number of pages increased.[3] Further modifications in the format were made from time to time. John S. Wright continued as publisher until the paper was sold in 1857; [4] he retained the editorship until 1852, but J. Ambrose Wight carried most of this responsibility.

[1] *Union Agriculturist*, I (Oct., 1840), 1.

[2] *Ibid.*, I (Nov., 1840), 14.

[3] *Prairie Farmer*, III (Jan., 1843), 1. Wright had had no practical experience in farming when he assumed his editorial duties. Wright, *Chicago: Relations to the Great Interior and Continent*, p. 290.

[4] Wright sold the *Prairie Farmer* to James C. Medill and William H.

The character and content of the *Prairie Farmer* are thus described by the editor in the prospectus of Volume IV (1844): [5]

Its character has become established as being an eminently practical paper, owing chiefly to the large proportion of matter supplied by its able correspondents, most of whom are themselves farmers, nearly three hundred in number, and residing in all parts of the West. Almost the entire western press pronounce it, for a farmer in the West, the best agricultural paper published.

The contents in general are as follows: Original correspondence; Editorial articles; Review of leading agricultural papers, presenting their more important parts; Mechanical department of two pages; Educational department of about two pages; departments entitled "Household Affairs"; "Orchard and Garden"; "Veterinary"; and the last two pages will be occupied with prices current, reviews of the Chicago, eastern, southern, and foreign markets, and with miscellany.

The editor realized that a large proportion of the farmers of the West had been unaccustomed to agricultural reading, therefore little emphasis was placed on the highly "scientific" phases of farming. The paper had a reputation among its contemporaries as being "spicy and interesting." Today it is considered a better historical source than any other early Western agricultural journal.[6]

The fact that John S. Wright was a conspicuous leader in the educational life of Illinois accounts for the emphasis on education,[7]

---

Medill in 1857; they in turn sold it to Emery and Company the following year. In 1858, up to the October issue, the title was *Emery's Journal of Agriculture*; it then became *Emery's Journal of Agriculture and Prairie Farmer.* The following year the title again became *Prairie Farmer*. During 1858–60, the editors were Henry D. Emery and Charles D. Bragdon, with Charles Kennicott acting as horticultural editor. The *Prairie Farmer* became a weekly in 1856.

[5] *Prairie Farmer*, IV (Jan., 1844), 32.

[6] *Ibid.*, VIII (Jan., 1848), 10; X (April, 1850), 133; Bidwell and Falconer, *History of Agriculture in the Northern United States*, p. 470.

[7] In 1835 Wright personally advanced funds for the first public school building erected in Chicago. It has been said that the "most influential school journal, until the appearance of the Illinois Teacher in 1854, was the Prairie Farmer." Belting, *The Development of the Free Public High School in Illinois to 1860*, p. 128; *Dictionary of American Biography*, XX, 557.

In 1849 Wright announced that the profits of the *Prairie Farmer*, aside

especially common school education. "Common Schools [wrote the editor in 1846] are too much neglected in the West, and two or three pages are occupied with disseminating the most important information concerning them." [8] The "Educational Department" of the *Prairie Farmer* was well conducted, and this column was widely copied by other farm periodicals.

Agricultural machinery received marked attention. In 1843 a "Mechanical Department" became a feature of this paper, with John Gage as the first editor. Wright was interested in machinery and in the early fifties entered into a short-lived partnership with Obed Hussey, the reaper manufacturer.[9] In 1852 he began to manufacture Jearum Atkins's Self-Raker, also called Atkins's Automaton, which developed into a thriving business.[10] Although Wright's editorial duties ceased in 1852, correspondents in the following year complained that the editor of the *Prairie Farmer* was financially interested in the machines discussed in its pages. Editor Wight felt called upon to answer:

> It is proper for us to say that the editor of this paper has no connection with any Store, Warehouse, Wool Depot, or Cabinet Shop; nor has he any interest in any Mower, Reaper, Thresher, Dog or Cat Churn, Seed Drill, Rat Trap, Cucumber Washer, or patent jewsharp, or any other machine or invention of any sort whatever. He is solely an editor. He is therefore without any undue bias to any one of these things above another.[11]

The *Prairie Farmer* was also noted for its work in horticulture.

---

from $1,200 retained by the editor each year, would be devoted to the advancement of education in the West. *Prairie Farmer*, IX (May, 1849), 167.

[8] *Prairie Farmer*, VI (Dec., 1846), 361.

[9] Hutchinson, *Cyrus Hall McCormick*, I, 413.

[10] However, in 1856, because of a shortage of seasoned timber, Wright used green wood in the manufacture of these machines, which resulted in their break-down when exposed to the summer sun. Hundreds of these self-rakers were returned, and in making good his guarantee, he was forced into bankruptcy. (Twenty years previously, Wright had made a fortune in real estate, which was swept away in the panic of 1837.) *Dictionary of American Biography*, XX, 557–58; I, 405–6.

[11] *Prairie Farmer*, XIII (Jan., 1853), 38.

Perhaps the greatest contributor on this subject, and certainly the best horticultural editor, was John A. Kennicott, M.D., long a student and teacher of botany.[12] His first extensive series of articles, "Fruits in the Lake Region," appeared in the *Farmer* in 1847. Finding the practice of medicine too laborious for his frail constitution, he gradually withdrew from this profession and more and more devoted himself to horticultural matters. He managed a large orchard and nursery at his home, "The Grove," about eighteen miles from Chicago. After serving only a short time as the chief editor Doctor Kennicott assumed the editorship of the "Horticultural Department" in 1853 and retained his connections with this journal in various capacities for a number of years.[13] While active in the affairs of the National Agricultural Society, Kennicott also served as secretary of the Illinois State Agricultural Society and president of the Illinois Horticultural Society.[14]

A special feature of the *Prairie Farmer* in the late fifties was literary articles, in particular, short stories. Such writers as Frances D. Gage and Frances E. Willard (her fame as a reformer came later) contributed stories written especially for the *Farmer*. "The City Belle" and "Jennie and John, or False Appearances" are examples of the works of the youthful Miss Willard.[15]

The magazine was directed principally to the Westerner, and the *Country Gentleman* advised all those who contemplated moving West to subscribe to it.[16] Wright was amused at his Eastern contemporaries who tried to stay the tide of emigration Westward.

[12] John A. Kennicott, (1800?–1863) was born in Montgomery County in New York. About 1823 he went to Buffalo, where he taught in a district school and served as a clerk in a drugstore. His winters were spent in the Medical College at Fairfield, Herkimer County, New York. Before going to Illinois in 1836, Kennicott had practiced medicine, traveled widely, taught school for a number of years at New Orleans, edited a "literary, scientific, and religious paper," and lectured widely on botany and other subjects. *Prairie Farmer*, n.s., XI (June 13, 1863), 369 ff.; n.s., XI (June 20, 1863), 389.

[13] *Prairie Farmer*, XIII (Jan., 1853), 16.

[14] *Ibid.*, n.s., XI (June 13, 1863), 369 ff.; n.s., XI (June 20, 1863), 389.

[15] *Emery's Journal of Agriculture*, XVIII (Oct. 7, 1858), 236; *Prairie Farmer*, XIX, (Feb. 3, 1859), 76–77.

[16] *Country Gentleman*, V (Jan. 1, 1855), 24.

He described the great opportunities in the new country and pointed out that the mass of the Western population was "contented, prosperous, growing in intelligence, and making great social progress."[17]

Beginning in 1841 with about 500 subscribers, the *Prairie Farmer* reached a total of 5,280 in 1845 and by 1860 boasted of a circulation the "largest of any paper of its class in the West and North-West." [18] Notwithstanding, financial difficulties became evident in 1857 when subscription arrears mounted to more than $10,000.[19] This journal, however, has weathered many storms and is today one of the leading farm papers in the United States.

[17] *Prairie Farmer*, XVII (May 14, 1857), 158.

[18] *Union Agriculturist*, I (July, 1841), 49; *Prairie Farmer*, V (Aug., 1845), 188; XXII (Sept. 27, 1860), 208.

[19] *Prairie Farmer*, XVII (March 5, 1857), 78.

# *The Michigan Farmer* (1843– )

Jackson, Detroit, and Niles, Michigan

THREE YEARS after the inauguration of the *Union Agriculturist* at Chicago, the first important agricultural journal in Michigan (*Michigan Farmer*) was established at Jackson by D. D. T. Moore, who was previously connected with the *State Gazette*.[1] This semi-monthly of eight small folio pages sold for $1.00 per annum and was devoted principally to the interests of Western agriculturists.[2] Within two years of its inauguration, Moore, who had served as both editor and proprietor, was obliged because of impaired health, to transfer the *Farmer* to W. F. Storey and R. S. Cheney, the former, an experienced publisher, and the latter a well-known cultivator.[3] Early in 1845 Henry Hurlbut, a practical and theoretical farmer, was engaged as editor. In December, 1847, Warren Isham purchased the paper and, under his guidance, the *Michigan Farmer* rose rapidly in popularity and influence.

Isham conducted his periodical in an informal, friendly manner,

[1] Moore purchased the subscription list of the *Western Farmer*, a periodical founded by Josiah Snow in 1841 and published in Detroit. This paper, probably the first farm journal in Michigan, had never proved profitable. Moore's new journal was known as the *Michigan Farmer and Western Agriculturist* for the first year, the *Michigan Farmer and Western Horticulturist* during the second year, and thereafter published under the simple title of *Michigan Farmer*. In 1847 the place of publication was removed to Detroit and in 1849 the *Farmer* was published simultaneously at Detroit and Niles.

[2] Modifications in size, price, and frequency of publication occurred occasionally in this period together with numerous changes in editors and proprietors.

[3] Moore subsequently became the publisher of the *Genesee Farmer* and later the popular editor of *Moore's Rural New-Yorker*.

and his dynamic personality was reflected on every page. His practical common sense approach to the problems of the frontier region is evident in this early editorial.

We intend to traverse the State, and find out what sort of farmers we have got here in Michigan—how they are getting along under the difficulties and hardships and privations of a new country—what sort of farms they are carving out for themselves, and what sort of keeping they allow them—what sort of houses and barns and orchards, and gardens, and fences, and horses, and oxen, and cows, and sheep, and pigs they have got—what sort of crops of wheat, and corn, and oats, and buckwheat, and clover, and timothy, and potatoes, and turnips, and carrots, and beans, and peas they are raising—whether these products of the soil give evidence of being *fullfed*, or being *starved*, checked in their growth, and *dwarfed*, for lack of the proper elements of nutrition in mother earth—noting special instances of successful or unsuccessful management—and to make the whole thing go off well and profitably to all parties, we intend as we go along, to deliver lectures on agricultural chemistry.[4]

The editor further stated that he intended "by hook or crook," to have an agricultural warehouse and seed store inaugurated at Detroit for the accommodation of the farmers throughout the state.

Probably no editor spent more time than Isham traveling about the country collecting a wide variety of interesting data on farm conditions and progress. His agricultural tours became justly famous in his columns, as "Notes by the Way." In 1851 he sailed to Europe to attend the World's Fair at London and eventually extended his trip to Ireland, France, Germany, and other Old World countries. For eighteen months his reports relating to agriculture and other interesting topics from all parts of Europe were printed in the *Michigan Farmer*.[5] While these observations proved valuable to his readers, the editor found on his return that the journal had been

[4] *Michigan Farmer*, VI (Jan. 15, 1848), 24.

[5] While in London Isham exchanged back numbers of his paper with the editor of the *London Agricultural Magazine*, the leading farm journal of England.

In 1853 he wrote *The Mud Cabin; or, the Character and Tendency of British Institutions*, a book covering his observations in Great Britain.

neglected and had experienced severe financial embarrassment during his absence.[6]

For many years Isham's editorials were eagerly awaited by thousands of farmers throughout the country, particularly in the Western states. Never straddling an issue, he took a definite stand on all contemporary problems of interest to cultivators. The California gold rush of 1849 was discussed under such headings as "Worse than the Cholera" and "Going Mad." Michigan farmers who had "visions of gold" and were disposing of their farms at cheap prices he dismissed by saying: "These men are in the delirious stage, and may as well be given over as hopeless cases." [7] In frequent editorials he pointed out the advantages of deep plowing and attempted to rid the farmers of the popular notion that this practice exhausted the soil. Occasionally his ideas were in error. For example, in the chess controversy, he supported the doctrine of transmutation; that is, he maintained that under certain circumstances wheat would "degenerate" into a weed called chess.[8] When wire fences were introduced into the West, Isham took pains to point out the impracticability of their use. In particular, he objected to their "appearance, or rather *want* of appearance," which he claimed was a disadvantage, as cattle would not be conscious of the obstruction until greeted with "a most unceremonious rebuff." [9] In general, however, his editorials were sound and always vigorous.

In 1853 he sold the *Michigan Farmer* to Robert F. Johnstone and William S. Duncklee and although Isham served as corresponding editor, Johnstone assumed editorial charge for most of the period to the time of the Civil War. After Isham's withdrawal from the editor's chair more than three hundred letters received at the office of the *Farmer*, expressing regret at his departure, bore eloquent evidence of his wide following.[10]

The new editor laid special stress on livestock, but in general conducted the magazine along the lines of his predecessor. He followed the wool market closely and in 1858, seeing a short clip,

[6] *Michigan Farmer*, X (Dec., 1852), 368.

[7] *Ibid.*, VII (Jan. 1, 1849), 11; VII (Feb. 15, 1849), 57; VII (March 15, 1849), 89; VII (Nov. 15, 1849), 340.

[8] *Ibid.*, VIII (Feb., 1850), 48.

[9] *Ibid.*, VII (Feb. 15, 1849), 57.

[10] *Ibid.*, XI (June, 1853), 176.

Johnstone, contrary to most "authorities," advised the farmers to hold for higher prices, a recommendation which made thousands of dollars for his readers.[11] Always a strong advocate for agricultural colleges, he supported the early efforts to obtain Federal aid for such institutions. He devoted much attention to the Michigan State Agricultural College and in 1859 accepted the position of general superintendent of its farm, retaining at the same time his editorship. From this time the *Farmer* contained detailed accounts of experiments at the college.

From its inauguration in the early forties, the *Michigan Farmer* was noted for a number of outstanding departments. For many years the "Horticultural Department" was edited by J. C. Holmes and later by S. B. Noble, both well-known horticulturists in the Western country. When Isham became editor, his interest in education resulted in a department which received the enthusiastic support of the State Superintendent of Public Instruction.[12] The "Ladies' Department," long under the supervision of male editors was taken over in the middle fifties by Mrs. L. B. Adams.[13] In her competent hands this department became a leader of its kind in the country. She took an "advanced" stand on contemporary issues, particularly on the woman's rights question.

Finances were always a precarious problem. During the first six years the paper struggled for existence with a small list of subscribers, many of whom refused to pay their obligations.[14] Although 2,500 copies of the first number were issued, many of these were given away as samples.[15]

[11] *Ibid.*, n.s., I (Jan. 8, 1859), 13; n.s., I (Feb. 12, 1859), 53.

[12] In 1848 the superintendent recommended that every district school and Township library be supplied with bound volumes of the *Farmer. Michigan Farmer*, n.s., I (Jan. 1, 1859), 5.

[13] Mrs. Adams, a native of New York state (born Oct. 17, 1817) moved to Michigan at an early age. She married James Randall Adams in 1841 and after his death in 1848 turned to teaching. Soon she became a regular contributor to the *Michigan Farmer* and other papers. In 1856 she took a proprietory interest in the *Michigan Farmer* and devoted all her time and talents to its literary and business affairs. Coggeshall, *The Poets and Poetry of the West*, p. 328.

[14] *Michigan Farmer*, VII (July 15, 1849), 220.

[15] *Ibid.*, I (Feb. 15, 1843), 8.

Unexpected encouragement came to the editor when the Michigan Senate passed a resolution requesting the delivery of the *Michigan Farmer* to every member of that body during the session of the legislature.[16] In an attempt to bolster the financial status of the paper, the editor announced that all kinds of produce would be received in payment of subscription dues.[17] Shortly after the middle of the century a hopeful tone was evident and by 1853 the *Farmer* claimed a circulation of 7,000 subscribers.[18] The following year this figure was increased by 3,000 and in 1857 it totalled nearly 12,-000.[19] This periodical has proved to be one of the most influential of the Western papers.

[16] *Ibid.*, I (Feb. 1, 1844), 188.

[17] *Ibid.*, III (May, 1845), 32.

[18] *Ibid.*, XI (Aug., 1853), 242.

[19] *Ibid.*, XII (April, 1854), 97; XV (Feb., 1857), 33.

Unpaid subscriptions continued to be a major problem throughout the period. In 1852 almost half of the obligations for the two preceding years remained unpaid and in 1857 there was more than $5,000 outstanding. *Michigan Farmer*, X (Oct., 1852), 305; XV (April, 1857), 120.

# *Ohio Cultivator* (1845–1864)

Columbus, Ohio

THE LAST journal to be considered is the *Ohio Cultivator* which was inaugurated on the first of January, 1845, in the city of Columbus, by M. B. Bateham, who was also its editor.[1] Bateham was not a stranger to this section of the country; as editor of the *New Genesee Farmer* of Rochester, New York, he had reached more than three thousand subscribers in Ohio and had made numerous agricultural tours to the Buckeye State.[2]

The first number outlined the editor's plans. He announced that the *Ohio Cultivator* would aim to impart such knowledge of the principles and practice of improved agriculture as would enable farmers to increase the value and productions of their lands and obtain greater return for their capital and labor. Descriptions of the different breeds of domestic animals with remarks on comparative value, management, and diseases were to be given. Also, improved agricultural implements, labor-saving inventions and machinery, farm buildings, and fences were to receive attention. A considerable portion of his time would be spent in visiting the farmers of Ohio, thus enabling him to adapt his editorial labors to the greatest advantage for his subscribers. Bateham's confidence in the future of Ohio is evident in his prophecy that it was destined to be the "greatest agricultural state in the Union." [3]

The *Ohio Cultivator,* like most of the Western journals, was

[1] This semimonthly quarto of eight pages sold for $1.00 a year. Bateham was born in England, September 13, 1813, and died in Parnsville, Ohio, August 5, 1880. Bailey, *Cyclopedia of American Agriculture,* IV, 555.

[2] *Ohio Cultivator,* I (Jan. 1, 1845), 1.

[3] *Ibid.,* I (Jan. 1, 1845), 1, 4.

written in a popular style and intended to appeal to a wide range of readers. Commenting on this periodical, the *Genesee Farmer* said it was "rich and racy—a real treat, not to the agriculturist alone, but to the philosopher, the student of human nature, the lover of literary oddities . . . and beyond all to the ladies."[4] Another contemporary wrote:

> The Cultivator is one of those journals which, like some portly, red-faced, jolly-hearted men, can get up a laugh on any occasion, and crack a joke with a pretty good explosion, whether the joke be big or little. Its editor aims, and rightly we think, to give his journal a popular rather than a scientific character; so that it shall reach and benefit the masses, rather than please a few highly cultivated minds.[5]

As the editor was vitally interested in a wide variety of subjects such as horticulture, education, agricultural societies, the woman's rights movement, and the peace crusade, it was natural that these topics should receive a large share of attention. The "Ladies' Department" of the *Ohio Cultivator,* edited by Mrs. Bateham was the most famous of its kind and is discussed elsewhere.[6] The *Cultivator* was influential in securing the passage of a law which established the Ohio State Board of Agriculture in 1846, and Bateham served as its secretary.[7] This journal was also helpful in obtaining state legislative aid for the encouragement of county agricultural societies in Ohio.[8] In 1852 Bateham played an important role in organizing the Ohio Pomological Society, gave it wide publicity in his paper, and served as one of its officials.[9]

The editor purchased a small farm outside of Columbus where he put into practice some of the lessons he attempted to teach. Within a year he was awarded the premium for the best Irish potatoes at an agricultural fair. The prize in this instance, happened to be a volume of his own magazine.[10]

[4] *Genesee Farmer,* VI (Feb., 1845), 22.

[5] *Prairie Farmer,* VII (Feb., 1847), 69.

[6] See pp. 162 ff.

[7] *Ohio Cultivator,* I (Feb. 15, 1845), 29; I (March 15, 1845), 41; I (April 1, 1845), 49, 53; II (March 1, 1846), 36.

[8] *Ibid.,* XI (Dec. 15, 1855), 375.

[9] *Ohio Farmer,* CLI (May 6, 1923), 4; Bailey, *Cyclopedia of American Agriculture,* IV, 555.

[10] *Ohio Cultivator,* II (Jan. 15, 1846), 12; II (Sept. 15, 1846), 140.

Early in 1851 Editor and Mrs. Bateham went to Europe to attend the World's Fair and Peace Conference at London.[11] The former was an official of the Ohio State Peace Society and both were delegates representing this organization at the conference.[12] The *Cultivator* reported in detail all phases of their trip. Bateham traveled widely throughout Europe, purchasing choice seeds for his clients and reporting the condition and progress of agriculture.[13]

In 1851 Sullivan D. Harris became associate and traveling editor of the *Ohio Cultivator.*[14] Four years later Bateham sold the periodical to Harris, who in turn took over the editorial responsibilities. In his valedictory the late proprietor stated that the condition of his health forbade the amount of mental labor requisite for conducting a paper of this kind in an "age of agricultural progress and of *quackery.*" Bateham returned to his favorite pursuit, horticulture, but continued as an active contributor to the agricultural press, which had already received fifteen years of his service.[15]

The new editor, Colonel Harris, as he was familiarly known, did not materially change the plan of the *Cultivator.*[16] He devoted his attention principally to general agriculture and livestock and

[11] A few months later, Mrs. Hannah M. Tracy, one of the leading contributors to the "Ladies' Department" of the *Cultivator* sailed to Europe for the same purpose, as well as to "take a look at European society." *Ohio Cultivator,* VII (Jan. 15, 1852), 216.

[12] *Ohio Cultivator,* VII (April 1, 1851), 109.

[13] *Ibid.,* VII (April 15, 1851), 121; VII (May 1, 1851), 136, 141; VII (June 1, 1851), 173; VII (July 5, 1851), 209; VII (Oct. 15, 1851), 309. Bateham was engaged in the nursery business in Columbus.

[14] *Ibid.,* VII (March 1, 1851), 65.

[15] *Ibid.,* XI (Dec., 1855), 374.

[16] Sullivan Dwight Harris was born at Middlebury, Vermont, January 21, 1812, and was reared a farmer. In 1836 he moved to Ohio where he was variously occupied as a farmer, printer, and teacher. As a youth he had gained a local reputation as a poet, a reputation sufficiently retained in middle life as to entitle him to be included in a book on poets of the West. "With Mr. Harris, poetry was an early and cherished passion, but the writing of verse was only a casual amusement, which he reckons among his juvenile indiscretions, and has abandoned for the more pressing duties of practical literature, only to be indulged in at the solicitation of personal friends whom he is too good-natured to refuse." Coggeshall, *op. cit.,* p. 401; *Ohio Cultivator,* XIV (Sept. 1, 1858), 264.

Bateham was induced to write on fruits and gardening. The departments of scientific agriculture, chemistry, and geology were in the hands of Professor W. W. Mather, formerly agricultural chemist of the Ohio State Board. Mrs. Bateham continued to contribute articles of interest to the fair sex. Of particular note was Harris's attitude on agricultural education. While a strong supporter of the common schools, he bitterly opposed state and federal supported agricultural colleges which he considered class legislation. Individual enterprise, he felt, would do "twice the work with half the money, and do it better every time." He editorialized as follows:

> We are opposed to an agricultural college or any other colleges, built up and supported by the State. The State has no money but what it takes from the people, and has no right to take money from one class of people to bestow it upon another class of people, except for purposes of general charity or absolute necessity. The State is a dear educator and a worse financial manager of any concern. The success of Morrill's College Land Bill would have been to build up the most stupendous literary hospital for political invalids and sap-rooted theorists, the world ever saw.[17]

The circulation of the *Ohio Cultivator*, while concentrated in the main in Ohio, Indiana, Illinois, Iowa, and Kentucky, reached almost every state and territory in the Union. With a consistency unusual in the farm press, each edition of the *Cultivator* approximated ten thousand copies during most of the years prior to the Civil War. The figure was bettered by two thousand in 1859.[18] This important and interesting periodical was one of the few journals to prove financially successful.[19] The *Ohio Cultivator* and the *Prairie Farmer* were probably the most influential agricultural papers in the West.[20]

[17] *Ohio Cultivator*, XV (May 1, 1859), 137. See also *ibid.*, XIV (Jan. 1, 1858), 9; XIV (Dec. 15, 1858), 377.

[18] *Ibid.*, II (Jan. 15, 1846), 9; V (Dec. 1, 1849), 368; VI (Jan. 15, 1850), 17; IX (Dec. 15, 1853), 379; XV (April 1, 1859), 112.

[19] *Ibid.*, XVI (Oct. 15, 1860), 312.

[20] Hutchinson, *op. cit.*, I, 234; Bidwell and Falconer, *op. cit.*, p. 470.

# BIBLIOGRAPHY

# BIBLIOGRAPHY

## PRIMARY SOURCES

### AGRICULTURAL PERIODICALS

*Agricultural Intelligencer*, Jan.–July, 1820 (Boston, Mass.).

*Agricultural Museum, The*, 1810–1812 (Georgetown, D.C.).

*Agriculturist, The*, 1840–1845 (Nashville, Tenn.).

*American Agriculturist, The*, 1842– (New York, N.Y.). Published as *The Plow* during 1852.

*American Cotton Planter, The*, 1853–1861 (Montgomery, Ala.). Title varies.

*American Farmer, The*, 1819–1834; 1839–1897 (Baltimore, Md., and Washington, D.C.). Called *Farmer and Gardener*, 1834–1839.

*American Farmers' Magazine*, 1857–1859 (New York, N.Y.). Merged with *American Agriculturist.*

*American Journal of Agriculture and Science*, 1845–1848 (Albany and New York, N.Y.). Vols. I–IV carry the title *American Quarterly Journal of Agriculture and Science.*

*American Quarterly Journal of Agriculture and Science*, 1845–1848 (Albany and New York, N.Y.). Vols. V–VII carry the title *American Journal of Agriculture and Science.*

*American Silk-Grower and Agriculturist*, 1836–1840 (Keene, N.H.). In 1838 title changed to *Cheshire Farmer* and a new series inaugurated; united with *Farmer's Monthly Visitor*, 1840.

*Arator, The*, 1855–1857? (Raleigh, N.C.).

*Boston Cultivator*, 1839–1876 (Boston, Mass.). Title varies. Combined with *New England Rural Home* and continued as *American Cultivator.*

*California Culturist, The,* 1858–1863? (San Francisco, Calif.).

*California Farmer, The,* 1854–1889? (San Francisco and Sacramento, Calif.).

*Carolina Planter, The,* monthly, 1844–1845 (Columbia, S.C.).

*Carolina Planter,* weekly, 1840–1841 (Columbia, S.C.). United with *Farmers' Register,* 1841.

*Central New York Farmer,* 1842–1844 (Rome and Oneida, N.Y.).

*Cheshire Farmer,* 1838–1840 (Keene, N.H.). See *American Silk-Grower and Agriculturist.*

*Cincinnatus, The,* 1856–1861 (College Hill, Ohio). Title varies.

*Connecticut Valley Farmer and Mechanic, The,* 1853–1855? (Springfield and Amherst, Mass.). In January, 1855, the title changed to *The Farmer.* Title varies. Merged with the second *New England Farmer.*

*Country Gentleman, The,* 1853– (Albany, N.Y., and Philadelphia, Pa.). Title varies.

*Cultivator, The,* 1834–1865 (Albany, N.Y.). Merged with *Country Gentleman,* 1866.

*Dollar Farmer, The,* 1842–1846 (Louisville, Ky.).

*Emery's Journal of Agriculture.* See *Prairie Farmer.*

*Farm and Garden,* 1853 (New York, N.Y.).

*Farmer, The.* See *Connecticut Valley Farmer.*

*Farmer and Artizan,* 1852–1854? (Portland, Me.).

*Farmer and Gardener, The,* 1834–1839 (Baltimore, Md.). See *American Farmer.*

*Farmer and Planter, The,* 1850–1860 (Pendleton and Columbia, S.C.).

*Farmer's Advocate, The,* 1847–1848? (Burlington, Iowa).

*Farmers' Cabinet, The,* 1836–1848 (Philadelphia, Pa.). Merged with *American Agriculturist,* 1849.

*Farmer's Companion and Horticultural Gazette, The,* 1853–1854 (Detroit, Mich.). Merged with *Michigan Farmer,* 1854.

*Farmer's Journal, The,* 1852–1855? (Bath and Raleigh, N.C.).

*Farmer's Monthly Visitor, The* (Concord), 1839–1849. (Concord, N.H.).

*Farmer's Monthly Visitor, The* (Manchester), 1852–1854

(Manchester, N.H.). Entirely distinct from the Concord *Farmer's Monthly Visitor.* Merged with the *Granite Farmer.*

*Farmers' Register, The,* 1833–1842 (Shellbanks and Petersburg, Va.).

*Franklin Farmer, The,* 1837–1840 (Frankfort and Lexington, Ky.). See *Kentucky Farmer.*

*Genesee Farmer, The,* 1831–1839 (Rochester, N.Y.). Consolidated with *The Cultivator,* 1839.

*Goodsell's Genesee Farmer,* 1833–1834 (Rochester, N.Y.).

*Granite Farmer, The,* 1850–1852? (Manchester, N.H.).

*Homestead, The,* 1855–1861 (Hartford, Conn.).

*Indiana Farmer, The,* 1845– (Indianapolis, Ind.). Title varies.

*Indiana Farmer and Gardener, The,* 1845–1847? (Indianapolis, Ind.). Called *Western Farmer and Gardener* after the first year.

*Iowa Farmer and Horticulturist, The,* 1853–1858 (Burlington, Fairfield, and Mount Pleasant, Iowa). Continued as *Pioneer Farmer and Iowa Home Visitor.*

*Journal of Agriculture, The,* 1851–1854 (Boston, Mass.).

*Kennebec Farmer.* See *Maine Farmer.*

*Kentucky Cultivator,* 1852–1853 (Cynthiana, Covington, and Louisville, Ky.). Title varies.

*Kentucky Farmer,* 1840–1842 (Frankfort and Lexington, Ky.). Continuation of *Franklin Farmer.*

*Kentucky Farmer,* monthly, 1842–? (Lexington, Ky.). Organized as a consolidation of the *Franklin Farmer* and *Kentucky Cultivator.*

*Maine Farmer, The,* 1833–1924 (Winthrop, Hallowell, and Augusta, Me.). Inaugurated as *Kennebec Farmer.* Title changed to *Maine Farmer* after second month.

*Michigan Farmer, The,* 1843– (Jackson, Detroit, and Niles, Mich.). Title varies.

*Moore's Rural New-Yorker,* 1849– (Rochester, N.Y.). Later *Rural New-Yorker.*

*New England Cultivator, The,* 1852–1854? (Boston, Mass.).

*New England Farmer, The,* 1822–1846 (Boston, Mass.).

*New England Farmer, The,* 1848–1913 (Boston, Mass.). Title varies. Unrelated to *New England Farmer* of 1822.

*New-England Farmers' and Mechanics' Journal, The,* 1828 (Gardiner, Me.).

*New Genesee Farmer, The,* 1840–1865 (Rochester, N.Y.). Merged with *American Agriculturist,* 1866.

*New Jersey Farmer, The,* 1855–1861 (Freehold and Trenton, N.J.).

*New Jersey and Pennsylvania Agricultural Monthly Intelligencer and Farmers' Magazine, The,* 1825–1826. (Camden, N.J.).

*New-York Farmer,* 1828–1839 (New York, N.Y.). Title varies.

*New York Farmer and Mechanic,* 1844–1852 (New York, N.Y.). Title varies.

*North-Carolina Planter, The,* 1858–1861? (Raleigh, N.C.).

*Northern Farmer, The* (Newport), 1832–1834 (Newport, N.H.).

*Northern Farmer, The* (Utica), 1852–1861? (Utica and Clinton, N.Y.).

*Northwestern Farmer,* 1856–1862 (Dubuque, Iowa).

*Ohio Agriculturist,* 1851–1852 (Tiffin, Ohio).

*Ohio Cultivator,* 1845–1864 (Columbus, Ohio). Merged with *Ohio Farmer,* 1864.

*Ohio Farmer, The,* 1852– (Cleveland, Ohio).

*Ohio Valley Farmer, The,* 1856–1861 (Cincinnati, Ohio).

*Oregon Farmer, The,* 1858–1861? (Portland, Ore.).

*Pennsylvania Farm Journal,* 1851–1857 (Lancaster, West Chester, and Philadelphia, Pa.). Title varies. Merged with *American Agriculturist,* 1857.

*Plough Boy, The,* 1819–1823 (Albany, N.Y.).

*Plough, the Loom, and the Anvil, The,* 1848–1857 (Philadelphia and New York). Title changed to *American Farmers' Magazine* in 1857–1859. Merged with the *American Agriculturist,* 1859.

*Plow, The,* 1852 (New York, N.Y.). See *American Agriculturist.*

*Prairie Farmer, The,* 1840– (Chicago, Ill.). Title varies.

Title for first two years, *The Union Agriculturist.* Called *Emery's Journal of Agriculture,* etc., during 1858.

*Rural American,* 1856–1870 (Utica, Clinton, and New York, N.Y.; New Brunswick, N.J.). Title varies.

*Rural Register, The,* 1859–1863? (Baltimore, Md.).

*School Journal, and Vermont Agriculturist, The,* 1847–1850 (Windsor, Vt.).

*Soil of the South, The,* 1851–1856 (Columbus, Ga.). Title varies. Absorbed by *American Cotton Planter,* 1857.

*South Carolina Agriculturist, The,* 1856–? (Columbia, S.C.).

*Southern Agriculturist, The,* 1828–1846 (Charleston, S.C.). Title varies. See *Southern Cabinet of Agriculture . . . ,* 1840.

*Southern Cabinet of Agriculture, Horticulture, Rural and Domestic Economy,* 1840 (Charleston, S.C.). Title of *Southern Agriculturist* during 1840.

*Southern Cultivator, The,* 1843–1935 (Augusta, Athens, and Atlanta, Ga.). Title varies.

*Southern Planter, The* (Mississippi), 1842 (Natchez and Washington, Miss.).

*Southern Planter* (Virginia), 1841– (Richmond, Va.). Title varies.

*South-Western Farmer, The,* 1842–1845? (Raymond, Miss.).

*Tennessee Farmer,* 1834–1840 (Jonesborough, Tenn.). Absorbed by the *Agriculturist.*

*Tippecanoe Farmer,* 1854–1855 (Lafayette, Ind.).

*Union Agriculturist, The,* 1840–1842 (Chicago, Ill.). Title varies. Became *Prairie Farmer,* 1843.

*United States Agriculturist and Farmer's Reporter,* 1830–1831 (Washington and Cincinnati, Ohio).

*Valley Farmer, The,* 1849–1864 (St. Louis, Mo., and Louisville, Ky.). Succeeded by *Colman's Rural World,* 1865.

*Western Agriculturist, The,* 1851–1852? (Columbus, Ohio).

*Western Farmer, The* (Cincinnati), 1839–1845 (Cincinnati, Ohio). Called *Western Farmer and Gardener* after first year.

*Western Farmer* (Detroit), 1841–1843 (Detroit, Mich.).

*Western Farmer and Gardener* (Cincinnati), 1839–1845 (Cincinnati, Ohio). Called *Western Farmer* for the first year.

*Western Farmer and Gardener* (Indianapolis), 1845–1847. (Indianapolis, Ind.). Called *Indiana Farmer and Gardener* for the first year.

*Western Plow-Boy, The,* 1853–1856? (Fort Wayne, Ind.).

*Western Reserve Magazine of Agriculture and Horticulture,* 1845–1846? (Cleveland, Ohio).

*Western Tiller,* 1828–1831? (Cincinnati, Ohio).

*Wisconsin Farmer, The,* 1849–1874 (Racine, Dubuque, Janesville, and Madison, Wis.). Title varies. Merged with *Western Rural,* 1875.

*Wool Grower and Magazine of Agriculture and Horticulture, The,* 1849–1856 (Buffalo and Rochester, N.Y.). Title changed to *The Wool Grower and Stock Register* with Vol. IV.

*Working Farmer, The,* 1849–1875 (New York, N.Y.). Title varies.

*Yankee Farmer, The,* 1835–1842 (Cornish and Portland, Me.; Boston, Mass.). Title varies. United with *Massachusetts Ploughman,* 1842.

*Yankee Farmer and News Letter.* Another title for *Yankee Farmer.*

*Yankee Farmer and New England Cultivator.* Another title for *Yankee Farmer.*

## OTHER SOURCES OF A PRIMARY NATURE

*Agricultural Almanack, The,* 1818 (Philadelphia, Pa.).

*American Turf Register and Sporting Magazine,* 1829–1844 (Baltimore, Md.).

Blegen, Theodore C., ed., Minnesota Farmers' Dairies (William R. Brown, 1845–1846; Mitchell Y. Jackson, 1852–1863). Publications of the Minnesota Historical Society, No. III, 1939.

Boyd, William K., ed., William Byrd's Histories of the Dividing Line betwixt Virginia and North Carolina. Raleigh, N.C., 1929.

Buel, Jesse, The Farmer's Companion. Boston, 1847. The first edition was published in New York in 1839.

Downing, Andrew Jackson, The Architecture of Country Houses. New York, 1850.

——— Rural Essays. New York, 1853.

Eliot, Jared, Essays upon Field Husbandry in New England. Boston, 1760.

Essex Agricultural Society, *Transactions*, 1820. Salem, 1821.

Fac Similes of Letters from George Washington to Sir John Sinclair. Philadelphia, 1844.

*Fessenden's Practical Farmer and Silk Manual*, 1835–1837 (Boston, Mass.).

Hawthorne, Nathaniel, Works. 12 vols., Boston, 1883.

*Horticulturist, The*, 1846–1875 (Albany, N.Y.).

Journal of the United States Agricultural Society. 10 vols., Washington and Boston, 1852–1862. Title varies.

Kellar, H. A., ed., Solon Robinson, Pioneer and Agriculturist; Selected Writings, 1825–1851. 2 vols., Indianapolis, 1936.

Kennedy, Joseph C., Preliminary Report of the Eighth Census, 1860. Washington, 1862.

*Lady's Home Magazine*, 1857–1860 (Philadelphia, Pa.). Inaugurated as *Home Magazine.*

Legare, John D., Account of the Medical Properties of the Grey Sulphur Springs, Virginia. Charleston, S.C., 1836.

*Maryland Gazette*, 1790, 1795, 1796, 1797, 1798. Annapolis, Md.

*Maryland Journal and Baltimore Advertiser* (Baltimore, 1790).

New York Agricultural Society, *Transactions*, 1792–1799.

*Niles Weekly Register*, 1811–1849 (Baltimore, Md., Washington, D.C., Philadelphia, Pa.). Title changed to *Niles National Register*, Vols. 53–75.

Philips, Martin W., "Diary of a Mississippi Planter." Edited by Franklin L. Ripey. Publications of the Mississippi Historical Society, No. X, 1909.

*Porter's Spirit of the Times*, 1856–1860. New York, N.Y.

Post-Office Law with Instructions and Forms, The. Washington, 1808.

*Richmond Enquirer*, 1804, 1805, 1806, 1816, 1818. Richmond, Va.

Robinson, Solon, "Account Book, 1840–1853." Manuscript in possession of McCormick Historical Association, Chicago.

Ruffin, Edmund, "Diary." Manuscript in the Library of Congress.

Stephenson, Wendell Holmes, ed., "A Quarter-Century of a Mississippi Plantation: Eli J. Campbell of 'Pleasant Hill,'" *Mississippi Valley Historical Review*, XXIII (Dec., 1936).

Strickland, William, Observations on the Agriculture of the United States of America. London, 1801.

Taylor, John, Arator. Georgetown, D.C., 1813.

*Virginia Herald*, 1815, 1816, 1817. Fredericksburg, Va.

Washington, George, Diaries, 1748–1799. Edited by John C. Fitzpatrick. 4 vols., New York, 1925.

Watson, Elkanah, History of Agricultural Societies, on the Modern Berkshire System. Albany, N.Y., 1820.

## SECONDARY SOURCES

*Agricultural History*, 1927–.

Allibone, S. Austin, Allibone's Dictionary of Authors. 5 vols., Philadelphia, 1870–1892.

*American Economic Review*, 1911–.

*American Historical Review*, 1896–.

American Library Association, Bulletins, 1907–.

Appletons' Cyclopaedia of American Biography. 7 vols., New York, 1888–1900.

Bailey, L. H., ed., Cyclopedia of American Agriculture. 4 vols., New York, 1907–1909.

——— ed., Standard Cyclopedia of Horticulture. 6 vols., New York, 1914–1917.

Barnett, Claribel R., "The Agricultural Museum; an Early American Periodical," *Agricultural History, II* (April, 1928).

Beal, William J., History of the Michigan Agricultural College. East Lansing, Mich., 1915.

Belting, Paul E., The Development of the Free Public High School in Illinois to 1860. Springfield, Ill., 1918.

Bidwell, Percy Wells, "The Agricultural Revolution in New England," *American Historical Review*, XXVI (July, 1921).

Bidwell, Percy Wells, "Rural Economy in New England at the Beginning of the Nineteenth Century," *Transactions*, Connecticut Academy of Arts and Sciences, XX (April, 1916).

——— and John I. Falconer, History of Agriculture in the Northern United States, 1620–1860. Washington, 1925.

Bowers, Claude C., Party Battles of the Jackson Period. Boston, 1922.

Bradley, Cyrus P., Biography of Isaac Hill of New Hampshire. Concord, N.H., 1835.

Branch, E. Douglas, The Sentimental Years, 1836–1860. New York, 1934.

Bruce, Kathleen, "Materials for Virginia Agricultural History," *Agricultural History*, IV (Jan., 1930).

Burkett, Charles W., History of Ohio Agriculture. Concord, N.H., 1900.

Burnham, George P., The History of the Hen Fever. Boston, 1855.

Cambridge History of American Literature, The. Edited by William P. Trent and others. 3 vols., New York, 1933.

Carman, Harry J., "Jesse Buel, Albany County Agriculturist," *New York History*, XIV (July, 1933).

——— Social and Economic History of the United States. 2 vols., New York, 1933.

——— and Rexford G. Tugwell, "The Significance of American Agricultural History," *Agricultural History*, XII (April, 1938).

Carrier, Lyman, Beginnings of Agriculture in America. New York, 1923.

——— "The United States Agricultural Society, 1852–1860," *Agricultural History*, XI (Oct., 1937).

Coggeshall, William T., ed., The Poets and Poetry of the West. Columbus, 1860.

Cole, Arthur Charles, The Irrepressible Conflict. New York, 1934.

Cole, Arthur H., "Agricultural Crazes," *American Economic Review*, XVI (Dec., 1926).

Connecticut Academy of Arts and Sciences, *Transactions*. 32 vols., New Haven, 1866–1937.

Cook, O. F., "The American Origin of Agriculture," *Popular Science Monthly*, LXI (Oct., 1902).

Craven, Avery O., Soil Exhaustion as a Factor in the Agricultural History of Virginia and Maryland, 1606–1860. Urbana, Ill., 1926.

——— "The Agricultural Reformers of the Ante-Bellum South," *American Historical Review*, XXXIII (Jan., 1928).

——— Edmund Ruffin, Southerner. New York, 1932.

——— "John Taylor and Southern Agriculture," *Journal of Southern History*, IV (May, 1938).

Crozier, A. A., Popular Errors about Plants. Ann Arbor, Mich., 1891.

Cutter, W. P., "A Pioneer in Agricultural Science," U.S. Dept. of Agriculture, *Yearbook* (1895), pp. 493-502.

Darlington, William, American Weeds and Useful Plants. Revised with additions by George Thuber. New York, 1859.

Dictionary of American Biography. 22 vols., New York, 1928–1936.

Dondore, Dorothy Anne, The Prairie and the Making of Middle America: Four Centuries of Description. Cedar Rapids, Iowa, 1926.

Dulles, Foster R., America Learns to Play. New York, 1940.

Edwards, Everett E., A Bibliography of the History of Agriculture in the United States. U.S. Dept. of Agriculture, Miscellaneous Publications, No. 84. Washington, D.C., 1930.

——— References on Agricultural History as a Field for Research. U.S. Dept. of Agriculture, Library, Bibliographical Contributions, No. 32. Washington, D.C., 1937.

——— "Some Sources for Northwest History; Agricultural Periodicals," *Minnesota History*, XVIII (Dec., 1937).

——— "Agricultural Records; Their Nature and Value for Research," *Agricultural History*, XIII (Jan., 1939).

Encyclopaedia Britannica, 14th ed., 24 vols., Cambridge, Eng., and New York, 1929–1938.

Encyclopedia Americana. 30 vols., New York, 1936.

Faulkner, Harold Underwood, American Economic History. New York, 1924.

Faulkner, Harold Underwood, American Political and Social History. New York, 1937.

Fish, Carl Russell, The Rise of the Common Man. New York, 1927.

Flint, Charles L., Eighty Years Progress of the United States. Hartford, Conn., 1869.

Fox, Charles James, and Samuel D. D. Osgood, eds., The New Hampshire Book. Boston, 1842.

Ganoe, John T., "The Beginnings of Irrigation in the United States," *Mississippi Valley Historical Review*, XXV (June, 1938).

Garwood, Irving, American Periodicals from 1850 to 1860. Monmouth, Ill., 1931.

Gates, Paul Wallace, The Illinois Central Railroad and Its Colonization Work. Cambridge, Mass., 1934.

——— "The Promotion of Agriculture by the Illinois Central Railroad, 1855–1870," *Agricultural History*, V (April, 1931).

Gras, Norman Scott Brien, A History of Agriculture in Europe and America. New York, 1925.

Gray, Lewis Cecil, History of Agriculture in the Southern United States to 1860. 2 vols., Washington, 1933.

Griffin, Joseph, History of the Press of Maine. Brunswick, Me., 1872.

Griswold, Rufus Wilmot, The Female Poets of America. New York, 1849.

Hamilton, Milton W., The Country Printer, New York State, 1785–1830. New York, 1936.

Harman, T. D., Jr., The History and Development of the Agricultural Press. State College, Pa., 1911.

Haworth, Paul Leland, George Washington, Country Gentleman. Indianapolis, 1925.

Hedrick, Ulysses Prentiss, A History of Agriculture in the State of New York. New York, 1933.

Helfrich, Myrtle, "A Baltimore Pioneer of Farm and Turf," *The Sun* (magazine section), Baltimore, Feb. 17, 1935.

Hill, Robert W., "John Pitkin Norton's Visit to England, 1844," *Agricultural History*, VIII (Oct., 1934).

Hudson, Frederic, Journalism in the United States from 1690 to 1872. New York, 1873.

Humphrey, Edward Frank, An Economic History of the United States. New York, 1931.

Hurd, D. Hamilton, comp., History of Hillsborough County, New Hampshire. Philadelphia, 1885.

Hutchinson, William T., Cyrus Hall McCormick. 2 vols., New York, 1930.

Johnstone, Paul H., "Turnips and Romanticism," *Agricultural History*, XII (July, 1938).

*Journal of Southern History*, 1935–.

King, Dan, Quackery Unmasked. New York, 1858.

Kirkland, Edward C., A History of American Economic Life. New York, 1932.

Kittredge, George Lyman, The Old Farmer and His Almanack. Boston, 1904.

Krout, John A., Annals of American Sport. Hartford, 1929.

Landon, Fred, "The Agricultural Journals of Upper Canada (Ontario)," *Agricultural History*, IX (Oct., 1935).

Leavitt, Charles T., "Attempts to Improve Cattle Breeds in the United States, 1790–1860," *Agricultural History*, VII (April, 1933).

——— "Transportation and the Livestock Industry of the Middle West to 1860," *Agricultural History*, VIII (Jan., 1934).

Lee, James Melvin, History of American Journalism. New York, 1917.

Lewton, Frederick L., "The Servant in the House," *Annual Report*, Smithsonian Institution, 1929.

Loehr, Rodney C., "The Influence of English Agriculture on American Agriculture," *Agricultural History*, II (Jan., 1937).

McCall, A. G., "The Development of Soil Science," *Agricultural History*, V (April, 1931).

McMillen [Article on the *Southern Cultivator*], American Library Association, *Bulletin*, XXVII (Dec. 15, 1933).

Major, Howard, The Domestic Architecture of the Early American Republic. Philadelphia, 1926.

Manning, Robert, ed., History of the Massachusetts Horticultural Society. Boston, 1880.

May, Caroline, The American Female Poets. Philadelphia, 1850.

*Minnesota History*, 1915–.

Mississippi Historical Society, Publications. 14 vols., Oxford, 1898–1914.

*Mississippi Valley Historical Review*, 1914–.

Mott, Frank L., A History of American Magazines, 1741–1850. New York, 1930.

——— A History of American Magazines, 1850–1865. Cambridge, Mass., 1938.

Mumford, F. B., "A Century of Missouri Agriculture," *Missouri Historical Review*, XV (Jan., 1921).

National Cyclopaedia of American Biography. 28 vols., New York, 1897–1940.

*National Horticultural Magazine*, 1922–.

Neely, Wayne Caldwell, The Agricultural Fair. New York, 1935.

*New York History*, 1919–.

Ogilvie, William E., Pioneer Agricultural Journalists. Privately printed, Chicago, 1927.

Parton, James, and others, Eminent Women of the Age. Hartford, 1868.

Perrin, Porter Gale, Thomas Green Fessenden. Orono, Me., 1925.

Phillips, Ulrich B., Life and Labor in the Old South. Boston, 1929.

Pickering, Octavius, The Life of Timothy Pickering. Boston, 1867.

Poore, Ben. Perley, "Biographical Notice of John S. Skinner," *Plough, the Loom, and the Anvil*, VII (July, 1854).

*Popular Science Monthly*, 1872–.

Power, Richard L., "Wet Lands and the Hoosier Stereotype," *Mississippi Valley Historical Review*, XXII (June, 1935).

Presbrey, Frank, The History and Development of Advertising. New York, 1929.

Proctor, John Clagett, Columbia Historical Society *Records*, XXVII, 1925.

Rippy, J. Fred, Joel R. Poinsett, Versatile American. Durham, N.C., 1935.

Roberts, Isaac Phillips, The Fertility of the Land. New York, 1907.

Rogin, Leo, The Introduction of Farm Machinery in its relation to the Productivity of Labor in the Agriculture of the United States During the Nineteenth Century. University of Calif., Publications in Economics, No. IX. Berkeley, 1931.

Ross, Earle D., "The 'Father' of the Land Grant College," *Agricultural History*, XII (April, 1938).

Ryerson, Knowles A., "History and Significance of the Foreign Plant Introduction Work of the United States Department of Agriculture," *Agricultural History*, VII (July, 1933).

Sanford, Albert H., The Story of Agriculture in the United States. New York, 1916.

Schafer, Joseph, The Social History of American Agriculture. New York, 1936.

Schmidt, Louis Bernard, "The Agricultural Revolution in the Prairies and the Great Plains of the United States," *Agricultural History*, VIII (Oct., 1934).

——— and Earle Dudley Ross, Readings in the Economic History of American Agriculture. New York, 1925.

Shafer, Henry B., The American Medical Profession, 1783–1850. New York, 1936.

Shryock, Richard H., The Development of Modern Medicine. Philadelphia, 1936.

Simms, Henry H., Life of John Taylor. Richmond, Va., 1932.

Smith, Harry Worcester, A Sporting Family of the Old South. Albany, N.Y., 1936.

Spaeth, Sigmund, Read 'Em and Weep. New York, 1926.

Stackpole, Everett S., History of Winthrop, Maine. Auburn, Me., 1925.

Stearns, Ezra S., ed., Genealogical and Family History of the State of New Hampshire. 4 vols., New York, 1908.

Stilwell, Lewis D., Migration from Vermont (1776–1860). Vermont Historical Society, Montpelier, Vt., 1937.

Stuntz, Stephen Conrad, "Compilation of Agricultural Journals" (unpublished and incomplete manuscript in U.S. Dept. of Agriculture Library, Washington, 1918). Cited as the Stuntz List.

Swem, Earl Gregg, An Analysis of Ruffin's Farmers' Register, with a Bibliography of Edmund Ruffin. Virginia State Library Bulletin, No. XI (July–Oct., 1918).

Tallmadge, Thomas E., The Story of Architecture in America. New York, 1927.

Traub, Hamilton Paul, "The Development of American Horticultural Literature, chiefly between 1800–1850," *National Horticultural Magazine*, VII (July, 1928).

True, Alfred Charles, A History of Agricultural Education in the United States, 1785–1925. U.S. Dept. of Agriculture, Miscellaneous Publications, No. 36. Washington, 1929.

True, N. T., "Biographical Sketch of Ezekiel Holmes," *Tenth Annual Report*, Maine Board of Agriculture, 1865.

True, Rodney H., "Jared Eliot, Minister, Physician, Farmer," *Agricultural History*, II (Oct., 1928).

——— "The Early Development of Agricultural Societies in the United States," *Annual Report*, American Historical Association, 1920.

Tucker, Gilbert M., American Agricultural Periodicals: An Historical Sketch. Privately printed. Albany, N.Y., 1909.

United States Department of Agriculture, *Yearbook*. Washington, 1895–.

United States Patent Office, Annual Reports of the Commissioner of Patents: Agriculture (1849–1862). Washington, D.C.

Wiest, Edward, Agricultural Organization in the United States. Lexington, Ky., 1923.

Wilson, Harold Fisher, The Hill Country of Northern New England. New York, 1936.

Woodward, Carl R., The Development of Agriculture in New Jersey, 1640–1880. New Brunswick, N.J., 1927.

Woodward, Carl R., "The Agricultural Magazine," *Journal* of the Rutgers University Library, III (Dec., 1939).

Wright, John S., Chicago: Relations to the Great Interior and Continent. Chicago, 1870.

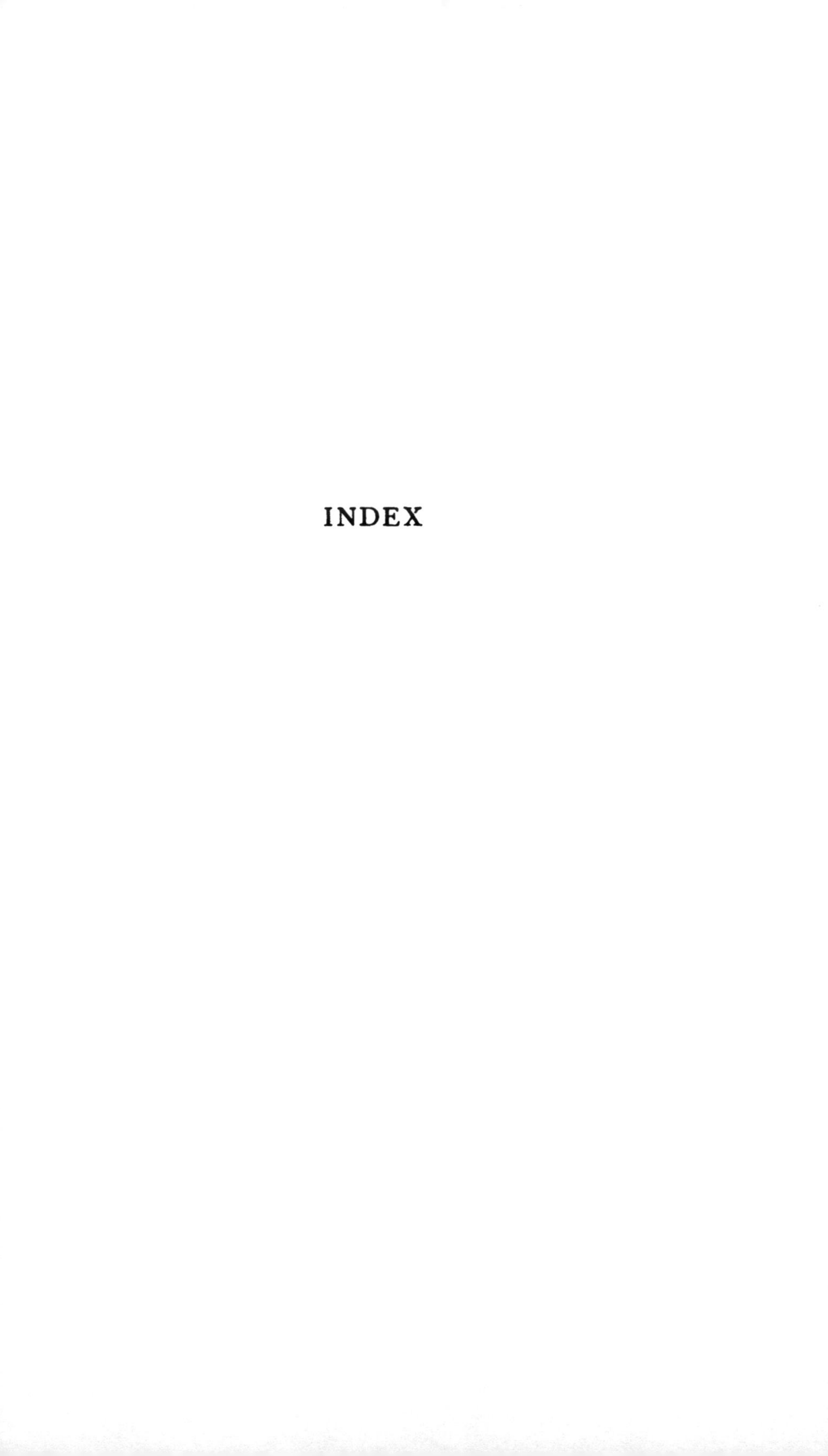

# INDEX

# INDEX

Bei Fragen zur Produktsicherheit wenden Sie sich bitte an:
If you have any questions regarding product safety,
please contact:

Walter de Gruyter GmbH
Genthiner Straße 13
10785 Berlin
productsafety@degruyterbrill.com